Study and Solutions Guide for

CALCULUS

SIXTH EDITION

Larson / Hostetler / Edwards

Volume I

David E. Heyd
The Pennsylvania State University
The Behrend College

Houghton Mifflin Company   Boston   New York

Editor in Chief, Mathematics: Charles Hartford
Managing Editor: Cathy Cantin
Senior Associate Editor: Maureen Brooks
Associate Editor: Michael Richards
Assistant Editor: Carolyn Johnson
Supervising Editor: Karen Carter
Art Supervisor: Gary Crespo
Marketing Manager: Sara Whittern
Associate Marketing Manager: Ros Kane
Marketing Assistant: Carrie Lipscomb
Design: Henry Rachlin
Composition and Art: Meridian Creative Group

Calculator Key font used with permission of Texas Instruments Incorporated. Copyright 1990, 1993, 1996.

Printed in the U.S.A.

ISBN: 0-395-88767-4

23456789-B-01 00 99 98

Preface

This *Study and Solutions Guide* is designed as a supplement to *Calculus,* Sixth Edition, by Roland E. Larson, Robert P. Hostetler, and Bruce H. Edwards. All references to chapters, theorems, and exercises relate to the main text. Although this supplement is not a substitute for good study habits, it can be valuable when incorporated into a well-planned course of study. The following suggestions may assist you in the use of the text, your lecture notes, and this *Guide*.

- *Read the section in the text for general content before class.* You will be surprised at how much more you will absorb from the lecture if you are aware of the objectives of the section and the types of problems that will be solved. If you are familiar with the topic, you will understand more of the lecture, and you will be able to take fewer (and better) notes.

- *As soon after class as possible, work problems from the exercise set.* The exercise sets in the text are divided into groups of similar problems and are presented in approximately the same order as the section topics. Try to get an overall picture of the various types of problems in the set. As you work your way through the exercise set, reread your class notes and the portion of the section that covers each type of problem. Pay particular attention to the solved examples.

- *Learning calculus takes practice.* You cannot learn calculus merely by reading, any more than you can learn to play the piano or to bowl merely by reading. Only after you have practiced the techniques of a section and have discovered your weak points can you make good use of the supplementary solutions in this *Guide*.

- *Technology.* Graphing utilities and symbolic algebra systems are now readily available. The computer and calculator are merely tools. Your ability to use these tools effectively requires that you continually sharpen your problem-solving skills and your understanding of fundamental mathematical principles.

During many years of teaching I have found that good study habits are essential for success in mathematics. My students have found the following additional suggestions to be helpful in making the best use of their time.

- *Write neatly in pencil.* A notebook filled with unorganized scribbling has little value.

- *Work at a deliberate and methodical pace, without skipping steps.* When you hurry through a problem you are more apt to make careless arithmetic or algebraic errors that, in the long run, waste time.

- *Keep up with the work.* This suggestion is crucial because calculus is a very structured topic. If you cannot do the problems in one section, you are not likely to be able to do the problems in the next. The night before a quiz or test is not the time to start working problems. In some instances cramming may help you pass an examination, but it is an inferior way to learn and retain essential concepts.

- After working some of the assigned exercises with access to the examples and answers, *try at least one of each type of exercise with the book closed.* This will increase your confidence on quizzes and tests.

- Do not be overly concerned with finding the most efficient way to solve a problem. *Your first goal is to find one way that works.* Short cuts and clever methods come later.

- If you have trouble with the algebra of calculus, *refer to the algebra review at the beginning of this guide.*

I wish to acknowledge several people whose help and encouragement were invaluable in the production of this *Guide*. First, I am grateful to Roland E. Larson, Robert P. Hostetler, and Bruce H. Edwards for the privilege of working with them on the main text. I also wish to thank the staffs at Larson Texts, Inc. and Houghton Mifflin Company. I am also grateful to my wife, Jean, and our children, Ed, Ruth, and Andy, for their love and support.

David E. Heyd

Contents

Algebra Review

0.1   Monomial Factors

Factor as indicated:

(a) $3x^4 + 4x^3 - x^2 = x^2(\quad)$

(b) $2\sqrt{x} + 6x^{3/2} = 2\sqrt{x}(\quad)$

(c) $e^{-x} - xe^{-x} + 2x^2e^{-x} = e^{-x}(\quad)$

(d) $x^{-1} - 2 + x = x^{-1} = x^{-1}(\quad)$

(e) $\dfrac{x}{2} - 6x^2 = \dfrac{x}{2}(\quad)$

(f) $\sin x + \tan x = \sin x(\quad)$

(g) $\dfrac{1}{2x^2 + 4x} = \dfrac{1}{2x}(\quad)$

Solution:

(a) $3x^4 + 4x^3 - x^2 = x^2(3x^2 + 4x - 1)$

(b) $2\sqrt{x} + 6x^{3/2} = 2\sqrt{x}(1 + 3x)$

(c) $e^{-x} - xe^{-x} + 2x^2e^{-x} = e^{-x}(1 - x + 2x^2)$

(d) $x^{-1} - 2 + x = x^{-1} = x^{-1}(1 - 2x + x^2)$

(e) $\dfrac{x}{2} - 6x^2 = \dfrac{x}{2}(1 - 12x)$

(f) $\sin x + \tan x = \sin x + \dfrac{\sin x}{\cos x} = \sin x\left(1 + \dfrac{1}{\cos x}\right)$

$$= \sin x(1 + \sec x)$$

(g) $\dfrac{1}{2x^2 + 4x} = \dfrac{1}{2x}\left(\dfrac{1}{x + 2}\right)$

0.2   Binomial Factors

Factor as indicated:

(a) $(x - 1)^2(x) - (x - 1) = (x - 1)(\quad)$

(b) $3(x^2 + 4)(x^2 + 1) + 6(x^2 + 4)^2 = 3(x^2 + 4)(\quad)$

(c) $\sqrt{x^2 + 1} - \dfrac{x^2}{\sqrt{x^2 + 1}} = \dfrac{1}{\sqrt{x^2 + 1}}(\quad)$

(d) $(x - 3)^3(x + 2) - 2(x - 3)^2(x + 2)^2 = (x - 3)^2(x + 2)(\quad)$

(e) $(2x + 1)^{3/2}(x^{1/2}) + (2x + 1)^{5/2}(x^{-1/2}) = (2x + 1)^{3/2}(x^{-1/2})(\quad)$

Solution:

(a) $(x - 1)^2(x) - (x - 1) = (x - 1)[(x - 1)x - 1]$

$$= (x - 1)(x^2 - x - 1)$$

(b) $3(x^2 + 4)(x^2 + 1) + 6(x^2 + 4)^2 = 3(x^2 + 4)[(x^2 + 1) + 2(x^2 + 4)]$

$$= 3(x^2 + 4)(3x^2 + 9)$$

—CONTINUED—

1

—CONTINUED—

(c) $\sqrt{x^2 + 1} - \dfrac{x^2}{\sqrt{x^2 + 1}} = (x^2 + 1)^{1/2} - x^2(x^2 + 1)^{-1/2}$

$$= (x^2 + 1)^{-1/2}[(x^2 + 1) - x^2]$$

$$= \dfrac{1}{\sqrt{x^2 + 1}}$$

(d) $(x - 3)^3(x + 2) - 2(x - 3)^2(x + 2)^2 = (x - 3)^2(x + 2)[(x - 3) - 2(x + 2)]$

$$= (x - 3)^2(x + 2)(-x - 7)$$

(e) $(2x + 1)^{3/2}(x^{1/2}) + (2x + 1)^{5/2}(x^{-1/2}) = (2x + 1)^{3/2}(x^{-1/2})[x + (2x + 1)]$

$$= (2x + 1)^{3/2}(x^{-1/2})(3x + 1)$$

0.3   Factoring Quadratic Expressions

Factor as indicated:

(a) $x^2 - 3x + 2 = ($ $)($ $)$

(b) $x^2 - 9 = ($ $)($ $)$

(c) $x^2 + 5x - 6 = ($ $)($ $)$

(d) $x^2 + 5x + 6 = ($ $)($ $)$

(e) $2x^2 + 5x - 3 = ($ $)($ $)$

(f) $e^{2x} + 2 + e^{-2x} = ($ $)^2$

(g) $x^4 - 7x^2 + 12 = ($ $)($ $)($ $)$

(h) $1 - \sin^2 x = ($ $)($ $)$

Solution:

(a) $x^2 - 3x + 2 = (x - 2)(x - 1)$

(b) $x^2 - 9 = (x + 3)(x - 3)$

(c) $x^2 + 5x - 6 = (x + 6)(x - 1)$

(d) $x^2 + 5x + 6 = (x + 2)(x + 3)$

(e) $2x^2 + 5x - 3 = (2x - 1)(x + 3)$

(f) $e^{2x} + 2 + e^{-2x} = (e^x + e^{-x})^2$

(g) $x^4 - 7x^2 + 12 = (x^2 - 3)(x^2 - 4) = (x^2 - 3)(x + 2)(x - 2)$

(h) $1 - \sin^2 x = (1 + \sin x)(1 - \sin x)$

0.4   Cancellation

Reduce each expression to lowest terms:

(a) $\dfrac{3x + 9}{6x}$

(b) $\dfrac{x^2}{x^{1/2}}$

(c) $\dfrac{(x + 1)^3(x - 2) + 3(x + 1)^2}{(x + 1)^4}$

(d) $\dfrac{x^{1/2} - x^{1/3}}{x^{1/6}}$

(e) $\dfrac{\sqrt{x - 1} + (x - 1)^{3/2}}{\sqrt{x - 1}}$

(f) $\dfrac{1 - (\sin x + \cos x)^2}{2 \sin x}$

Solution:

(a) $\dfrac{3x + 9}{6x} = \dfrac{3(x + 3)}{3(2x)} = \dfrac{x + 3}{2x}$

(b) $\dfrac{x^2}{x^{1/2}} = \dfrac{(x^{1/2})(x^{3/2})}{x^{1/2}} = x^{3/2}$

—CONTINUED—

—CONTINUED—

(c) $\dfrac{(x + 1)^3(x - 2) + 3(x + 1)^2}{(x + 1)^4} = \dfrac{(x + 1)^2[(x + 1)(x - 2) + 3]}{(x + 1)^4}$

$$= \dfrac{x^2 - x + 1}{(x + 1)^2}$$

(d) $\dfrac{x^{1/2} - x^{1/3}}{x^{1/6}} = \dfrac{x^{1/6}(x^{2/6} - x^{1/6})}{x^{1/6}} = x^{1/3} - x^{1/6}$

(e) $\dfrac{\sqrt{x - 1} + (x - 1)^{3/2}}{\sqrt{x - 1}} = \dfrac{\sqrt{x - 1}[1 + (x - 1)]}{\sqrt{x - 1}} = x$

(f) $\dfrac{1 - (\sin x + \cos x)^2}{2 \sin x} = \dfrac{1 - (\sin^2 x + 2 \sin x \cos x + \cos^2 x)}{2 \sin x}$

$$= \dfrac{1 - (\sin^2 x + \cos^2 x) - 2 \sin x \cos x}{2 \sin x}$$

$$= \dfrac{1 - 1 - 2 \sin x \cos x}{2 \sin x} = -\cos x$$

0.5   Quadratic Formula

Equation	*Solve for*
(a) $x^2 - 4x - 1 = 0$	x
(b) $2x^2 + x - 3 = 0$	x
(c) $\cos^2 x + 3 \cos x + 2 = 0$	$\cos x$
(d) $x^2 - xy - (1 + y^2) = 0$	x
(e) $x^4 - 4x^2 + 2 = 0$	x^2

Solution:

(a) $x = \dfrac{4 \pm \sqrt{16 + 4}}{2} = \dfrac{4 \pm \sqrt{20}}{2} = \dfrac{4 \pm 2\sqrt{5}}{2} = 2 \pm \sqrt{5}$

(b) $x = \dfrac{-1 \pm \sqrt{1 + 24}}{4} = \dfrac{-1 \pm 5}{4}$

$\quad x = \dfrac{4}{4} = 1 \quad \text{or} \quad x = -\dfrac{6}{4} = -\dfrac{3}{2}$

(c) $\cos x = \dfrac{-3 \pm \sqrt{9 - 8}}{2} = \dfrac{-3 \pm 1}{2}$

$\quad \cos x = -\dfrac{2}{2} = -1 \quad \text{or} \quad \cos x = -\dfrac{4}{2} = -2$

(d) $x = \dfrac{y \pm \sqrt{y^2 + 4(1 + y^2)}}{2} = \dfrac{y \pm \sqrt{y^2 + 4 + 4y^2}}{2}$

$$= \dfrac{y \pm \sqrt{5y^2 + 4}}{2}$$

(e) $x^2 = \dfrac{4 \pm \sqrt{16 - 8}}{2} = \dfrac{4 \pm \sqrt{8}}{2} = \dfrac{4 \pm 2\sqrt{2}}{2} = 2 \pm \sqrt{2}$

0.6 Synthetic Division

Using synthetic division to factor as indicated:

(a) $x^3 - 4x^2 + 2x + 1 = (x - 1)(\quad)$

(b) $2x^3 + 5x + 7 = (x + 1)(\quad)$

(c) $x^4 - 3x^3 + x^2 + x + 2 = (x - 2)(\quad)$

(d) $4x^4 + 3x^2 - 1 = (2x - 1)(\quad)$

Solution:

(a) $x^3 - 4x^2 + 2x + 1$

$$
\begin{array}{r|rrrr}
1 & 1 & -4 & 2 & 1 \\
  &   & 1 & -3 & -1 \\
\hline
  & 1 & -3 & -1 & 0
\end{array}
$$

$x^3 - 4x^2 + 2x + 1 = (x - 1)(x^2 - 3x - 1)$

(b) $2x^3 + 5x + 7$

$$
\begin{array}{r|rrrr}
-1 & 2 & 0 & 5 & 7 \\
   &   & -2 & 2 & -7 \\
\hline
   & 2 & -2 & 7 & 0
\end{array}
$$

$2x^3 - 5x + 7 = (x + 1)(2x^2 - 2x + 7)$

(c) $x^4 - 3x^3 + x^2 + x + 2$

$$
\begin{array}{r|rrrrr}
2 & 1 & -3 & 1 & 1 & 2 \\
  &   & 2 & -2 & -2 & -2 \\
\hline
  & 1 & -1 & -1 & -1 & 0
\end{array}
$$

$x^4 - 3x^3 + x^2 + x + 2 = (x - 2)(x^3 - x^2 - x - 1)$

(d) $4x^4 + 3x^2 - 1$

$$
\begin{array}{r|rrrrr}
\frac{1}{2} & 4 & 0 & 3 & 0 & -1 \\
  &   & 2 & 1 & 2 & 1 \\
\hline
  & 4 & 2 & 4 & 2 & 0
\end{array}
$$

$4x^4 + 3x^2 - 1 = \left(x - \dfrac{1}{2}\right)(4x^3 + 2x^2 + 4x + 2)$

$\qquad\qquad\quad = (2x - 1)(2x^3 + x^2 + 2x + 1)$

0.7 Special Products

Factor completely (into linear or irreducible quadratic factors):

(a) $x^3 - 27$

(b) $x^3 - 3x^2 + 3x - 1$

(c) $x^3 + 6x^2 + 12x + 8$

(d) $x^4 - 25$

(e) $x^4 - 8x^3 + 24x^2 - 32x + 16$

Solution:

(a) $x^3 - 27 = (x - 3)(x^2 + 3x + 9)$

(b) $x^3 - 3x^2 + 3x - 1 = (x - 1)^3$

(c) $x^3 + 6x^2 + 12x + 8 = x^3 + 3(2)x^2 + 3(2^2)x + 2^3 = (x + 2)^3$

(d) $x^4 - 25 = (x^2 + 5)(x^2 - 5) = (x^2 + 5)(x + \sqrt{5})(x - \sqrt{5})$

(e) $x^4 - 8x^3 + 24x^2 - 32x + 16 = x^4 - 4(2)x^3 + 6(2^2)x^2 - 4(2^3)x + 2^4 = (x - 2)^4$

0.8 Factoring by Grouping

Factor completely (into linear or irreducible quadratic factors):

(a) $x^3 + 4x^2 - 2x - 8$

(b) $x^3 + 2x^2 + 3x + 6$

(c) $5\cos^2 x - 5\sin^2 x + \sin x + \cos x$

(d) $\cos^2 x + 4\cos x + 4 - \tan^2 x$

Solution:

(a) $x^3 + 4x^2 - 2x - 8 = x^2(x + 4) - 2(x + 4)$

$\qquad\qquad\qquad\quad = (x^2 - 2)(x + 4)$

$\qquad\qquad\qquad\quad = \left(x + \sqrt{2}\right)\left(x - \sqrt{2}\right)(x + 4)$

—CONTINUED—

—CONTINUED—

(b) $x^3 + 2x^2 + 3x + 6 = x^2(x + 2) + 3(x + 2)$

$$= (x^2 + 3)(x + 2)$$

(c) $5\cos^2 x - 5\sin^2 x + \sin x + \cos x = 5(\cos^2 x - \sin^2 x) + (\sin x + \cos x)$

$$= 5(\cos x - \sin x)(\cos x - \sin x) + (\cos x + \sin x)$$

$$= (\cos x + \sin x)[5(\cos x - \sin x) + 1]$$

(d) $\cos^2 x + 4\cos x + 4 - \tan^2 x = (\cos x + 2)^2 - \tan^2 x$

$$= (\cos x + 2 + \tan x)(\cos x + 2 - \tan x)$$

0.9   Simplifying

Rewrite each of the following in simplest form:

(a) $\dfrac{(x - 1)(x + 3) - (x + 1)^2}{x + 1}$

(b) $\dfrac{\sqrt{x^2 + 1} - \dfrac{1}{\sqrt{x^2 + 1}}}{x^2 + 1}.$

(c) $\dfrac{x^2 - 5x + 6}{x^2 - 4x + 4}$

(d) $\dfrac{1}{x + 1} - \dfrac{1}{x - 1} - \dfrac{2}{x^2 - 1}$

(e) $\dfrac{x(-2x)}{2\sqrt{1 - x^2}} + \sqrt{1 - x^2} + \dfrac{1}{\sqrt{1 - x^2}}$

Solution:

(a) $\dfrac{(x - 1)(x + 3) - (x + 1)^2}{x + 1} = \dfrac{(x^2 + 2x - 3) - (x^2 + 2x + 1)}{x + 1} = \dfrac{-4}{x + 1}$

(b) $\dfrac{\sqrt{x^2 + 1} - \dfrac{1}{\sqrt{x^2 + 1}}}{x^2 + 1} = \dfrac{\dfrac{1}{\sqrt{x^2 + 1}}(x^2 + 1 - 1)}{x^2 + 1} = \dfrac{x^2 + 1 - 1}{\sqrt{x^2 + 1}(x^2 + 1)} = \dfrac{x^2}{(x^2 + 1)^{3/2}}$

(c) $\dfrac{x^2 - 5x + 6}{x^2 - 4x + 4} = \dfrac{(x - 2)(x - 3)}{(x - 2)^2} = \dfrac{x - 3}{x - 2}$

(d) $\dfrac{1}{x + 1} - \dfrac{1}{x - 1} - \dfrac{2}{x^2 - 1} = \dfrac{(x - 1) - (x + 1) - 2}{x^2 - 1} = \dfrac{-4}{x^2 - 1}$

(e) $\dfrac{x(-2x)}{2\sqrt{1 - x^2}} + \sqrt{1 - x^2} + \dfrac{1}{\sqrt{1 - x^2}} = \dfrac{-x^2}{\sqrt{1 - x^2}} + \dfrac{1 - x^2}{\sqrt{1 - x^2}} + \dfrac{1}{\sqrt{1 - x^2}} = \dfrac{2 - 2x^2}{\sqrt{1 - x^2}}$

$$= \dfrac{2(1 - x^2)}{\sqrt{1 - x^2}} = 2\sqrt{1 - x^2}$$

0.10   Rationalizing

Remove the sum or difference from the denominator by multiplying the numerator and denominator by the conjugate of the denominator.

(a) $\dfrac{1}{1 - \cos x}$

(b) $\dfrac{x}{1 - \sqrt{x^2 + 1}}$

(c) $\dfrac{2}{x + \sqrt{x^2 + 1}}$

Solution:

(a) $\dfrac{1}{1 - \cos x} = \left(\dfrac{1}{1 - \cos x}\right)\left(\dfrac{1 + \cos x}{1 + \cos x}\right)$

$$= \dfrac{1 + \cos x}{1 - \cos^2 x} = \dfrac{1 + \cos x}{\sin^2 x}$$

—CONTINUED—

—CONTINUED—

(b) $\left(\dfrac{x}{1 - \sqrt{x^2 + 1}}\right)\left(\dfrac{1 + \sqrt{x^2 + 1}}{1 + \sqrt{x^2 + 1}}\right) = \dfrac{x\left(1 + \sqrt{x^2 + 1}\right)}{1 - (x^2 + 1)}$

$\qquad\qquad = \dfrac{x\left(1 + \sqrt{x^2 + 1}\right)}{-x^2} = \dfrac{1 + \sqrt{x^2 + 1}}{-x}$

(c) $\left(\dfrac{2}{x + \sqrt{x^2 + 1}}\right)\left(\dfrac{x - \sqrt{x^2 + 1}}{x - \sqrt{x^2 + 1}}\right) = \dfrac{2\left(x - \sqrt{x^2 + 1}\right)}{x^2 - (x^2 + 1)} = -2\left(x - \sqrt{x^2 + 1}\right)$

0.11   Algebraic Errors to Avoid

Error	Correct form	Comments
$a - (x - b) \neq a - x - b$	$a - (x - b) = a - x + b$	Change all signs when distribution negative through parentheses.
$(a + b)^2 \neq a^2 + b^2$	$(a + b)^2 = a^2 + 2ab + b^2$	Don't forget middle term when squaring binomials.
$\left(\frac{1}{2}a\right)\left(\frac{1}{2}b\right) \neq \frac{1}{2}ab$	$\left(\frac{1}{2}a\right)\left(\frac{1}{2}b\right) = \frac{1}{4}(ab)$	1/2 occurs twice as a factor.
$\dfrac{a}{x + b} \neq \dfrac{a}{x} + \dfrac{a}{b}$	Leave as $\dfrac{a}{x + b}$	Don't add denominators when adding fractions.
$\dfrac{1}{a} + \dfrac{1}{b} \neq \dfrac{1}{a + b}$	$\dfrac{1}{a} + \dfrac{1}{b} = \dfrac{a + b}{ab}$	Use definition for adding fractions.
$\dfrac{\frac{x}{a}}{b} \neq \dfrac{bx}{a}$	$\dfrac{\frac{x}{a}}{b} = \left(\dfrac{x}{a}\right)\left(\dfrac{1}{b}\right) = \dfrac{x}{ab}$	Multiply by reciprocal of the denominator.
$\dfrac{1}{3x} \neq \dfrac{1}{3}x$	$\dfrac{1}{3x} = \dfrac{1}{3} \cdot \dfrac{1}{x}$	Use definition for multiplying fractions.
$1/x + 2 \neq \dfrac{1}{x + 2}$	$1/x + 2 = \dfrac{1}{x} + 2$	Be careful when using a slash to denote division.
$(x^2)^3 \neq x^5$	$(x^2)^3 = x^{2 \cdot 3} = x^6$	Multiply exponents when an exponential form is raised to a power.
$2x^3 \neq (2x)^3$	$2x^3 = 2(x^3)$	Exponents have priority over coefficients.
$\dfrac{1}{x^2 + x^3} \neq x^{-2} + x^{-3}$	Leave as $\dfrac{1}{x^2 + x^3}$	Don't shift term-by-term from denominator to numerator.
$\sqrt{5x} \neq 5\sqrt{x}$	$\sqrt{5x} = \sqrt{5}\sqrt{x}$	Radicals apply to every factor inside radical.
$\sqrt{x^2 + a^2} \neq x + a$	Leave as $\sqrt{x^2 + a^2}$	Don't apply radicals term-by-term.
$\dfrac{a + bx}{a} \neq 1 + bx$	$\dfrac{a + bx}{a} = 1 + \dfrac{b}{a}x$	Cancel common factor, *not* common terms.
$\dfrac{a + ax}{a} \neq a + x$	$\dfrac{a + ax}{a} = 1 + x$	Factor *before* canceling.

CHAPTER P
Preparation for Calculus

CHAPTER P
Preparation for Calculus

Section P.1   Graphs and Models
Solutions to Selected Odd-Numbered Exercises

5. To find the x-intercepts, let $y = 0$. Then

$$x^2 + x - 2 = 0$$

and by factoring (or by the quadratic formula),

$$(x + 2)(x - 1) = 0.$$

Therefore, $y = 0$ when $x = -2$ or $x = 1$ and the x-intercepts are $(-2, 0)$ and $(1, 0)$. To find the y-intercepts, let $x = 0$. Then $y = 0^2 + 0 - 2 = -2$, and the y-intercept is $(0, -2)$.

9. To find the x-intercepts you let $y = 0$. Then

$$x^2(0) - x^2 + 4(0) = -x^2 = 0,$$

which implies that $x = 0$. Letting $y = 0$ yields the same result and the only intercept is $(0, 0)$.

13. There is *no symmetry* about the y-axis since replacing x with $-x$ in the equation yields

$$y^2 = (-x)^3 - 4(-x) = -x^3 + 4x$$

which is *not* equivalent to the original equation.

There is *symmetry* about the x-axis since replacing y with $-y$ in the equation yields

$$(-y)^2 = x^3 - 4x \quad \text{or} \quad y^2 = x^3 - 4x$$

which *is* equivalent to the original equation.

There is *no symmetry* about the origin since replacing x with $-x$ and y with $-y$ in the equation yields

$$(-y)^2 = (-x)^3 - 4(-x) \quad \text{or} \quad y^2 = -x^3 + 4x$$

which is *not* equivalent to the original equation.

17. There is *no symmetry* about the y-axis since replacing x with $-x$ in the equation yields

$$y = \frac{-x}{(-x)^2 + 1} = \frac{-x}{x^2 + 1}$$

which is *not* equivalent to the original equation.

There is *no symmetry* about the x-axis since replacing y with $-y$ in the equation yields

$$-y = \frac{x}{x^2 + 1}$$

which is *not* equivalent to the original equation.

There is *symmetry* with respect to the origin since replacing x with $-x$ and y with $-y$ in the equation yields

$$-y = \frac{-x}{(-x)^2 + 1} \quad \text{or} \quad y = \frac{x}{x^2 + 1}$$

which *is* equivalent to the original equation.

23. To find the x-intercepts, let $y = 0$. Then

$$0 = 1 - x^2$$

and by factoring

$$0 = (1 + x)(1 - x).$$

Therefore, the x-intercepts are $(-1, 0)$ and $(1, 0)$. To find any y-intercept, let $x = 0$. Then $y = 1$ and the y-intercept is $(0, 1)$.

There is *symmetry* with respect to the y-axis since replacing x with $-x$ in the equation yields

$$y = 1 - (-x)^2 \quad \text{or} \quad y = 1 - x^2$$

which is equivalent to the original equation. Some solution points are given in the table.

x	0	± 1	± 2
y	1	0	-3

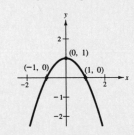

8

29. The graph has no intercepts since the equation has no solution if either $x = 0$ or $y = 0$. There is *symmetry* with respect to the origin since replacing x with $-x$ and y with $-y$ in the equation yields

$$-y = \frac{1}{-x}$$

which is equivalent to the original equation. Some solution points are given in the table.

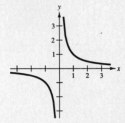

x	$\frac{1}{5}$	$\frac{1}{2}$	1	2	3	4	5
y	5	2	1	$\frac{1}{2}$	$\frac{1}{3}$	$\frac{1}{4}$	$\frac{1}{5}$

From the table we see that y increases without bound as x approaches 0, and y approaches 0 as x increases without bound. Using this information and symmetry, we obtain the graph in the figure.

39. The graph of the equation passes through the points $(-2, 0)$, $(4, 0)$, and $(6, 0)$. If the equation is a polynomial equation, then factors of the equation are $(x + 2)$, $(x - 4)$, and $(x - 6)$. Therefore, one possible equation is

$$y = (x + 2)(x - 4)(x - 6) = x^3 - 8x^2 + 4x + 48.$$

The graph of the equation is shown in the figure.

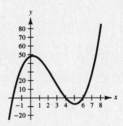

45. To solve the two equations

$$x + y = 7$$
$$3x - 2y = 11$$

simultaneously, multiply the first equation by 2 and add. Thus,

$$
\begin{aligned}
2x + 2y &= 14 \\
(+)\ 3x - 2y &= 11 \\
\hline
5x\quad\ &= 25 \\
x &= 5.
\end{aligned}
$$

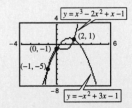

Substituting $x = 5$ into the first equation, we have

$$5 + y = 7 \quad \text{or} \quad y = 2.$$

Thus, the point of intersection is $(5, 2)$.

47. To solve the two equations

$$x^2 + y^2 = 5$$
$$x - y = 1$$

simultaneously, solve the second equation for x and obtain $x = y + 1$.

Substituting into the first equation yields

$$
\begin{aligned}
(y + 1)^2 + y^2 &= 5 \\
(y^2 + 2y + 1) + y^2 &= 5 \\
2y^2 + 2y - 4 &= 0 \\
2(y + 2)(y - 1) &= 0
\end{aligned}
$$

which implies that $y = -2$ or $y = 1$. Therefore,

$$x = (-2) + 1 = -1 \quad \text{or} \quad x = 1 + 1 = 2$$

and the points of intersection are $(-1, -2)$ and $(2, 1)$.

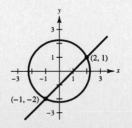

51. The graphs are shown in the figure. Equating the two expressions for y yields

$$x^3 - 2x^2 + x - 1 = -x^2 + 3x - 1$$
$$x^3 - 2x^2 + x^2 + x - 3x - 1 + 1 = 0$$
$$x^3 - x^2 - 2x = 0$$
$$x(x - 2)(x + 1) = 0.$$

Thus, $x = 0$, 2, or -1, and substituting these values into the second (or first) equation, we have

$$y = -(0)^2 + 3(0) - 1 = -1$$
$$y = -(2)^2 + 3(2) - 1 = 1$$
$$y = -(-1)^2 + 3(-1) - 1 = -5.$$

Therefore, the points of intersection are $(0, -1)$, $(2, 1)$, and $(-1, -5)$.

57. The graph of

$$y = \frac{10{,}770}{x^2} - 0.37$$

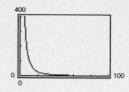

for $5 \le x \le 100$ is shown in the figure. If x is replaced by $2x$, we have

$$y = \frac{10{,}770}{(2x)^2} - 0.37 = \frac{10{,}770}{4x^2} - 0.37.$$

Therefore, the resistance is changed by approximately a factor of $1/4$.

Section P.2   Linear Models and Rates of Change

9. Let $(3, -4) = (x_1, y_1)$ and $(5, 2) = (x_2, y_2)$. The slope of the line passing through (x_1, y_1) and (x_2, y_2) is

$$m = \frac{y_2 - y_1}{x_2 - x_1} = \frac{2 - (-4)}{5 - 3} = \frac{6}{2} = 3.$$

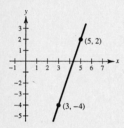

15. Since the slope is -3, it follows that

$$\frac{\Delta y}{\Delta x} = \frac{-3}{1}.$$

Therefore, for each positive one-unit increase in x, y decreases three units. Starting at the point $(1, 7)$, we have

$$(1 + 1, 7 - 3) = (2, 4),$$

$$(2 + 1, 4 - 3) = (3, 1), \quad \text{and}$$

$$(3 + 1, 1 - 3) = (4, -2).$$

Since the slope is -3, it also follows that

$$\frac{\Delta y}{\Delta x} = \frac{3}{-1}.$$

Therefore, for each one-unit decrease in x, y increases three units. Starting at the point $(1, 7)$, we have

$$(1 - 1, 7 + 3) = (0, 10),$$

$$(0 - 1, 10 + 3) = (-1, 13), \quad \text{and}$$

$$(-1 - 1, 13 + 3) = (-2, 16).$$

25. Let $(2, 1) = (x_1, y_1)$ and $(0, -3) = (x_2, y_2)$. The slope of the line passing through (x_1, y_1) and (x_2, y_2) is

$$m = \frac{y_2 - y_1}{x_2 - x_1} = \frac{-3 - 1}{0 - 2} = 2.$$

Using the point-slope form of the equation of a line, we have

$$y - y_1 = m(x - x_1)$$

$$y - 1 = 2(x - 2)$$

$$y - 1 = 2x - 4$$

$$0 = 2x - y - 3. \quad \text{General form}$$

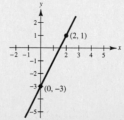

31. Using the slope-intercept form of the equation of a line, we have

$$y = mx + b$$

$$y = \tfrac{3}{4}x + 3$$

$$4y = 3x + 12$$

$$0 = 3x - 4y + 12.$$

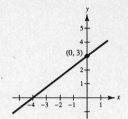

43. The line given by $4x - 2y = 3$ has a slope of 2 since

$$4x - 2y = 3$$

$$-2y = -4x + 3$$

$$y = 2x - \tfrac{3}{2} = mx + b.$$

(a) The line through $(2, 1)$ parallel to $4x - 2y = 3$ must also have a slope of 2. Thus its equation must be

$$y - y_1 = m(x - x_1)$$

$$y - 1 = 2(x - 2)$$

$$-2x + y + 3 = 0$$

$$2x - y - 3 = 0.$$

(b) The line through $(2, 1)$ perpendicular to $4x - 2y = 3$ must have a slope of $m = -\tfrac{1}{2}$. Thus its equation must be

$$y - y_1 = m(x - x_1)$$

$$y - 1 = -\tfrac{1}{2}(x - 2)$$

$$2y - 2 = -x + 2$$

$$x + 2y - 4 = 0.$$

51. Writing the equation in slope-intercept form we have

$$y = 2x - 3.$$

Therefore, the slope is $m = 2$ and the y-intercept is -3. The line passes through the points $(0, 3)$ and $(4, 5)$. Plotting the two points and drawing a line through them yields the line shown in the figure.

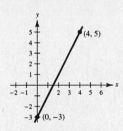

63. Let $(-2, 1) = (x_1, y_1)$, $(-1, 0) = (x_2, y_2)$, and $(2, -2) = (x, y)$. The point (x, y) lies on the line passing through (x_1, y_1) and (x_2, y_2) if and only if

$$\frac{y - y_1}{x - x_1} = m = \frac{y - y_2}{x - x_2}.$$

Since

$$\frac{-2 - 1}{2 - (-2)} = -\frac{3}{4} \neq -\frac{2}{3} = \frac{-2 - 0}{2 - (-1)},$$

the three points are *not* collinear.

73. (a) Two solution points to the linear equation are $(x_1 \, p_1) = (50, 580)$ and $(x_2, p_2) = (47, 625)$. Therefore, the slope is

$$m = \frac{625 - 580}{47 - 50} = -15,$$

and the equation of the line is

$$p - 580 = -15(x - 50)$$

$$p = -15x + 1330 \text{ or } x = \frac{1}{15}(1330 - p).$$

(b) The graph of the line is shown in the figure.

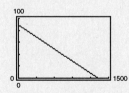

When $p = \$655$, $x = \dfrac{1}{15}(1330 - 655) = 45$ units.

(c) When $p = \$595$, $x = \dfrac{1}{15}(1330 - 595) = 49$ units.

79. A point on the line $x + y = 1$ is $(2, -1)$. The distance between the given parallel lines is equal to the distance from $(2, -1)$ to the line $x + y = 5$. Letting $(2, -1) = (x_1, y_1)$ and $x + y - 5 = Ax + By + C = 0$, we have

$$d = \frac{|Ax_1 + By_1 + C|}{\sqrt{A^2 + B^2}} = \frac{|1(2) + 1(-1) - 5|}{\sqrt{1^2 + 1^2}} = \frac{|-4|}{\sqrt{2}} = \frac{4}{\sqrt{2}} = 2\sqrt{2}.$$

Section P.3   Functions and Their Graphs

3. $f(x) = \begin{cases} 2x + 1, & x < 0 \\ 2x + 2, & x \geq 0 \end{cases}$

    (a) $f(-1) = 2(-1) + 1 = -1$

    (b) $f(0) = 2(0) + 2 = 2$

    (c) $f(2) = 2(2) + 2 = 6$

    (d) Since $t^2 + 1 > 0$, we have

$$f(t^2 + 1) = 2(t^2 + 1) + 2 = 2t^2 + 4.$$

7. $f(x) = x^3$

$$\frac{f(x + \Delta x) - f(x)}{\Delta x} = \frac{(x + \Delta x)^3 - x^3}{\Delta x}$$

$$= \frac{x^3 + 3x^2\Delta x + 3x(\Delta x)^2 + (\Delta x)^3 - x^3}{\Delta x}$$

$$= \frac{\Delta x[3x^2 + 3x\Delta x + (\Delta x)^2]}{\Delta x}$$

$$= 3x^2 + 3x\Delta x + (\Delta x)^2$$

15. $f(x) = \sqrt{9 - x^2}$

Since $9 - x^2$ must be nonnegative $(9 - x^2 \geq 0)$, the domain is $[-3, 3]$. The range is $[0, 3]$. There is symmetry with respect to the y-axis since

$$y = \sqrt{9 - (-x)^2} = \sqrt{9 - x^2}$$

is equivalent to the original equation. Squaring both members of the equation, you have

$$y^2 = 9 - x^2 \quad \text{or} \quad x^2 + y^2 = 3^2.$$

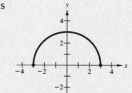

This is the standard form of an equation of a circle with center $(0, 0)$ and radius 3. Therefore, the graph of $f(x) = \sqrt{9 - x^2}$ is a semicircle in the first and second quadrants with center $(0, 0)$ and radius 3.

25. Solving the equation for y, yields

$$y^2 = x^2 - 1$$

$$y = \pm\sqrt{x^2 - 1}, \quad |x| \geq 1.$$

For each value of the independent variable x such that $|x| > 1$, there corresponds two values of the dependent variable y. Therefore, y is not a function of x.

37. The graph of $f(x) = \sqrt{x}$ is shown in the figure.

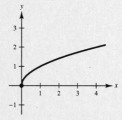

    (a) The graph of $y = \sqrt{x} + 2$ is a vertical shift of f 2 units upward.

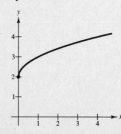

    (b) The graph of $y = -\sqrt{x}$ is a reflection of f in the x-axis.

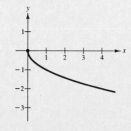

    (c) The graph of $y = \sqrt{x - 2}$ is a horizontal shift of f 2 units to the right.

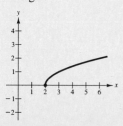

41. Given the functions $f(x) = x^2$ and $g(x) = \sqrt{x}$, we have

$$(f \circ g)(x) = f[g(x)] = f(\sqrt{x}) = (\sqrt{x})^2 = x \quad \text{and} \quad (g \circ f)(x) = g[f(x)] = g(x^2) = \sqrt{x^2} = |x|.$$

The domain of $(f \circ g)$ is $[0, \infty)$ and the domain of $(g \circ f)$ is $(-\infty, \infty)$. If $x \geq 0$, then $(f \circ g) = (g \circ f)$.

49. Since $\cos(-x) = \cos x$, we have

$$f(-x) = (-x)\cos(-x) = -x \cos x = -f(x).$$

Therefore, f is an odd function. The graph is shown in the figure.

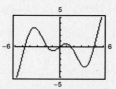

55. Let $F(x) = f(x)g(x)$ where f and g are even functions. Thus, $f(-x) = f(x)$ and $g(-x) = g(x)$. Therefore,

$$F(-x) = f(-x)g(-x) = f(x)g(x) = F(x)$$

which implies that F is an even function.

Let $F(x) = f(x)g(x)$ where f and g are odd functions. Therefore, $f(-x) = -f(x)$ and $g(-x) = -g(x)$. Hence,

$$F(-x) = f(-x)g(-x) = [-f(x)][-g(x)] = f(x)g(x) = F(x)$$

which implies that F is an even function.

61. (a) Using the table feature of a graphing utility yields the following.

Height, x	Length and width	Volume, V
1	$24 - 2(1)$	$1[24 - 2(1)]^2 = 484$
2	$24 - 2(2)$	$2[24 - 2(2)]^2 = 800$
3	$24 - 2(3)$	$3[24 - 2(3)]^2 = 972$
4	$24 - 2(4)$	$4[24 - 2(4)]^2 = 1024$
5	$24 - 2(5)$	$5[24 - 2(5)]^2 = 980$
6	$24 - 2(6)$	$6[24 - 2(6)]^2 = 864$

The estimate of the maximum volume is 1024 cubic centimeters.

(c) Using the form of the expressions in the table of part (a) and the variables assigned to the dimensions of the box shown in the text, the volume is given by

$$V = x(24 - 2x)^2 = x[2(12 - x)]^2 = 4x(12 - x)^2.$$

The size of the material for constructing the box constrains the values of x to the domain $(0, 12)$.

(b) For each value of x there corresponds exactly one value of V. Therefore, V is a function of x. The graph of the points (x, V) are shown in the figure.

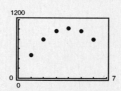

(d) The graph of the function for the volume is given in the figure. The volume is maximum when $x \approx 4$. Therefore, the approximate dimensions of the box are $4 \times 16 \times 16$ centimeters.

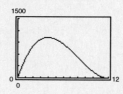

Section P.4   Fitting Models to Data

9. (a) Begin by entering the data in the statistics lists of your graphing utility. Access the statistic calculation menu and select linear regression ($y = ax + b$). Fitting the linear model to the data yields

$$y = 2.62x + 2.66.$$

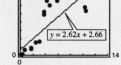

(b) Enter the regression model in the $\boxed{Y=}$ screen, and turn on the statistical plot capabilities. The plot of the data and the graph of the model are shown in the figure.

(c) The slope of the model indicates that the per capita gross national product increases an average of $2620 for each 1000 kilograms of coal equivalent per capita energy consumption. Denmark, Finland, France, Italy, and Japan appear to not follow the linear pattern. Their per capita gross national product appears above average for the amount of energy usage.

17. (a) y is a function of t since there corresponds one and only one value of y for each value of t.

(b) The average of the y-values at the maximum displacements from equilibrium is

$$\frac{2.35 + 1.65}{2} = 2.$$

Therefore, the weight is at equilibrium when $y = 2$. The amplitude is the maximum displacement from equilibrium which is 0.35. The period p is twice the distance between the t-values for the consecutive maximum displacements from equilibrium shown in the figure. Thus,

$$p = 2(0.375 - 0.125) = 0.5.$$

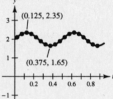

(c) Using the model $y = a\sin(bt) + c$, we have $a = 0.35$, $c = 2$, and $0.5 = 2\pi/b$ or $b = 4\pi$. Therefore,

$$y = 0.35\sin(4\pi t) + 2.$$

(d) The graph of the model is shown in the figure.

Review Exercises for Chapter P

3. To find the x-intercepts, let $y = 0$. Then

$$0 = \frac{x - 1}{x - 2}.$$

Multiply both sides of the equation by $(x - 2)$ yields $x - 1 = 0$ or $x = 1$. Therefore, $y = 0$ when $x = 1$ and the x-intercept is $(1, 0)$. To find the y-intercepts, let $x = 0$. Then

$$y = \frac{0 - 1}{0 - 2} = \frac{1}{2},$$

and the y-intercept is $\left(0, \frac{1}{2}\right)$.

13. Since $5 - x \geq 0$ when $x \leq 5$, the domain of the function $y\sqrt{5 - x}$ is $(-\infty, 5]$. To find the x-intercepts, let $y = 0$. Then

$$0 = \sqrt{5 - x}$$

$$0 = 5 - x \quad \text{or} \quad x = 5.$$

Therefore, the x-intercept is $(5, 0)$. To find the y-intercepts, let $x = 0$. Then $y = \sqrt{5 - 0} = \sqrt{5}$. Therefore, the y-intercept is $\left(0, \sqrt{5}\right)$. Some solution points are:

x	-4	1	4	5
y	3	2	1	0

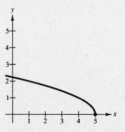

19. Since $x = -2$ and $x = 2$ are intercepts of the graph of the equation, the factors of a polynomial equation with the given intercepts are $(x + 2)$ and $(x - 2)$. The polynomial equation is $y = (x + 2)(x - 2) = x^2 - 4$. However, the graph of this equation is symmetric to the y-axis (an even function). An even function divided by and odd function (symmetric to the origin) is an odd function. Dividing by the odd function $y = x$ yields an equation satisfying the given requirements.

$$y = \frac{x^2 - 4}{x}$$ (There are many correct answers.)

25. (a) Using the point-slope form for the equation of a line we have

$$y - y_1 = m(x - x_1)$$

$$y - 4 = \frac{7}{16}[x - (-2)]$$

$$y = \frac{7}{16}x + \frac{7}{8} + 4$$

$$y = \frac{7}{16}x + \frac{39}{8}.$$

(c) The slope of the line is given by

$$m = \frac{4 - 0}{-2 - 0} = -2.$$

Using the slope-intercept form for the equation of a line we have

$$y = mx + b$$

$$y = -2x + 0 \quad \text{or} \quad y = -2x.$$

(b) Rewriting the equation $5x - 3y = 3$ in slope-intercept form we have $y = \frac{5}{3}x - 1$. The slope of the given line is $\frac{5}{3}$ and the required equation is

$$y - y_1 = m(x - x_1)$$

$$y - 4 = \frac{5}{3}[x - (-2)]$$

$$y = \frac{5}{3}x + \frac{10}{3} + 4$$

$$y = \frac{5}{3}x + \frac{22}{3}.$$

(d) The standard form of the equation of a vertical line is $x = a$ where a is the x-intercept. Therefore, the equation of the line is

$$x = -2.$$

33. $f(x) = \dfrac{1}{x}$

(a) $f(0) = \dfrac{1}{0}$. Division by 0 is undefined.

(b) $\dfrac{f(x + \Delta x) - f(1)}{\Delta x} = \dfrac{\left(\dfrac{1}{1 + \Delta x} - 1\right)}{\Delta x}$

$$= \frac{\left(\dfrac{1}{1 + \Delta x} - 1\right)(1 + \Delta x)}{\Delta x(1 + \Delta x)}$$

$$= \frac{1 - (1 + \Delta x)}{\Delta x(1 + \Delta x)}$$

$$= \frac{-\Delta x}{\Delta x(1 + \Delta x)} = \frac{-1}{1 + \Delta x}, \Delta x \neq -1, 0$$

39. (a) If x and y are the lengths of the shortest and longest sides of the rectangle (see figure), respectively, then

$$2x + 2y = 24 \quad \text{or} \quad y = 12 - x.$$

Therefore, the area A as a function of x is

$$A = xy = x(12 - x).$$

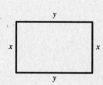

(b) Because of the constraint that the wire has length 24 inches, the domain is $(0, 12)$. The graph of the area function is shown in the figure.

(c) Using the graph from part (b), the maximum area is approximately 36 square inches which occurs when the rectangle is a square with sides 6 inches.

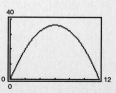

CHAPTER 1
Limits and Their Properties

CHAPTER 1
Limits and Their Properties

Section 1.1   A Preview of Calculus
Solutions to Selected Odd-Numbered Exercises

5. The shaded region is a triangle with base 5 units and height 3 units. From the precalculus formula for the area of a triangle we have

$$A = \tfrac{1}{2}bh = \tfrac{1}{2}(5)(3) = \tfrac{15}{2}.$$

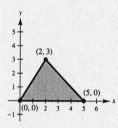

11. (a) Using the Distance Formula to find the straight line distance between the points $(1, 5)$ and $(5, 1)$, we have the following approximation of the length L of the curve.

$$L \approx \sqrt{(5 - 1)^2 + (1 - 5)^2} = \sqrt{32} = 4\sqrt{2} \approx 5.66$$

(b) The intermediate points on the graph of the function are $\left(2, \tfrac{5}{2}\right)$, $\left(3, \tfrac{5}{3}\right)$, and $\left(4, \tfrac{5}{4}\right)$. Therefore, an approximation of the length L of the curve is given by the sum of the lengths of four line segments.

$$L \approx \sqrt{(2 - 1)^2 + \left(\tfrac{5}{2} - 5\right)^2} + \sqrt{(3 - 2)^2 + \left(\tfrac{5}{3} - \tfrac{5}{2}\right)^2} + \sqrt{(4 - 3)^2 + \left(\tfrac{5}{4} - \tfrac{5}{3}\right)^2} + \sqrt{(5 - 4)^2 + \left(1 - \tfrac{5}{4}\right)^2} \approx 6.11$$

(c) To improve the accuracy of the approximation increase the number of line segments. In Section 6.4, we will use calculus to find that the length of the curve (accurate to two decimal places) is 6.14.

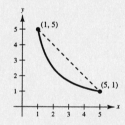
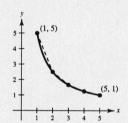

Section 1.2   Finding Limits Graphically and Numerically

7. The table lists values of $f(x) = \dfrac{\dfrac{1}{x + 1} - \dfrac{1}{4}}{x - 3}$ at several x-values near 3.

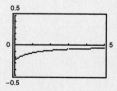

x	2.9	2.99	2.999	3.001	3.01	3.1
$f(x)$	-0.0641	-0.0627	-0.0625	-0.0625	-0.0623	-0.0610

As x approaches 3 from the left and from the right $f(x)$ approaches -0.0625. Therefore, we estimate the limit to be $-\tfrac{1}{16}$.

17. From the graph we observe that as x approaches $\pi/2$ from the left, $f(x)$ increases without bound, and as x approaches $\pi/2$ from the right, $f(x)$ decreases without bound. Since $f(x)$ is not approaching a real number L as x approaches $\pi/2$, we say that the limit does *not* exist.

21. We use the definition of limit to verify that

$$\lim_{x \to 2} (3x + 2) = 8.$$

We are required to show that there exists a δ such that $|f(x) - L| < 0.01$ whenever $0 < |x - 2| < \delta$.

$$|f(x) - L| < 0.01$$

$$|(3x + 2) - 8| < 0.01$$

$$|3x - 6| < 0.01$$

$$3|x - 2| < 0.01$$

$$0 < |x - 2| < \frac{0.01}{3}$$

Therefore, $\delta = 0.01/3$.

31. We use the definition of limit to verify that

$$\lim_{x \to 0} \sqrt[3]{x} = 0.$$

Given $\epsilon > 0$,

$$\left|\sqrt[3]{x} - 0\right| < \epsilon$$

$$\left(\left|\sqrt[3]{x}\right|\right)^3 < \epsilon^3$$

$$|x| < \epsilon^3 = \delta$$

39. The graph of

$$f(x) = \frac{x - 9}{\sqrt{x} - 3}$$

is shown in the figure. From the graph it appears as though the limit exists, and $\lim_{x \to 9} f(x) = 6$.

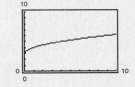

The domain of the function is $[0, 9) \cup (9, \infty)$. The graph generated by the graphing utility does not show that the function is indeterminate at $x = 9$. Therefore, it is important to analyze a function analytically as well as graphically.

49. If $\lim_{x \to c} f(x) = L_1$ and $\lim_{x \to c} f(x) = L_2$, then for every $\epsilon > 0$, there exists $\delta_1 > 0$ and $\delta_2 > 0$ such that

$$|x - c| < \delta_1 \implies |f(x) - L_1| < \epsilon$$

and

$$|x - c| < \delta_2 \implies |f(x) - L_2| < \epsilon.$$

Let δ equal the smaller of δ_1 and δ_2. Then for $|x - c| < \delta$, we have

$$|L_1 - L_2| = |L_1 - f(x) + f(x) - L_2|$$

$$\leq |L_1 - f(x)| + |f(x) - L_2| < \epsilon + \epsilon.$$

Therefore, $|L_1 - L_2| < 2\epsilon$. Since $\epsilon > 0$ is arbitrary, it follows that $L_1 = L_2$.

Section 1.3   Evaluating Limits Analytically

3. The graph of the function

$$f(x) = x \cos x$$

is shown in the figure. From the figure we estimate the limits.

(a) $\lim_{x \to 0} f(x) \approx 0$ (b) $\lim_{x \to \pi/3} f(x) \approx 0.52$

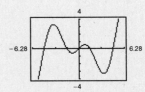

11. $\lim_{x \to 3} \sqrt{x + 1} = \sqrt{3 + 1}$

$$= \sqrt{4} = 2$$

19. $\lim_{x \to \pi/2} \sin x = \sin \dfrac{\pi}{2} = 1$

27. $\lim_{x \to 3} \tan \dfrac{\pi x}{4} = \tan \dfrac{3\pi}{4} = -1$

31. (a) $\lim_{x \to c} [f(x)]^3 = \left[\lim_{x \to c} f(x)\right]^3 = 4^3 = 64$

(b) $\lim_{x \to c} \sqrt{f(x)} = \sqrt{\lim_{x \to c} f(x)} = \sqrt{4} = 2$

(c) $\lim_{x \to c} [3 f(x)] = 3 \lim_{x \to c} f(x) = 3(4) = 12$

(d) $\lim_{x \to c} [f(x)]^{3/2} = \left[\lim_{x \to c} f(x)\right]^{3/2} = 4^{3/2} = 8$

37. Simplification of the given function yields

$$\frac{x^2 - 1}{x + 1} = \frac{(x + 1)(x - 1)}{x + 1} = x - 1.$$

If we let $f(x) = (x^2 - 1)/(x + 1)$ and $g(x) = x - 1$, then $f = g$ for all $x \neq -1$. Therefore,

$$\lim_{x \to -1} f(x) = \lim_{x \to -1} g(x) = -2.$$

The graph of f is given in the figure. Note that the graph generated by the graphing utility does not show that f is indeterminate at $x = -1$. Therefore, it is important to analyze a function analytically as well as graphically.

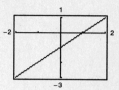

47. $\displaystyle\lim_{x \to 0} \frac{\dfrac{1}{2 + x} - \dfrac{1}{2}}{x} = \lim_{x \to 0} \frac{\dfrac{2 - (2 + x)}{2(2 + x)}}{x}$

$$= \lim_{x \to 0} \frac{-x}{x(2)(2 + x)}$$

$$= \lim_{x \to 0} \frac{-1}{2(2 + x)} = -\frac{1}{4}$$

51. $\displaystyle\lim_{\Delta x \to 0} \frac{(x + \Delta x)^2 - 2(x + \Delta x) + 1 - (x^2 - 2x + 1)}{\Delta x} = \lim_{\Delta x \to 0} \frac{x^2 + 2x\Delta x + (\Delta x)^2 - 2x - 2\Delta x + 1 - x^2 + 2x - 1}{\Delta x}$

$$= \lim_{\Delta x \to 0} \frac{2x\Delta x + (\Delta x)^2 - 2\Delta x}{\Delta x}$$

$$= \lim_{\Delta x \to 0} \frac{\Delta x(2x + \Delta x - 2)}{\Delta x}$$

$$= \lim_{\Delta x \to 0} (2x + \Delta x - 2) = 2x - 2$$

53. The graph of

$$f(x) = \frac{\sqrt{x + 2} - \sqrt{2}}{x}$$

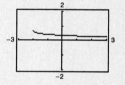

is shown in the figure. From the graph we estimate the limit to be 0.35. The table lists values of $f(x)$ at several x-values near 0.

x	-0.1	-0.01	-0.001	0.001	0.01	0.1
$f(x)$	0.3581	0.3540	0.3536	0.3535	0.3531	0.3492

As x approaches 0 from the left and from the right, $f(x)$ approached 0.35. Finding the limit analytically, we have the following.

$$\lim_{x \to 0} \frac{\sqrt{x + 2} - \sqrt{2}}{x} = \lim_{x \to 0} \left(\frac{\sqrt{x + 2} - \sqrt{2}}{x} \right)\left(\frac{\sqrt{x + 2} + \sqrt{2}}{\sqrt{x + 2} + \sqrt{2}} \right)$$

$$= \lim_{x \to 0} \frac{x + 2 - 2}{x(\sqrt{x + 2} + \sqrt{2})}$$

$$= \lim_{x \to 0} \frac{1}{\sqrt{x + 2} + \sqrt{2}} = \frac{1}{2\sqrt{2}} = \frac{\sqrt{2}}{4}$$

57. $\displaystyle\lim_{x \to \pi/2} \frac{\sin x}{5x} = \lim_{x \to \pi/2} \left[\left(\frac{1}{5} \right)\left(\frac{\sin x}{x} \right) \right]$

$$= \left(\frac{1}{5} \right) 1 = \frac{1}{5}$$

65. $\displaystyle\lim_{x \to \pi/2} \frac{\cos x}{\cot x} = \lim_{x \to \pi/2} \frac{\cos x}{\cos x/\sin x}$

$$= \lim_{x \to \pi/2} \sin x = 1$$

75. $\displaystyle\lim_{h\to0}\frac{f(x+h)-f(x)}{h}=\lim_{h\to0}\frac{\dfrac{4}{x+h}-\dfrac{4}{x}}{h}$

$$=\lim_{h\to0}\left[\frac{\dfrac{4}{x+h}-\dfrac{4}{x}}{h}\cdot\frac{x(x+h)}{x(x+h)}\right]$$

$$=\lim_{h\to0}\frac{4x-4(x+h)}{hx(x+h)}$$

$$=\lim_{h\to0}\frac{-4\cancel{h}}{\cancel{h}x(x+h)}$$

$$=\lim_{h\to0}\frac{-4}{x(x+h)}=-\frac{4}{x^2}$$

83. The graphs of $y=x$, $y=-x$, and $y=x\sin(1/x)$ are shown in the figure. What follows is a direct application of the Squeeze Theorem to find the required limit. Since

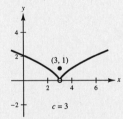

$$\lim_{x\to0}\left(-|x|\right)=0\quad\text{and}\quad\lim_{x\to0}|x|=0\quad\text{and}\quad-|x|\le x\sin\frac{1}{x}\le|x|,$$

we have

$$0=\lim_{x\to0}\left(-|x|\right)\le\lim_{x\to0}x\sin\frac{1}{x}\le\lim_{x\to0}|x|=0.$$

Therefore, by the Squeeze Theorem $\displaystyle\lim_{x\to0}x\sin\frac{1}{x}=0.$

89. $\displaystyle\lim_{t\to3}\frac{s(3)-s(t)}{3-t}=\lim_{t\to3}\frac{105.9-(-4.9t^2+150)}{3-t}$

$$=\lim_{t\to3}\frac{4.9t^2-44.1}{3-t}$$

$$=\lim_{t\to3}\frac{-4.9(3+1)(3-t)}{3-t}$$

$$=\lim_{x\to3}\left[-4.9(3+t)\right]=-29.4\text{ meters per second}$$

Section 1.4 Continuity and One-Sided Limits

7. (a) $\displaystyle\lim_{x\to c^+}f(x)=0$

(b) $\displaystyle\lim_{x\to c^-}f(x)=0$

(c) Since

$$\lim_{x\to c^+}f(x)=0=\lim_{x\to c^-}f(x),$$

$\displaystyle\lim_{x\to c}f(x)=0.$ [Note that $f(c)\ne\lim_{x\to c}f(x)$.]

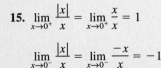

15. $\displaystyle\lim_{x\to0^+}\frac{|x|}{x}=\lim_{x\to0^+}\frac{x}{x}=1$

$\displaystyle\lim_{x\to0^-}\frac{|x|}{x}=\lim_{x\to0^-}\frac{-x}{x}=-1$

Since the limit from the left is *not equal* to the limit from the right, the limit does *not* exist.

19. $\displaystyle\lim_{x\to3^-}f(x)=\lim_{x\to3^-}\frac{x+2}{5}=\frac{5}{2}$

$\displaystyle\lim_{x\to3^+}f(x)=\lim_{x\to3^+}\frac{12-2x}{3}=2$

Since the limit from the left is *not equal* to the limit from the right, the limit does *not* exist.

21. $\lim\limits_{x\to 1^-} f(x) = \lim\limits_{x\to 1^-} (x^3 + 1) = 2$

$\lim\limits_{x\to 1^+} f(x) = \lim\limits_{x\to 1^+} (x + 1) = 2$

Since the limit from the left is *equal* to the limit from the right, we have $\lim\limits_{x\to 1} f(x) = 2$.

29. As shown in the figure, the limit as x approaches a given integer from the left and from the right are not equal because of the greatest integer function. For any integer n, we have the following limits.

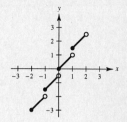

$$\lim\limits_{x\to n^+} \left(\tfrac{1}{2}[\![x]\!] + x\right) = \tfrac{1}{2}(n) + n = \tfrac{3}{2}n$$

$$\lim\limits_{x\to n^-} \left(\tfrac{1}{2}[\![x]\!] + x\right) = \tfrac{1}{2}(n - 1) + n = \tfrac{3}{2}n - \tfrac{1}{2}$$

Therefore, f has a nonremovable discontinuity at any integer n.

35. From Theorem 1.11 we know that

$$f(x) = \frac{1}{x - 1}$$

is continuous for all x other than $x = 1$. At $x = 1$ the function is discontinuous and the discontinuity is nonremovable since

$$\lim\limits_{x\to 1^-} \frac{1}{x - 1} = -\infty \quad \text{and} \quad \lim\limits_{x\to 1^+} \frac{1}{x - 1} = \infty.$$

39. Since $x^2 - 3x - 10 = (x - 5)(x + 2)$, $x = 5$, and $x = -2$ are not in the domain of

$$f(x) = \frac{x + 2}{x^2 - 3x - 10}.$$

By Theorem 1.11, f is continuous for all x other than $x = 5$ or $x = -2$. At $x = 5$ the function is discontinuous and the discontinuity is nonremovable since

$$\lim\limits_{x\to 5^-} \frac{x + 2}{(x - 5)(x + 2)} = \lim\limits_{x\to 5^-} \frac{1}{x - 5} = -\infty$$

$$\lim\limits_{x\to 5^+} \frac{x + 2}{(x - 5)(x + 2)} = \lim\limits_{x\to 5^+} \frac{1}{x - 5} = \infty.$$

At $x = -2$ the function is discontinuous but it is removable since

$$\lim\limits_{x\to -2} \frac{x + 2}{x^2 - 3x - 10} = \lim\limits_{x\to -2} \frac{1}{x - 5} = -\frac{1}{7}.$$

45. Since f is linear to the right and left of $x = 2$, it is continuous for all x other than possibly at $x = 2$. At $x = 2$,

$$\lim\limits_{x\to 2^-} f(x) = \lim\limits_{x\to 2^-} \left(\frac{x}{2} + 1\right) = 2$$

$$\lim\limits_{x\to 2^+} f(x) = \lim\limits_{x\to 2^+} (3 - x) = 1$$

Thus, f is discontinuous at $x = 2$ and this discontinuity is nonremovable since the limit from the left is *not equal* to the limit from the right.

55. Since each part of the piecewise-defined function is a polynomial, f is continuous except possibly at $x = 2$. Since $f(2) = 8$, find a so that

$$\lim\limits_{x\to 2^+} ax^2 = 8.$$

$$\lim\limits_{x\to 2^+} ax^2 = a(2^2) = 8 \implies a = \frac{8}{2^2} = 2$$

65. Since each part of the piecewise-defined function is a polynomial, f is continuous except possibly at $x = 3$. The graph of f is shown in the figure. From the graph, we see that

$$\lim\limits_{x\to 3^-} f(x) \neq \lim\limits_{x\to 3^+} f(x).$$

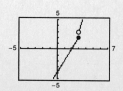

Therefore, f has a nonremovable discontinuity at $x = 3$.

71. The graph of the function

$$f(x) = \frac{\sin x}{x}$$

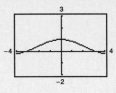

over the interval $[-4, 4]$ is shown in the figure. The graph was produced by a graphing utility and appears continuous. Since $f(0)$ is not defined, the function is not continuous at $x = 0$. This example shows that a removable continuity usually cannot be observed on a graph produced by a graphing utility. Therefore, it is important to examine a function analytically as well as graphically.

81. Since

$$f(x) = x^3 - x^2 + x - 2$$

is a polynomial function, it is continuous on the entire real line and in particular on the interval $[0, 3]$. Also,

$$f(0) = -2 < 19 = f(3).$$

Hence, the Intermediate Value Theorem applies. To find c, we solve the following equation.

$$x^3 - x^2 + x - 2 = 4$$
$$x^3 - x^2 + x - 6 = 0$$
$$(x - 2)(x^2 + x + 3) = 0$$

Since $x^2 + x + 3 = 0$ has no real solution, the only real solution is $x = 2$ and it is in the specified interval $[0, 3]$. Therefore, $c = 2$ and $f(2) = 4$.

85. Since the number of units in inventory is given by

$$N(t) = 25\left(2\left[\!\left[\frac{t + 2}{2}\right]\!\right] - t\right) = 25\left(2\left[\!\left[\frac{t}{2} + 1\right]\!\right] - t\right),$$

we observe that the greatest integer function is discontinuous at every positive even integer. Therefore, $N(t)$ is discontinuous at every positive even integer. We demonstrate this by evaluating the function for several values of t and plotting the points to generate the figure.

$$N(0) = 25(2[\![1]\!] - 0) = 25[2(1) - 0] = 50$$
$$N(1) = 25(2[\![1.5]\!] - 1) = 25[2(1) - 1] = 25$$
$$N(1.8) = 25(2[\![1.9]\!] - 1.8) = 25[2(1) - 1.8] = 5$$
$$N(2) = 25(2[\![2]\!] - 2) = 25[2(2) - 2] = 50$$
$$N(3) = 25(2[\![2.5]\!] - 3) = 25[2(2) - 3] = 25$$
$$N(3.8) = 25(2[\![2.9]\!] - 3.8) = 25[2(2) - 3.8] = 5$$
$$N(4) = 25(2[\![3]\!] - 4) = 25[2(3) - 4] = 50$$

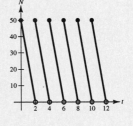

The company must replenish his inventory every two months.

99. To find the domain begin by observing that f is indeterminate at $x = 0$. It is also necessary that $x + c^2$ be nonnegative which implies that $x \geq -c^2$. Therefore, the domain is $[-c^2, 0) \cup (0, \infty)$.

$$\lim_{x \to 0} \frac{\sqrt{x + c^2} - c}{x} = \lim_{x \to 0} \left[\frac{\sqrt{x + c^2} - c}{x} \cdot \frac{\sqrt{x + c^2} + c}{\sqrt{x + c^2} + c}\right]$$

$$= \lim_{x \to 0} \frac{(x + c^2) - c^2}{x\left[\sqrt{x + x^2} + c\right]}$$

$$= \lim_{x \to 0} \frac{1}{\sqrt{x + c^2} + c} = \frac{1}{2c}$$

Define $(0) = 1/(2c)$ to make f continuous at $x = 0$.

Section 1.5   Infinite Limits

5. When $x = 3$, the values of the numerator and denominator of the function

$$f(x) = \frac{1}{x^2 - 9}$$

are 1 and 0, respectively. Therefore, $x = -3$ is a vertical asymptote. The behavior of f and x approaches -3 from the left and from the right can be seen from the table:

x	-3.5	-3.1	-3.01	-3.001	-2.999	-2.99	-2.9	-2.5
$f(x)$	0.31	1.64	16.64	166.64	-166.69	-16.69	-1.69	-0.36

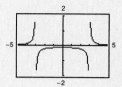

Therefore,

$$\lim_{x \to -3^-} \frac{1}{x^2 - 9} = \infty \quad \text{or} \quad \lim_{x \to -3^+} \frac{1}{x^2 - 9} = -\infty. \text{ The graph is shown in the figure.}$$

17. By rewriting the equation for this function as the ratio of two polynomials, we have

$$T(t) = 1 - \frac{4}{t^2} = \frac{t^2 - 4}{t^2}$$

When $t = 0$, the values of the numerator and denominator of the function are -4 and 0, respectively. Therefore, $t = 0$ is the only vertical asymptote.

19. By factoring the denominator we have

$$f(x) = \frac{x}{x^2 + x - 2} = \frac{x}{(x - 1)(x + 2)}.$$

Therefore, the function has two vertical asymptotes, and they are $x = 1$ and $x = -2$.

27. At $x = -1$ the numerator of the function is not zero and the denominator is equal to zero. Therefore, the graph of f has a vertical asymptote at $x = -1$. This result is also verified by the following one-sided limits.

$$\lim_{x \to -1^+} \frac{x^2 + 1}{x + 1} = \infty$$

$$\lim_{x \to -1^-} \frac{x^2 + 1}{x + 1} = -\infty$$

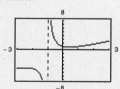

The graph of the function is shown in the figure.

37. Since

$$\lim_{x \to 0^+} 2 = 2 \quad \text{or} \quad \lim_{x \to 0^+} \sin x = 0,$$

it follows that

$$\lim_{x \to 0^+} \frac{2}{\sin x} = \infty.$$

47. (a) When $x = 7$, we have

$$r = \frac{2(7)}{\sqrt{625 - 7^2}} = \frac{14}{\sqrt{576}} = \frac{7}{12} \text{ feet per second.}$$

(b) When $x = 15$, we have

$$r = \frac{2(15)}{\sqrt{625 - 15^2}} = \frac{30}{\sqrt{400}} = \frac{3}{2} \text{ feet per second.}$$

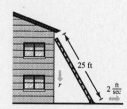

(c) As x approaches 25 from the left, we have

$$\lim_{x \to 25^-} \frac{2x}{\sqrt{625 - x^2}} = \infty.$$

53. (a) The linear velocities of the two pulleys must be the same. Linear velocity is given by $v = r\omega$ where r is the radius of a pulley and ω is its angular velocity in radians per unit time. Letting R be the number of revolutions per minute of the saw, we have the following.

$$10\left(\frac{1700}{2\pi}\right) = 20\left(\frac{R}{2\pi}\right)$$

The solution of the equation is $R = 850$ revolutions per minute.

(b) The direction of rotation of the saw is reversed.

(c) Observe the figure for the assignment of the variables x and y to the straight sections of the belt and for the magnitudes of the angles required in the following calculations.

Begin by recalling that the length of an arc of a circle is $s = r\theta$ where r is the radius of the circle and θ is magnitude of the central angle. Therefore, the lengths of the belt in contact with the pulleys are the following.

Small pulley: $10(\pi + 2\phi)$

Large pulley: $20(\pi + 2\phi)$

The lengths x and y of the straight sections of the belt are the following.

$$\cot \phi = \frac{y}{10} \implies y = 10 \cot \phi$$

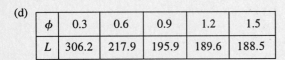

$$\cot \phi = \frac{x}{20} \implies x = 20 \cos \phi$$

Therefore, the total length L of the belt is

$$L = 2x + 2y + 10(\pi + 2\phi) + 20(\pi + 2\phi)$$

$$= 60 \cot \phi + 30(\pi + 2\phi).$$

The domain of the function is $\left(0, \frac{\pi}{2}\right)$.

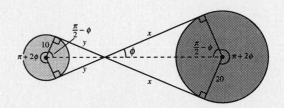

(d)

ϕ	0.3	0.6	0.9	1.2	1.5
L	306.2	217.9	195.9	189.6	188.5

(e) The graph of L is shown in the figure.

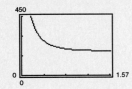

(f) $\displaystyle\lim_{\phi \to (\pi/2)^-} L = 60 \cot \frac{\pi}{2} + 30\left(\pi + 2 \cdot \frac{\pi}{2}\right)$

$$= 0 + 60\pi = 60\pi$$

As ϕ approaches $\pi/2$ from the left, the distance between the pulleys decreases and the belt tends to wrap farther around each pulley. Therefore, the length of the belt approaches the sum of the two circumferences of the pulleys.

(g) $\displaystyle\lim_{\phi \to 0^+} L = \infty.$

Review Exercises for Chapter 1

15. $\lim\limits_{x \to 0} \dfrac{\dfrac{1}{x+1} - 1}{x} = \lim\limits_{x \to 0} \dfrac{\dfrac{1 - (x+1)}{x+1}}{x}$

$$= \lim\limits_{x \to 0} \dfrac{1 - x - 1}{x(x+1)}$$

$$= \lim\limits_{x \to 0} \dfrac{-1}{x+1} = -1$$

21. Since

$$\sin\left(\frac{\pi}{6} + \Delta x\right) = \sin\frac{\pi}{6}\cos \Delta x + \cos\frac{\pi}{6}\sin \Delta x,$$

we have

$$\lim\limits_{\Delta x \to 0} \dfrac{\sin\left(\dfrac{\pi}{6} + \Delta x\right) - \dfrac{1}{2}}{\Delta x} = \lim\limits_{\Delta x \to 0} \dfrac{\dfrac{1}{2}\cos \Delta x + \dfrac{\sqrt{3}}{2}\sin \Delta x - \dfrac{1}{2}}{\Delta x}$$

$$= \lim\limits_{\Delta x \to 0}\left[\dfrac{\sqrt{3}\sin \Delta x}{2\Delta x} + \dfrac{1}{2}\left(\dfrac{\cos \Delta x - 1}{\Delta x}\right)\right]$$

$$= \dfrac{\sqrt{3}}{2}.$$

25. $\lim\limits_{x \to -1^+} \dfrac{x+1}{x^3+1} = \lim\limits_{x \to -1^+} \dfrac{x+1}{(x+1)(x^2-x+1)}$

$$= \lim\limits_{x \to -1^+} \dfrac{1}{x^2-x+1} = \dfrac{1}{3}$$

35. $\lim\limits_{x \to k^+} [\![x+3]\!] = k + 3$ where k is an integer.

$\lim\limits_{x \to k^-} [\![x+3]\!] = k + 2$ where k is an integer.

Therefore, $f(x) = [\![x+3]\!]$ has a nonremovable discontinuity at each integer k, and is continuous on $(k, k+1)$ for all integers k.

45. $f(x) = \begin{cases} x + 3, & x \le 2 \\ cx + 6, & x > 2 \end{cases}$

Since f is linear to the right and left of 2, it is continuous for all values of x other than possibly at $x = 2$. Furthermore, since $f(2) = 2 + 3 = 5$, we can make f continuous at $x = 2$ by finding c such that

$$c(2) + 6 = 5$$

$$2c = -1$$

$$c = -\dfrac{1}{2}$$

Now we have

$$\lim\limits_{x \to 2^-} f(x) = \lim\limits_{x \to 2^-} (x + 3) = 2 + 3 = 5$$

and

$$\lim\limits_{x \to 2^+} f(x) = \lim\limits_{x \to 2^+} \left(-\frac{x}{2} + 6\right) = -1 + 6 = 5.$$

Thus, for $c = -1/2$, f is continuous at $x = 2$ and consequently f is continuous for all x.

57. $\lim\limits_{x \to 0^+} \dfrac{|x|}{x} = \lim\limits_{x \to 0^+} \dfrac{x}{x} = \lim\limits_{x \to 0^+} 1 = 1$

$\lim\limits_{x \to 0^-} \dfrac{|x|}{x} = \lim\limits_{x \to 0^-} \dfrac{-x}{x} = \lim\limits_{x \to 0^-} (-1) = -1$

Since the limit from the right and the limit from the left are *not* equal, the limit does *not* exist and the given statement is false.

C H A P T E R   2
Differentiation

CHAPTER 2
Differentiation

Section 2.1   The Derivative and the Tangent Line Problem

Solutions to Selected Odd-Numbered Exercises

9. $f'(x) = \lim\limits_{\Delta x \to 0} \dfrac{f(x + \Delta x) - f(x)}{\Delta x} = \lim\limits_{\Delta x \to 0} \dfrac{2(x + \Delta x)^2 + (x + \Delta x) - 1 - (2x^2 + x - 1)}{\Delta x}$

$\qquad = \lim\limits_{\Delta x \to 0} \dfrac{2x^2 + 4x\Delta x + 2(\Delta x)^2 + x + \Delta x - 1 - 2x^2 - x + 1}{\Delta x} = \lim\limits_{\Delta x \to 0} \dfrac{\Delta x(4x + 2\Delta x + 1)}{\Delta x}$

$\qquad = \lim\limits_{\Delta x \to 0} (4x + 2\Delta x + 1) = 4x + 1$

13. $f'(x) = \lim\limits_{\Delta x \to 0} \dfrac{f(x + \Delta x) - f(x)}{\Delta x} = \lim\limits_{\Delta x \to 0} \dfrac{\dfrac{1}{(x + \Delta x - 1)} - \dfrac{1}{x - 1}}{\Delta x}$

$\qquad = \lim\limits_{\Delta x \to 0} \dfrac{x - 1 - (x + \Delta x - 1)}{\Delta x(x + \Delta x - 1)(x - 1)} = \lim\limits_{\Delta x \to 0} \dfrac{-\Delta x}{\Delta x(x + \Delta x - 1)(x - 1)}$

$\qquad = \lim\limits_{\Delta x \to 0} \dfrac{-1}{(x + \Delta x - 1)(x - 1)} = \dfrac{-1}{(x - 1)(x - 1)} = \dfrac{-1}{(x - 1)^2}$

19. $f'(x) = \lim\limits_{\Delta x \to 0} \dfrac{f(x + \Delta x) - f(x)}{\Delta x}$

$\qquad = \lim\limits_{\Delta x \to 0} \dfrac{(x + \Delta x)^3 - x^3}{\Delta x}$

$\qquad = \lim\limits_{\Delta x \to 0} \dfrac{x^3 + 3x^2\Delta x + 3x(\Delta x)^2 + (\Delta x)^3 - x^3}{\Delta x}$

$\qquad = \lim\limits_{\Delta x \to 0} \dfrac{\Delta x[3x^2 + 3x\Delta x + (\Delta x)^2]}{\Delta x}$

$\qquad = \lim\limits_{\Delta x \to 0} [3x^2 + 3x\Delta x + (\Delta x)^2]$

$\qquad = 3x^2$

Thus, $f'(x) = 3x^2$ and the slope of the tangent line at $(2, 8)$ is $f'(2) = 3(2)^2 = 12$. Finally, by the point-slope form of the equation of a line, we have

$\qquad y - y_1 = m(x - x_1)$

$\qquad y - 8 = 12(x - 2)$

$\qquad y = 12x - 16.$

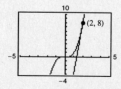

23. The slope of the line given by $3x - y + 1 = 0$ is 3 since

$\qquad 3x - y + 1 = 0$

$\qquad y = 3x + 1 = mx + b$

Thus, it is necessary to find a point (or points) on the graph of $y = x^3$ such that the tangent line at that point has a slope of 3. From Exercise 19, we know that $dy/dx = 3x^2$ is the slope at any point on the graph of $y = x^3$. Therefore,

$\qquad \dfrac{dy}{dx} = 3x^2 = 3$

$\qquad x^2 = 1 \quad \text{or} \quad x = \pm 1.$

Finally, we conclude that the slope of the graph of $y = x^3$ is 3 at the points $(1, 1)$ and $(-1, -1)$. The tangent lines at these two points are:

$\qquad y - 1 = 3(x - 1) \qquad y - (-1) = 3[x - (-1)]$

$\qquad\qquad y = 3x - 2 \qquad\qquad\qquad y = 3x + 2$

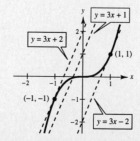

25. To begin, find $f'(x)$ as follows:

$$f'(x) = \lim_{\Delta x \to 0} \frac{f(x + \Delta x) - f(x)}{\Delta x}$$

$$= \lim_{\Delta x \to 0} \frac{4(x + \Delta x) - (x + \Delta x)^2 - (4x - x^2)}{\Delta x}$$

$$= \lim_{\Delta x \to 0} \frac{4x + 4\Delta x - x^2 - 2x\Delta x - (\Delta x)^2 - 4x + x^2}{\Delta x}$$

$$= \lim_{\Delta x \to 0} \frac{\Delta x(4 - 2x - \Delta x)}{\Delta x}$$

$$= \lim_{\Delta x \to 0} (4 - 2x - \Delta x)$$

$$= 4 - 2x$$

Now let (x, y) be a point on the graph of $y = 4x - x^2$. Since $dy/dx = 4 - 2x$, the slope of the tangent line at (x, y) is $m = 4 - 2x$. On the other hand, if the tangent line at (x, y) passes through the point $(2, 5)$, then its slope must be

$$m = \frac{y - 5}{x - 2} = \frac{4x - x^2 - 5}{x - 2}.$$

Equating these two expressions for m, we have

$$\frac{4x - x^2 - 5}{x - 2} = 4 - 2x$$

$$4x - x^2 - 5 = (4 - 2x)(x - 2)$$

$$4x - x^2 - 5 = 4x - 2x^2 - 8 + 4x$$

$$x^2 - 4x + 3 = 0$$

$$(x - 3)(x - 1) = 0$$

$$x = 3 \quad \text{or} \quad x = 1.$$

If $x = 3$, then $y = 4(3) - 3^2 = 3$ and $m = 4 - 2(3) = -2$. Therefore,

$$y - 3 = -2(x - 3) \quad \text{or} \quad y = -2x + 9.$$

If $x = 1$, then $y = 4(1) - 1^2 = 3$ and $m = 4 - 2(1) = 2$. Therefore,

$$y - 3 = 2(x - 1) \quad \text{or} \quad y = 2x + 1.$$

33. (a) $g'(0) = -3$

(b) $g'(3) = 0$

(c) Since $g'(1) = -\frac{8}{3}$, the slope of the tangent line to the graph of g is $-\frac{8}{3}$ when $x = 1$. This means the graph is moving downward to the right when $x = 1$.

(d) Since $g'(-4) = \frac{7}{3}$, the slope of the tangent line to the graph of g is $\frac{7}{3}$ when $x = -4$. This means the graph is moving upward to the right when $x = -4$.

(e) Since $g'(x) > 0$ on the interval $[3, 6]$, the graph of g is moving upward to the right for all x in the interval. Therefore, $g(6) > g(4)$ and $g(6) - g(4) > 0$.

(f) No. Knowing only $g'(2)$ is not sufficient information for finding $g(2)$. $g'(2)$ remains the same for any vertical translation of the graph of g.

39. $f(x) = 2x - x^2$, $g(x) = \dfrac{f(x + 0.01) - f(x)}{0.01}$

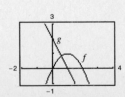

The graphs of f and g are shown in the figure. For any real number x, the difference quotient g gives the slope of the secant line through the points $(x, f(x))$ and $(x + 0.01, f(x + 0.01))$ on the graph of f. Since the two points are "close" to each other, the secant line approximates the slope of the tangent line to the graph of f at $(x, f(x))$. Therefore $g(x) \approx f'(x)$.

47. $f'(-2) = \lim\limits_{x \to -2} \dfrac{f(x) - f(-2)}{x - (-2)}$

$= \lim\limits_{x \to -2} \dfrac{(x^3 + 2x^2 + 1) - 1}{x + 2}$

$= \lim\limits_{x \to -2} \dfrac{x^2(x + 2)}{x + 2}$

$= \lim\limits_{x \to -2} x^2 = 4$

55. The graph of $f(x) = (x - 3)^{2/3}$ is continuous at $x = 3$. However, the one-sided limits

$\lim\limits_{x \to 3^+} \dfrac{f(x) - f(3)}{x - 3} = \lim\limits_{x \to 3^+} \dfrac{(x - 3)^{2/3}}{x - 3} = \lim\limits_{x \to 3^+} \dfrac{1}{(x - 3)^{1/3}} = \infty$

$\lim\limits_{x \to 3^-} \dfrac{f(x) - f(3)}{x - 3} = \lim\limits_{x \to 3^-} \dfrac{(x - 3)^{2/3}}{x - 3} = \lim\limits_{x \to 3^-} \dfrac{1}{(x - 3)^{1/3}} = -\infty$

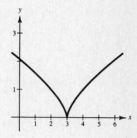

are not equal. Therefore, f is not differentiable at $x = 3$ and the point $(3, 0)$ is called a **cusp**. Hence, f is differentiable in the intervals $(-\infty, 3)$ and $(3, \infty)$.

65. The derivative from the left is

$\lim\limits_{x \to 2^-} \dfrac{f(x) - f(2)}{x - 2} = \lim\limits_{x \to 2^-} \dfrac{(x^2 + 1) - 5}{x - 2}$

$= \lim\limits_{x \to 2^-} \dfrac{x^2 - 4}{x - 2}$

$= \lim\limits_{x \to 2^-} (x + 2) = 4.$

The derivative from the right is

$\lim\limits_{x \to 2^+} \dfrac{f(x) - f(2)}{x - 2} = \lim\limits_{x \to 2^+} \dfrac{(4x - 3) - 5}{x - 2}$

$= \lim\limits_{x \to 2^+} \dfrac{4x - 8}{x - 2}$

$= \lim\limits_{x \to 2^+} 4 = 4.$

The one-sided limits are equal. Thus, f is differentiable at $x = 2$ and $f'(2) = 4$.

Section 2.2   Basic Differentiation Rules and Rates of Change

11. $s(t) = t^3 - 2t + 4$

$s'(t) = 3t^2 - 2$

15. $y = \dfrac{1}{x} - 3 \sin x = x^{-1} - 3 \sin x$

$y' = -x^{-2} - 3 \cos x$

$= -\dfrac{1}{x^2} - 3 \cos x$

21. Function: $f(x) = \dfrac{\sqrt{x}}{x}$

Rewrite: $f(x) = \dfrac{x^{1/2}}{x} = x^{-1/2}$

Derivative: $f'(x) = -\dfrac{1}{2}x^{-3/2}$

Simplify: $f'(x) = -\dfrac{1}{2x^{3/2}}$

27. $y = (2x + 1)^2 = 4x^2 + 4x + 1$

$y' = 8x + 4$

At the point $(0, 1)$ the derivative is

$y' = 8(0) + 4 = 4.$

33. $g(t) = t^2 - \dfrac{4}{t} = t^2 - 4t^{-1}$

$g'(t) = 2t - (-1)4t^{-2} = 2t + \dfrac{4}{t^2}$

35. $f(x) = \dfrac{x^3 - 3x^2 + 4}{x^2} = \dfrac{x^3}{x^2} - \dfrac{3x^2}{x^2} + \dfrac{4}{x^2} = x - 3 + 4x^{-2}$

$f'(x) = 1 + 4(-2)x^{-3} = 1 - \dfrac{8}{x^3} = \dfrac{x^3 - 8}{x^3}$

41. $f(x) = 4\sqrt{x} + 3\cos x = 4x^{1/2} + 3\cos x$

$f'(x) = 4\left(\dfrac{1}{2}\right)x^{-1/2} + 3(-\sin x) = \dfrac{2}{\sqrt{x}} - 3\sin x$

43. (a) $y = x^4 - 3x^2 + 2$

$y' = 4x^3 - 6x$

Thus the slope of the tangent line at $(1, 0)$ is

$y' = 4(1)^3 - 6(1) = 4 - 6 = -2.$

The equation of the tangent line at $(1, 0)$ is

$y - y_1 = m(x - x_1)$

$y - 0 = -2(x - 1)$

$2x + y - 2 = 0$

(b) The graph is shown in the figure.

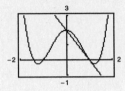

47. A tangent line is horizontal if the derivative (the slope) at the point of tangency is zero. Since $y' = 4x^3 - 16x = 4x(x^2 - 4)$, we must find all values of x that satisfy the equation $y' = 4x(x^2 - 4) = 4x(x + 2)(x - 2) = 0$. The solutions to this equation are $x = -2$, $x = 0$, and $x = 2$. At $x = \pm2$, we have

$y = (\pm2)^4 - 8(\pm2)^2 + 2 = -14.$

At $x = 0$, we have $y = 2$. Thus, the points of horizontal tangency are:

$(-2, -14), (0, 2),$ and $(2, -14)$

55. Let (x_1, y_1) and (x_2, y_2) be the points of tangency on the graphs of $y = x^2$ and $y = -x^2 + 6x - 5$, respectively. We know that $y_1 = x_1^2$ and $y_2 = -x_2^2 + 6x_2 - 5$. Let m be the slope of the tangent line. Since the line passes through (x_1, y_1) and (x_2, y_2), it follows that

(1) $m = \dfrac{y_2 - y_1}{x_2 - x_1} = \dfrac{-x_2^2 + 6x_2 - 5 - x_1^2}{x_2 - x_1}.$

Since the line is tangent to $y = x^2$ at (x_1, y_1) and the derivative of this curve is $y' = 2x$, it follows that $m = 2x_1$.

Since the line is tangent to $y = -x^2 + 6x - 5$ at (x_2, y_2) and the derivative of this curve is $y' = -2x_2 + 6$, it follows that

(2) $m = -2x_2 + 6.$

Thus, from the preceding two equations, we have: $m = 2x_1 = -2x_2 + 6$

(3) $x_1 = -x_2 + 3.$

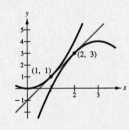

Using equations (1), (2), and (3) yields

$m = -2x_2 + 6 = \dfrac{-x_2^2 + 6x_2 - 5 - x_1^2}{x_2 - x_1}$

$-2x_2 + 6 = \dfrac{-x_2^2 + 6x_2 - 5 - (-x_2 + 3)^2}{x_2 - (-x_2 + 3)}$

$(2x_2 - 3)(-2x_2 + 6) = -x_2^2 + 6x_2 - 5 - (x_2^2 - 6x_2 + 9)$

$-4x_2^2 + 18x_2 - 18 = -2x_2^2 + 12x_2 - 14$

$-2x_2^2 + 6x_2 - 4 = 0$

$x_2^2 - 3x_2 + 2 = 0$

$(x_2 - 2)(x_2 - 1) = 0$

$x_2 = 1 \text{ or } x_2 = 2.$

—CONTINUED—

55. —CONTINUED—

If $x_2 = 1$, then $y_2 = -1^2 + 6(1) - 5 = 0$, $x_1 = -1 + 3 = 2$, and $y_1 = 2^2 = 4$. Thus the line containing $(1, 0)$ and $(2, 4)$ is tangent to both curves. The equation of this line is

$$y - 0 = \left(\frac{4 - 0}{2 - 1}\right)(x - 1) \quad \text{or} \quad y = 4x - 4.$$

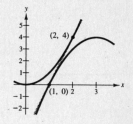

If $x_2 = 2$, then $y_2 = -2^2 + 6(2) - 5 = -4 + 12 - 5 = 3$, $x_1 = -2 + 3 = 1$, and $y_1 = 1^2 = 1$. Thus the line containing $(2, 3)$ and $(1, 1)$ is tangent to both curves. The equation of this line is

$$y - 1 = \left(\frac{3 - 1}{2 - 1}\right)(x - 1) = 2x - 2$$

$$y = 2x - 1.$$

57. From the figure we observe that the slope of the tangent line can be determined by the derivative of f and by the two-point formula for the slope of the line. Therefore,

$$f'(x) = \frac{y_0 - y}{x_0 - x}.$$

Since $f(x) = \sqrt{x}, f'(x) = 1/(2\sqrt{x})$, $y = \sqrt{x}$, $x_0 = -4$, and $y_0 = 0$, we have the following.

$$f'(x) = \frac{y_0 - y}{x_0 - x}$$

$$\frac{1}{2\sqrt{x}} = \frac{0 - \sqrt{x}}{-4 - x}$$

$$\frac{1}{2\sqrt{x}} = \frac{\sqrt{x}}{4 + x}$$

$$4 + x = 2x \implies x = 4$$

Hence the point of tangency is $(4, 2)$ and the slope of the tangent line is $f'(4) = \frac{1}{4}$. Now, using the point-slope form of the equation of a line, we can write

$$y - 2 = \frac{1}{4}(x - 4)$$

$$4y - 8 = x - 4$$

$$0 = x - 4y + 4.$$

67. The average rate of change is given by

$$\frac{\Delta y}{\Delta x} = \frac{f(2) - f(1)}{2 - 1} = \frac{-\frac{1}{2} - (-1)}{2 - 1} = \frac{1}{2}.$$

To find the instantaneous rate of change we first find $f'(x)$.

$$f(x) = -\frac{1}{x} = -x^{-1}$$

$$f'(x) = -(-1)x^{-2} = \frac{1}{x^2}$$

The instantaneous rate of change when $x = 1$ is

$$f'(1) = \frac{1}{1^2} = 1.$$

and the instantaneous rate of change when $x = 2$ is

$$f'(2) = \frac{1}{2^2} = \frac{1}{4}.$$

71. (a) Since the initial height of the coin is $s_0 = 1362$ feet and the initial velocity is $v_0 = 0$, we have

$$s(t) = -\frac{1}{2}gt^2 + v_0 t + s_0 = -16t^2 + 1362$$

and

$$v(t) = s'(t) = -32t.$$

(b) The average velocity on the interval $[1, 2]$ is

$$\frac{s(2) - s(1)}{2 - 1} = 1298 - 1346 = -48 \text{ ft/sec.}$$

(c) Since $v(t) = -32t$, the instantaneous velocity at time $t = 1$ is

$$s'(1) = -32(1) = -32 \text{ ft/sec}$$

and at time $t = 2$ is $s'(2) = -32(2) = -64$ ft/sec.

(d) The dollar will reach ground level when

$$s(t) = -16t^2 + 1362 = 0$$

$$t^2 = \frac{1362}{16}$$

$$t = \frac{1}{4}\sqrt{1362}$$

$$= \frac{15}{4} \approx 9.2 \text{ sec.}$$

(e) The velocity just before it hits the ground is

$$v\left(\frac{1}{4}\sqrt{1362}\right) = -32\left(\frac{1}{4}\sqrt{1362}\right) = -8\sqrt{1362}$$

$$\approx -295.2 \text{ ft/sec.}$$

79. (a) Begin by entering the data in the statistics lists of your graphing utility. Access the statistics calculation menu and select linear regression ($y = ax + b$). Fitting this linear model to the data yields

$$R(v) = 0.167v - 0.02.$$

(b) Access the statistics calculation menu and select quadratic regression. Fitting this quadratic model to the data yields

$$B(v) = 0.006v^2 - 0.024v + 0.460.$$

(c) $T(v) = R(v) + B(v)$

$$= (0.167v - 0.02) + (0.006v^2 - 0.024v + 0.460)$$

$$= 0.006v^2 + 0.143v + 0.440$$

(d) The graphs of the specified functions are shown in the figure.

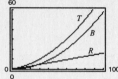

(e) $T(v) = 0.006v^2 + 0.143v + 0.440$

$T'(v) = 0.012v + 0.143$

$T'(40) = 0.012(40) + 0.143 = 0.623$ m/km/hr

$T'(80) = 0.012(80) + 0.143 = 1.103$ m/km/hr

$T'(100) = 0.012(100) + 0.143 = 1.343$ m/km/hr

(f) As the speed increases the total stopping distance increases at an increasing rate.

87. The parabola passes through the points $(0, 1)$ and $(1, 0)$ which implies that these points are solution points of the equation $y = ax^2 + bx + c$. By substitution we have

$(0, 1)$: $1 = a(0)^2 + b(0) + c \Rightarrow c = 1$

$(1, 0)$: $0 = a(1)^2 + b(1) + c \Rightarrow b = -a - c = -a - 1.$

The tangent line $y = x - 1$ has slope 1 which must equal the value of the derivative of the quadratic function at $x = 1$. Therefore, at $x = 1$ we have the following.

$$y = ax^2 + (-a - 1)x + 1$$

$$y' = 2ax + (-a - 1)$$

$$1 = 2a(1) + (-a - 1)$$

$$1 = a - 1 \Rightarrow a = 2$$

Hence, $a = 2, b = -a - 1 = -3, c = 1$, and $y = 2x^2 - 3x + 1$.

Section 2.3   The Product and Quotient Rules and Higher-Order Derivatives

3. $f(x) = (x^3 - 3x)(2x^2 + 3x + 5)$

$f'(x) = (x^3 - 3x)(4x + 3) + (2x^2 + 3x + 5)(3x^2 - 3)$

$\quad = 4x^4 + 3x^3 - 12x^2 - 9x + 6x^4 - 6x^2 + 9x^3 - 9x + 15x^2 - 15$

$\quad = 10x^4 + 12x^3 - 3x^2 - 18x - 15$

$f'(0) = -15$

9. Function: $y = \dfrac{7}{3x^3}$

Rewrite: $y = \dfrac{7}{3}x^{-3}$

Derivative: $y' = -3\left(\dfrac{7}{3}\right)x^{-4}$

Simplify: $y' = -\dfrac{7}{x^4}$

15. $f(x) = \dfrac{3 - 2x - x^2}{x^2 - 1}$

$\quad = -\dfrac{x^2 + 2x - 3}{x^2 - 1} = -\dfrac{(x + 3)(x - 1)}{(x + 1)(x - 1)} = -\dfrac{x + 3}{x + 1}$

$f'(x) = -\dfrac{(x + 1)(1) - (x + 3)(1)}{(x + 1)^2} = \dfrac{2}{(x + 1)^2}, x \neq -1$

23. We begin by using the Associative Law to group the first two factors and then use the Product Rule twice.

$$f(x) = [(3x^3 + 4x)(x - 5)](x + 1)$$

$$f'(x) = [(3x^3 + 4x)(x - 5)](1) + (x + 1)[(3x^3 + 4x)(1) + (x - 5)(9x^2 + 4)]$$

$$= (3x^3 + 4x)(x - 5) + (x + 1)(3x^3 + 4x) + (x + 1)(x - 5)(9x^2 + 4)$$

$$= 15x^4 - 48x^3 - 33x^2 - 32x - 20$$

29. $f(t) = \dfrac{\cos t}{t}$

$$f'(t) = \frac{t(-\sin t) - (\cos t)(1)}{t^2} = \frac{-(t \sin t + \cos t)}{t^2}$$

37. $y = -\csc x - \sin x$

$$y' = -(-\csc x \cot x) - \cos x = \frac{1}{\sin x}\left(\frac{\cos x}{\sin x}\right) - \cos x$$

$$= \cos x\left(\frac{1}{\sin^2 x} - 1\right) = \frac{\cos x}{\sin^2 x}(1 - \sin^2 x) = \cos x \cot^2 x$$

41. $f(x) = x^2 \tan x$

$$f'(x) = x^2 \sec^2 x + (\tan x)(2x) = x(x \sec^2 x + 2 \tan x)$$

51. (a) $f(x) = \dfrac{x}{x - 1}$

$$f'(x) = \frac{(x - 1)(1) - (x)(1)}{(x - 1)^2} = \frac{x - 1 - x}{(x - 1)^2} = \frac{-1}{(x - 1)^2}$$

Therefore, the slope of the tangent line at $(2, 2)$ is $f'(2) = -1/(2 - 1)^2 = -1$ and an equation of the tangent line at $(2, 2)$ is

$$y - 2 = -1(x - 2)$$

$$= -x + 2$$

$$y = -x + 4.$$

(b) The graphs are shown in the figure.

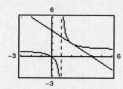

57. Any horizontal tangents to the graph of f will occur where $f'(x) = 0$.

$$f(x) = \frac{x^2}{x - 1}$$

$$f'(x) = \frac{(x - 1)(2x) - x^2(1)}{(x - 1)^2}$$

$$= \frac{x^2 - 2x}{(x - 1)^2} = \frac{x(x - 2)}{(x - 1)^2}$$

Therefore, $f'(x) = 0$ when $x = 0$ or $x = 2$. Since $f(0) = 0$ and $f(2) = 4$, the points of horizontal tangency are $(0, 0)$ and $(2, 4)$.

67. $P(t) = 500\left(1 + \dfrac{4t}{50 + t^2}\right)$

$$P'(t) = 500\left[\frac{(50 + t^2)(4) - (4t)(2t)}{(50 + t^2)^2}\right]$$

$$= \frac{500(200 + 4t^2 - 8t^2)}{(50 + t^2)^2} = \frac{2000(50 - t^2)}{(50 + t^2)^2}$$

Therefore, the rate of population growth when $t = 2$ is

$$P'(2) = \frac{2000(50 - 4)}{(50 + 4)^2} \approx 31.55 \text{ bacteria/hr.}$$

75. The first derivative of $f(x) = \dfrac{x}{x - 1}$ was found in Exercise 51.

$$f'(x) = \frac{-1}{(x - 1)^2} = \frac{-1}{x^2 - 2x + 1}$$

$$f''(x) = \frac{(x^2 - 2x + 1)(0) - (-1)(2x - 2)}{(x^2 - 2x + 1)^2}$$

$$= \frac{2(x - 1)}{[(x - 1)^2]^2} = \frac{2}{(x - 1)^3}$$

85. $f(x) = g(x)h(x)$

(a) $f'(x) = g(x)h'(x) + h(x)g'(x)$

   $f''(x) = g(x)h''(x) + g'(x)h'(x) + h(x)g''(x) + h'(x)g'(x)$

   $= g(x)h''(x) + 2g'(x)h'(x) + h(x)g''(x)$

   $f'''(x) = g(x)h'''(x) + g'(x)h''(x) + 2g'(x)h''(x) + 2g''(x)h'(x) + h(x)g'''(x) + h'(x)g''(x)$

   $= g(x)h'''(x) + 3g'(x)h''(x) + 3g''(x)h'(x) + g'''(x)h(x)$

   $f^{(4)}(x) = g(x)h^{(4)}(x) + g'(x)h'''(x) + 3g'(x)h'''(x) + 3g''(x)h''(x) + 3g''(x)h''(x) + 3g'''(x)h'(x) + g^{(4)}(x)h'(x) + g^{(4)}(x)h(x)$

   $= g(x)h^{(4)}(x) + 4g'(x)h'''(x) + 6g''(x)h''(x) + 4g'''(x)h'(x) + g^{(4)}(x)h(x)$

(b) $f^{(n)}(x) = g(x)h^{(n)}(x) + \dfrac{n(n-1)(n-2)\cdots(2)(1)}{1[(n-1)(n-2)\cdots(2)(1)]}g'(x)h^{(n-1)}(x) + \dfrac{n(n-1)(n-2)\cdots(2)(1)}{(2)(1)[(n-2)(n-3)\cdots(2)(1)]}g''(x)h^{(n-2)}(x)$

$+ \dfrac{n(n-1)(n-2)\cdots(2)(1)}{(3)(2)(1)[(n-3)(n-4)\cdots(2)(1)]}g'''(x)h^{(n-3)}(x) + \cdots$

$+ \dfrac{n(n-1)(n-2)\cdots(2)(1)}{[(n-1)(n-2)\cdots(2)(1)](1)}g^{(n-1)}(x)h'(x) + g^{(n)}(x)h(x)$

$= g(x)h^{(n)}(x) + \dfrac{n!}{1!(n-1)!}g'(x)h^{(n-1)}(x) + \dfrac{n!}{2!(n-2)!}g''(x)h^{(n-2)}(x) + \cdots + \dfrac{n!}{(n-1)!1!}g^{(n-1)}(x)h'(x) + g^{(n)}(x)h(x)$

Note: For a definition of $n!$ (read "n factorial"), see Section 8.1 of the text.

91. (a) $f(x) = \cos x$ $f\left(\dfrac{\pi}{3}\right) = \dfrac{1}{2}$

   $f'(x) = -\sin x$ $f'\left(\dfrac{\pi}{3}\right) = -\dfrac{\sqrt{3}}{2}$

   $f''(x) = -\cos x$ $f''(x)\left(\dfrac{\pi}{3}\right) = -\dfrac{1}{2}$

   $P_1(x) = f'(a)(x-a) + f(a)$

   $= -\dfrac{\sqrt{3}}{2}\left(x - \dfrac{\pi}{3}\right) + \dfrac{1}{2}$

   $P_2(x) = \dfrac{1}{2}f''(a)(x-a)^2 + f'(a)(x-a) + f(a)$

   $= -\dfrac{1}{4}\left(x - \dfrac{\pi}{3}\right)^2 - \dfrac{\sqrt{3}}{2}\left(x - \dfrac{\pi}{3}\right) + \dfrac{1}{2}$

(b) The graphs for f, P_1, and P_2 are shown in the figure.

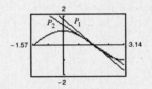

(c) P_2 is a better approximation.

(d) P_1 and P_2 become less accurate for values of x farther from $x = a$.

Section 2.4   The Chain Rule

5. If $u = g(x) = \csc x$ and $y = f(u) = u^3$, then

   $y = f(g(x)) = (\csc x)^3 = \csc^3 x.$

15. $y = \sqrt[3]{9x^2 + 4} = (9x^2 + 4)^{1/3}$

   $\dfrac{dy}{dx} = \overbrace{\left(\dfrac{1}{3}\right)}^{n}\overbrace{(9x^2 + 4)^{-2/3}}^{u^{n-1}}\overbrace{(18x)}^{u'} = \dfrac{6x}{(9x^2 + 4)^{2/3}}$

21. $f(t) = \dfrac{1}{(t-3)^2} = (t-3)^{-2}$

   $f'(x) = \overbrace{(-2)}^{n}\overbrace{(t-3)^{-3}}^{u^{n-1}}\overbrace{(1)}^{u'} = \dfrac{-2}{(t-3)^3}$

23. $y = \dfrac{1}{\sqrt{x+2}} = (x+2)^{-1/2}$

   $y' = \left(-\dfrac{1}{2}\right)(x+2)^{-3/2}(1) = \dfrac{-1}{2(x+2)^{3/2}}$

25. Applying the Product Rule and the General Power Rule produces

$$f(x) = x^2(x - 2)^4$$

$$f'(x) = x^2(4)(x - 2)^3(1) + (x - 2)^4(2x)$$

$$= 2x(x - 2)^3(2x + x - 2) = 2x(x - 2)^3(3x - 2).$$

33. Using a symbolic differentiation utility, first derivative of the function

$$g(t) = \frac{3t^2}{\sqrt{t^2 + 2t - 1}} \text{ is } g'(t) = \frac{3t(t^2 + 3t - 2)}{(t^2 + 2t - 1)^{3/2}}.$$

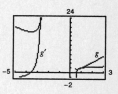

(Your differentiation utility may give the derivative in unsimplified form.) The graphs of the function and its derivative are given in the figure. The zeros of the derivative are approximately $t = -3.56$ and $t = 0.56$. These are the t-coordinates of the points where the graph of $g(t)$ has horizontal tangents.

43. $y = \cos 3x$

$$\frac{dy}{dx} = \overbrace{-\sin 3x}^{-\sin u} \overbrace{3}^{u'}$$

$$= -3 \sin 3x$$

47. $f(\theta) = \frac{1}{4}(\sin 2\theta)^2$

$$f'(\theta) = \frac{1}{4}(2)(\sin 2\theta)(\cos 2\theta)(2)$$

$$= \frac{1}{2}(2 \sin 2\theta \cos 2\theta) = \frac{1}{2} \sin 4\theta \quad \text{Double Angle Identity}$$

55. $f(x) = \dfrac{3}{x^3 - 4} = 3(x^3 - 4)^{-1}$

$$f'(x) = -3(x^3 - 4)^{-2}(3x^2) = -\frac{9x^2}{(x^3 - 4)^2}$$

$$f'(-1) = -\frac{9(-1)^2}{[(-1)^3 - 4]^2} = -\frac{9}{25}$$

Verify this result by using the numerical differentiation capabilities of your graphing utility.

63. (a) $f(x) = \sin 2x$

$$f'(x) = (\cos 2x)(2) = 2 \cos 2x$$

Therefore, the slope of the tangent line at $(\pi, 0)$ is
$f'(\pi) = 2 \cos 2\pi = 2$ and an equation of the tangent
line at $(\pi, 0)$ is

$$y - 0 = 2(x - \pi)$$

$$y = 2(x - \pi)$$

$$2x - y - 2\pi = 0$$

(b) The graphs are shown in the figure.

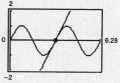

71. $f(x) = \sin x^2$

$$f'(x) = (\cos x^2)(2x) = 2x \cos x^2$$

$$f''(x) = 2[x(-\sin x^2)(2x) + (\cos x^2)(1)]$$

$$= 2(\cos x^2 - 2x^2 \sin x^2)$$

79. Using the Chain Rule and the fact that r is constant, we have

$$S = C(R^2 - r^2)$$

$$\frac{dS}{dt} = C\left(2R\frac{dR}{dt} - 0\right)$$

Substituting the given constants yields

$$\frac{dS}{dt} = (1.76 \times 10^5)[2(1.2 \times 10^{-2})(10^{-5})]$$

$$= 4.224 \times 10^{-2}.$$

87. If $y = |u|$, then

$$\frac{dy}{dx} = u' \frac{u}{|u|}, \, u \neq 0.$$

Since $g(x) = |u|$ where $u = 2x - 3$ and $u' = 2$, we have the following.

$$g'(x) = 2\left(\frac{2x - 3}{|2x - 3|}\right)$$

Section 2.5   Implicit Differentiation

5.
$$x^3 - xy + y^2 = 4$$

$$\frac{d}{dx}[x^3 - xy + y^2] = \frac{d}{dx}[4]$$

$$\frac{d}{dx}[x^3] - \frac{d}{dx}[xy] + \frac{d}{dx}[y^2] = \frac{d}{dx}[4]$$

$$3x^2 - \left[x\frac{dy}{dx} + y(1)\right] + 2y\frac{dy}{dx} = 0$$

$$3x^2 - x\frac{dy}{dx} - y + 2y\frac{dy}{dx} = 0$$

$$(2y - x)\frac{dy}{dx} = y - 3x^2$$

$$\frac{dy}{dx} = \frac{y - 3x^2}{2y - x}$$

11.
$$\sin x + 2\cos 2y = 1$$

$$\frac{d}{dx}[\sin x + 2\cos 2y] = \frac{d}{dx}[1]$$

$$\frac{d}{dx}[\sin x] + \frac{d}{dx}[2\cos 2y] = \frac{d}{dx}[1]$$

$$\cos x + 2(-\sin 2y)(2)\frac{dy}{dx} = 0$$

$$\cos x - 4\sin 2y\frac{dy}{dx} = 0$$

$$-4\sin 2y\frac{dy}{dx} = -\cos x$$

$$\frac{dy}{dx} = \frac{\cos x}{4\sin 2y}$$

17. Implicit differentiation of the equation $xy = 4$ yields

$$x\frac{dy}{dx} + y(1) = 0$$

$$x\frac{dy}{dx} = -y$$

$$\frac{dy}{dx} = -\frac{y}{x}$$

At $(-4, -1)$, $\frac{dy}{dx} = -\frac{-1}{-4} = -\frac{1}{4}$.

21. Implicit differentiation of the equation $x^{2/3} + y^{2/3} = 5$ yields

$$\frac{2}{3}x^{-1/3} + \frac{2}{3}y^{-1/3}\frac{dy}{dx} = 0$$

$$x^{-1/3} + y^{-1/3}\frac{dy}{dx} = 0$$

$$y^{-1/3}\frac{dy}{dx} = -x^{-1/3}$$

$$\frac{dy}{dx} = -\frac{x^{-1/3}}{y^{-1/3}} = -\sqrt[3]{\frac{y}{x}}$$

At $(8, 1)$, $\frac{dy}{dx} = -\sqrt[3]{\frac{1}{8}} = -\frac{1}{2}$.

23. Implicit differentiation of the equation $\tan(x + y) = x$ yields

$$\sec^2(x + y)\left(1 + \frac{dy}{dx}\right) = 1$$

$$\sec^2(x + y)\frac{dy}{dx} = 1 - \sec^2(x + y)$$

$$\frac{dy}{dx} = \frac{1 - \sec^2(x + y)}{\sec^2(x + y)}$$

$$= \frac{-\tan^2(x + y)}{\tan^2(x + y) + 1} = \frac{-x^2}{x^2 + 1}$$

At $(0, 0)$, $\frac{dy}{dx} = \frac{0}{1} = 0$.

33. (a) Solving the equation for y, yields

$$9x^2 + 16y^2 = 144$$

$$16y^2 = 144 - 9x^2 = 9(16 - x^2)$$

$$y^2 = \frac{9}{16}(16 - x^2)$$

$$y = \pm\frac{3}{4}\sqrt{16 - x^2}, \quad -4 \le x \le 4$$

(c) When the explicit functions are differentiated explicitly, we obtain

$$\frac{dy}{dx} = \pm\left(\frac{3}{4}\right)\left(\frac{1}{2}\right)(16 - x^2)^{-1/2}(-2x)$$

$$= \frac{\mp 3x}{4\sqrt{16 - x^2}} = \frac{-3(3x)}{16[\pm(3/4)]\sqrt{16 - x^2}} = \frac{-9x}{16y}$$

(b) The graph is given in the figure.

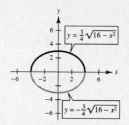

(d) Differentiating the equation $9x^2 + 16y^2 = 144$ implicitly yields

$$18x + 32y\frac{dy}{dx} = 0$$

$$\frac{dy}{dx} = \frac{-9x}{16y}$$

37. $x^2 - y^2 = 16$

$$2x - 2yy' = 0$$

$$-2yy' = -2x$$

$$y' = \frac{x}{y}$$

Differentiating implicitly again yields

$$y'' = \frac{y(1) - (x)y'}{y^2} = \frac{y - x(x/y)}{y^2} = \frac{y^2 - x^2}{y^3} = -\frac{16}{y^3}.$$

43. By implicit differentiation,

$$x^2 + y^2 = r^2$$

$$2x + 2yy' = 0$$

$$y' = -\frac{x}{y}.$$

Thus, if (x, y) is a point on the circle $x^2 + y^2 = r^2$, the slope of the tangent line at (x, y) is $-x/y$. On the other hand, the slope of the line passing through (x, y) and $(0, 0)$ is

$$m = \frac{y - 0}{x - 0} = \frac{y}{x}.$$

Since this slope is the negative reciprocal of y', the line passing through (x, y) and $(0, 0)$ must be perpendicular to the tangent line at (x, y).

47. To find the points of intersection, set $y^2 = 6 - 2x^2$ and $y^2 = 4x$ equal to each other.

$$4x = 6 - 2x^2$$

$$2x^2 + 4x - 6 = 0$$

$$x^2 + 2x - 3 = 0$$

$$(x + 3)(x - 1) = 0$$

$$x = -3 \quad \text{and} \quad x = 1$$

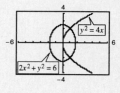

When $x = 1$, $y = \pm 2$, and when $x = -3$, y is undefined. Thus the two points of intersection are $(1, 2)$ and $(1, -2)$. Differentiating the equation $2x^2 + y^2 = 6$ implicitly yields

$$4x + 2yy' = 0$$

$$y' = -\frac{2x}{y}$$

and differentiating $y^2 = 4x$ implicitly yields

$$2yy' = 4$$

$$y' = \frac{2}{y}.$$

—CONTINUED—

47. —CONTINUED—

Thus at $(1, 2)$ the slopes of the two curves are

$$\frac{-2(1)}{2} = -1 \quad \text{and} \quad \frac{2}{2} = 1$$

which implies that the tangent lines at this point are perpendicular. Finally, at $(1, -2)$ the slopes of the two curves are

$$\frac{-2(1)}{-2} = 1 \quad \text{and} \quad \frac{2}{-2} = -1$$

and the tangent lines at this point are also perpendicular

57. (a) The graph of $x^4 = 4(4x^2 - y^2)$ is shown in the figure.

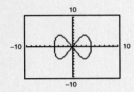

(b) $\quad x^4 = 4(4x^2 - y^2)$

$$4x^3 = 4\left(8x - 2y\frac{dy}{dx}\right)$$

$$2y\frac{dy}{dx} = 8x - x^3$$

$$\frac{dy}{dx} = \frac{x(8 - x^2)}{2y}$$

To find the x-coordinates of the points of tangency when $y = 3$, solve the given equation for x when $y = 3$.

$$x^4 = 4(4x^2 - 3^2)$$

$$x^4 - 16x^2 + 36 = 0$$

Let $t = x^2$ and obtain the quadratic equation $t^2 - 16t + 36 = 0$. Using the quadratic formula we have the following.

$$t = \frac{16 \pm \sqrt{(-16)^2 - 4(1)(36)}}{2}$$

$$t = x^2 = 2(4 \pm \sqrt{7})$$

$$x = \pm\sqrt{2(4 \pm \sqrt{7})} = \pm(1 \pm \sqrt{7})$$

Therefore, the four points of tangency and the slopes of the slopes of the tangent lines at each point are the following.

Point	*Slope*		
$(1 + \sqrt{7}, 3)$	$\dfrac{dy}{dx} = \dfrac{(1 + \sqrt{7})[8 - (1 + \sqrt{7})^2]}{2(3)}$	$= -\dfrac{1}{3}(7 + \sqrt{7})$	
$(1 - \sqrt{7}, 3)$	$\dfrac{dy}{dx} = \dfrac{(1 - \sqrt{7})[8 - (1 - \sqrt{7})^2]}{2(3)}$	$= -\dfrac{1}{3}(7 - \sqrt{7})$	
$(-1 + \sqrt{7}, 3)$	$\dfrac{dy}{dx} = \dfrac{(-1 + \sqrt{7})[8 - (-1 + \sqrt{7})^2]}{2(3)}$	$= \dfrac{1}{3}(7 - \sqrt{7})$	
$(-1 - \sqrt{7}, 3)$	$\dfrac{dy}{dx} = \dfrac{(-1 - \sqrt{7})[8 - (1 - \sqrt{7})^2]}{2(3)}$	$= \dfrac{1}{3}(7 + \sqrt{7})$	

—CONTINUED—

57. —CONTINUED—

Using each point of tangency and the corresponding slope of the tangent line given above, yields the following equations of the tangent lines.

$$y_1 = -\frac{1}{3}\left[(7 + \sqrt{7})x - (8\sqrt{7} + 23)\right]$$

$$y_2 = -\frac{1}{3}\left[(7 - \sqrt{7})x + (8\sqrt{7} - 23)\right]$$

$$y_3 = \frac{1}{3}\left[(7 - \sqrt{7})x - (8\sqrt{7} - 23)\right]$$

$$y_4 = \frac{1}{3}\left[(7 + \sqrt{7})x + (8\sqrt{7} + 23)\right]$$

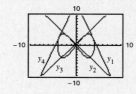

The graphs of the tangent lines are shown in the figure.

(c) From the figure in Part (b), we observe that y_1 and y_3 are the two tangent lines with points of tangency in the first quadrant. We now find the point of intersection of these two lines.

$$y_1 = y_3$$

$$-\frac{1}{3}\left[(7 + \sqrt{7})x - (8\sqrt{7} + 23)\right] = \frac{1}{3}\left[(7 - \sqrt{7})x - (8\sqrt{7} - 23)\right]$$

$$-(7 + \sqrt{7})x - (7 - \sqrt{7})x = -(8\sqrt{7} + 23) - (8\sqrt{7} - 23)$$

$$-14x = -16\sqrt{7} \implies x = \frac{8\sqrt{7}}{7}$$

To find the y-coordinate of the point of intersection substitute this value of x into the equation of one of the two lines. Using y_1 we have the following.

$$y_1 = -\frac{1}{3}\left[(7 + \sqrt{7})\left(\frac{8\sqrt{7}}{7}\right) - (8\sqrt{7} + 23)\right]$$

$$= -\frac{1}{3}(8\sqrt{7} + 8 - 8\sqrt{7} - 23) = 5$$

Therefore, the point of intersection is $\left(\frac{8\sqrt{7}}{7}, 5\right)$.

Section 2.6   Related Rates

3. Since x and y are differentiable functions of t, differentiating the equation $xy = 4$ with respect to t yields

(1) $\quad x\dfrac{dy}{dt} + y\dfrac{dx}{dt} = 0.$

(a) Solving (1) for dy/dt, and substituting $y = 1/2$ and $dx/dt = 10$ when $x = 8$, yields

$$\frac{dy}{dt} = -\frac{y}{x}\frac{dx}{dt} = -\frac{1/2}{8}(10) = -\frac{5}{8}.$$

(b) Solving (1) for dx/dt, and substituting $y = 4$ and $dy/dt = -6$ when $x = 1$, yields

$$\frac{dx}{dt} = -\frac{x}{y}\frac{dy}{dt} = -\frac{1}{4}(-6) = \frac{3}{2}.$$

13. *Area of a circle:* $A = \pi r^2$

Given rate: $\dfrac{dr}{dt} = 2$ centimeters per minute

Find: $\dfrac{dA}{dt}$

Differentiation *with respect to t* of the area formula produces

$$\frac{dA}{dt} = 2\pi r\frac{dr}{dt}.$$

(a) When $r = 6$ the rate of change of area is

$$\frac{dA}{dt} = 2\pi(6)(2) = 24\pi \text{ centimeters per minute.}$$

(b) When $r = 24$ the rate of change of area is

$$\frac{dA}{dt} = 2\pi(24)(2) = 96\pi \text{ centimeters per minute.}$$

21. Let

$$V = \text{volume of cone} = \frac{1}{3}\pi r^2 h.$$

Since the diameter of the base is approximately three times the altitude, we have

$$2r = 3h \quad \text{or} \quad r = \frac{3}{2}h$$

Therefore,

$$V = \frac{1}{3}\pi\left(\frac{3}{2}h\right)^2 h = \frac{3\pi}{4}h^3.$$

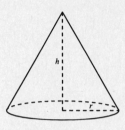

Differentiating with respect to t, yields

$$\frac{dV}{dt} = \frac{9\pi}{4}h^2\frac{dh}{dt} \quad \text{or} \quad \frac{4(dV/dt)}{9\pi h^2} = \frac{dh}{dt}.$$

Now, substituting $h = 15$ and $dV/dt = 10$, yields the following rate of change of the height of the conical pile.

$$\frac{dh}{dt} = \frac{4(10)}{9\pi(15)^2} = \frac{8}{405\pi} \text{ ft/min}.$$

23. From the figure we see that x and y are related by the equation

$$m = \frac{y - 0}{x - 0} = \frac{2 - 0}{12 - 0}$$

$$\frac{y}{x} = \frac{1}{6} \quad \text{or} \quad x = 6y.$$

The volume of the inclined portion of the pool is given by the product of the width of the pool and the area of the triangular cross section.

$$V_L = 6\left(\frac{1}{2}\right)xy = 3xy = 3(6y)y = 18y^2$$

When $y = 2$, the inclined portion of the pool has a volume of

$$V_L = 18(2^2) = 72 \text{ cubic meters}.$$

Since the upper rectangular portion of the pool has a volume of

$$V_U = 1(12)(6) = 72 \text{ cubic meters}$$

the total volume of the pool is

$$V = V_L + V_U = 72 + 72 = 144 \text{ cubic meters}.$$

(a) When $y = 1$, the ratio of the filled portion of the pool to the total volume is

$$\frac{18y^2}{V} = \frac{18(1^2)}{144} = 12.5\%.$$

(b) When $y = 1$, we can find dy/dt by differentiating $V_L = 18y^2$ and substituting $dV_L/dt = 1/4$.

$$V_L = 18y^2$$

$$\frac{dV_L}{dt} = 36y\frac{dy}{dt}$$

$$\frac{1}{4} = 36(1)\frac{dy}{dt}$$

$$\frac{dy}{dt} = \frac{1}{4(36)} = \frac{1}{144} \text{ meters per minute}$$

25. (a) From the figure it follows that x and y are related by the equation $x^2 + y^2 = 25^2$. Differentiating this equation with respect to t, yields

$$2x\frac{dx}{dt} + 2y\frac{dy}{dt} = 0$$

$$y\frac{dy}{dt} = -x\frac{dx}{dt}$$

$$\frac{dy}{dt} = -\frac{x}{y}\frac{dx}{dy}$$

Since $dx/dt = 2$, we have $dy/dt = -2(x/y)$.

When $x = 7$, $y = \sqrt{(25)^2 - 7^2} = \sqrt{576} = 24$.

$$\frac{dy}{dt} = -2\left(\frac{7}{24}\right) = -\frac{7}{12} \approx -0.583 \text{ ft/sec}$$

When $x = 15$, $y = \sqrt{(25)^2 - (15)^2} = \sqrt{400} = 20$.

$$\frac{dy}{dt} = -2\left(\frac{15}{20}\right) = -\frac{3}{2} = -1.5 \text{ ft/sec}$$

When $x = 24$, $y = \sqrt{(25)^2 - (24)^2} = \sqrt{49} = 7$.

$$\frac{dy}{dt} = -2\left(\frac{24}{7}\right) = -\frac{48}{7} \approx -6.857 \text{ ft/sec}$$

(b) The area of the triangle in the accompanying figure is $A = \frac{1}{2}xy = \frac{1}{2}x\sqrt{625 - x^2}$. Differentiating with respect to t yields

$$\frac{dA}{dt} = \frac{1}{2}\left[x\left(\frac{1}{2}\right)(625 - x^2)^{-1/2}\left(-2x\frac{dx}{dt}\right) + \sqrt{625 - x^2}\frac{dx}{dt}\right]$$

$$= \left(\frac{625 - 2x^2}{2\sqrt{625 - x^2}}\right)\frac{dx}{dt}.$$

The rate of change of the area when $x = 7$ and $dx/dt = 2$ is $\dfrac{dA}{dt} = \left(\dfrac{625 - 2(7)^2}{2\sqrt{625 - (7^2)}}\right)(2) = \dfrac{527}{24} \text{ ft}^2 \text{ sec}.$

(c) If θ is the angle between the top of the ladder and the wall of the house, then

$$\sin\theta = \frac{x}{25}$$

$$\cos\theta\frac{d\theta}{dt} = \frac{1}{25}\frac{dx}{dt}$$

$$\frac{d\theta}{dt} = \frac{\sec\theta}{25}\frac{dx}{dt}.$$

When $x = 7$, $\sin\theta = 7/25$ and $\sec\theta = 25/24$. Since $dx/dt = 2$, the rate of change of the angle θ is

$$\frac{d\theta}{dt} = \frac{25/24}{25}(2) = \frac{1}{12} \text{ rad/sec}.$$

27. The relationship between x and y in the figure is given by $x = \sqrt{12^2 - y^2}$ and the relationship among the variables x, y, and s is given by $s^2 = x^2 + (12 - y)^2$. Substituting for x in the second equation and differentiating with respect to t yields

$$s^2 = \left(\sqrt{12^2 - y^2}\right)^2 + (12 - y)^2 = 12^2 - y^2 + (12 - y)^2$$

$$2s\frac{ds}{dt} = -2y\frac{dy}{dt} + 2(12 - y)(-1)\frac{dy}{dt}$$

$$s\frac{ds}{dt} = -12\frac{dy}{dt}$$

$$-\frac{s}{12}\frac{ds}{dt} = \frac{dy}{dt}.$$

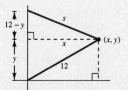

—CONTINUED—

27. **—CONTINUED—**

When $y = 6$, $s = \sqrt{(12^2 - 6^2) + (12 - 6)^2} = 12$. Substituting this value for s and -0.2 meters per second for ds/dt yields

$$\frac{dy}{dt} = -\frac{12}{12}(-0.2) = \frac{1}{5} \text{ meters per second.}$$

When $y = 6$, $x = \sqrt{12^2 - 6^2} = 6\sqrt{3}$. Differentiating the equation $s^2 = x^2 + (12 - y)^2$ with respect to t yields

$$2s\frac{ds}{dt} = 2x\frac{dx}{dt} + 2(12 - y)(-1)\frac{dy}{dt}.$$

Solving this equation for dx/dt and substituting the known values for x, y, dy/dt and ds/dt, yields

$$\frac{dx}{dt} = \frac{s(ds/dt) + (12 - y)(dy/dt)}{x}$$

$$= \frac{12(-0.2) + (12 - 6)(1/5)}{6\sqrt{3}}$$

$$= \frac{-2.4 + 1.2}{6\sqrt{3}} = -\frac{\sqrt{3}}{15} \text{ meters per second.}$$

33. From the figure we see that x and s are related by similar triangles in such a way that

$$\frac{s - x}{6} = \frac{s}{15}$$

$$15s - 15x = 6s$$

$$9s = 15x$$

$$s = \frac{5}{3}x.$$

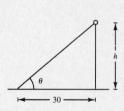

(a) To find ds/dt, given that $dx/dt = 5$, differentiate with respect to t as follows:

$$\frac{ds}{dt} = \frac{5}{3} \cdot \frac{dx}{dt} = \frac{5}{3}(5) = \frac{25}{3} \approx 8.3 \text{ ft/sec}$$

(b) The rate at which the shadow in increasing is

$$\frac{ds}{dt} - \frac{dx}{dt} = \frac{25}{3} - 5 = \frac{10}{3} \approx 3.3 \text{ ft/sec.}$$

(Note: The measurement 10 feet given in this problem is a "red herring" since the distance from the base of the light does not affect ds/dt.)

41. From the figure it follows that $\tan \theta = h/30$. Differentiating with respect to t yields the following.

$$\tan \theta = \frac{h}{30}$$

$$\sec^2 \theta \frac{d\theta}{dt} = \frac{1}{30}\frac{dh}{dt}$$

$$\frac{d\theta}{dt} = \frac{1}{30}\cos^2 \theta \frac{dh}{dt}$$

We are given that $dh/dt = 3$ meters per second. Also, when $h = 30$, $\theta = \pi/4$. Substituting into the derivative yields

$$\frac{d\theta}{dt} = \frac{1}{30}\cos^2 \frac{\pi}{4}(3)$$

$$= \frac{1}{10} \cdot \frac{1}{2}$$

$$= \frac{1}{20} \text{ radians per second}$$

$$\approx 2.86 \text{ degrees per second.}$$

45. (a) We are given that the wheel rotates at 10 revolutions per second in a clockwise direction. Therefore,

$$\frac{d\theta}{dt} = 10(2\pi) = 20\pi \text{ radians per second.}$$

From the figure we have

$$x = 30 \cos \theta$$

$$\frac{dx}{dt} = -30 \sin \theta \frac{d\theta}{dt}$$

$$= -30 \sin \theta(20\pi) = -600\pi \sin \theta.$$

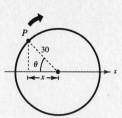

(b) The graph is given in the figure.

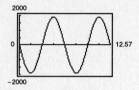

(c) The absolute value of dx/dt is greatest when absolute value of $\sin \theta$ is greatest. This occurs when

$$\theta = (2n - 1)\frac{\pi}{2}, \ n \text{ an integer.}$$

The absolute value of dx/dt is least when $\sin \theta$ is 0. This occurs when

$$\theta = n\pi, \quad n \text{ and integer.}$$

(d) $\theta = 30°$: $\dfrac{dx}{dt} = -600\pi\left(\dfrac{1}{2}\right) = -300\pi$ centimeters per second.

$\theta = 60$: $\dfrac{dx}{dt} = -600\pi\left(\dfrac{\sqrt{3}}{2}\right) = -300\sqrt{3}\pi$ centimeters per second.

Review Exercises for Chapter 2

9. One method of finding the derivative of f is by rewriting the function prior to taking the derivative.

$$f(x) = \frac{2x^3 - 1}{x^2} = 2x - x^{-2}$$

$$f'(x) = 2 + 2x^{-3}$$

$$= 2\left(1 + \frac{1}{x^3}\right) = \frac{2(x^3 + 1)}{x^3}$$

A second method of finding the derivative is by using the Quotient Rule.

$$f(x) = \frac{2x^3 - 1}{x^2}$$

$$f'(x) = \frac{x^2(6x^2) - (2x^3 - 1)(2x)}{x^4}$$

$$= \frac{6x^4 - 4x^4 + 2x}{x^4}$$

$$= \frac{2x^4 + 2x}{x^4} = \frac{2(x^3 + 1)}{x^3}$$

15. $f(x) = (3x^2 + 7)(x^2 - 2x + 3)$

$f'(x) = (3x^2 + 7)(2x - 2) + (x^2 - 2x + 3)(6x)$

$= 6x^3 - 6x^2 + 14x - 14 + 6x^3 - 12x^2 + 18x$

$= 2(6x^3 - 9x^2 + 16x - 7)$

19. $f(x) = \dfrac{x^2 + x - 1}{x^2 - 1}$

$$f'(x) = \frac{(x^2 - 1)(2x + 1) - (x^2 + x - 1)(2x)}{(x^2 - 1)^2}$$

$$= \frac{2x^3 + x^2 - 2x - 1 - 2x^3 - 2x^2 + 2x}{(x^2 - 1)^2}$$

$$= -\frac{x^2 + 1}{(x^2 - 1)^2}$$

23. $y = 3 \cos(3x + 1)$

$$\frac{dy}{dx} = 3[-\sin(3x + 1)]\frac{d}{dx}[3x + 1]$$

$$= -3 \sin(3x + 1)(3) = -9 \sin(3x + 1)$$

31. $f(x) = -x \tan x$

$f'(x) = -x \sec^2 x + (\tan x)(-1)$

$= -(x \sec^2 x + \tan x)$

35. Using a symbolic differentiation utility, the first derivative of the function $f(t) = t^2(t-1)^5$ is

$$f'(t) = t(t-1)^4(7t-2).$$

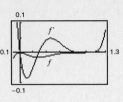

(Your differentiation utility may give the derivative in a different form.) The graphs of f and f' are given in the figure. The zeros of f' correspond to the points on the graph of f where the tangent line is horizontal.

47. $f(x) = \cot x$

$f'(x) = -\csc^2 x$

$f''(x) = -2\csc x(-\csc x \cot x)$

$\qquad = 2\csc^2 x \cot x$

53. $\qquad\qquad x^2 + 3xy + y^3 = 10$

$$\frac{d}{dx}[x^2 + 3xy + y^3] = \frac{d}{dx}[10]$$

$$\frac{d}{dx}[x^2] + \frac{d}{dx}[3xy] + \frac{d}{dx}[y^3] = \frac{d}{dx}[10]$$

$$2x + 3x\frac{dy}{dx} + 3y + 3y^2\frac{dy}{dx} = 0$$

$$(3x + 3y^2)\frac{dy}{dx} = -2x - 3y$$

$$\frac{dy}{dx} = \frac{-(2x + 3y)}{3(x + y^2)}$$

63. $y = \sqrt[3]{(x-2)^2} = (x-2)^{2/3}$

$$y' = \left(\frac{2}{3}\right)(x-2)^{-1/3} = \frac{2}{3(x-2)^{1/3}}$$

Thus the slope of the tangent line at $(3, 1)$ is

$$y' = \frac{2}{3(3-2)^{1/3}} = \frac{2}{3},$$

and the equation of the tangent line at $(3, 1)$ is

$$y - 1 = \frac{2}{3}(x - 3)$$

$$3y - 3 = 2x - 6$$

$$-2x + 3y + 3 = 0$$

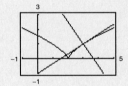

Since the slope of the tangent line is $2/3$, the slope of the normal line is $-3/2$ and the equation of the normal line is

$$y - 1 = -\frac{3}{2}(x - 3)$$

$$2y - 2 = -3x + 9$$

$$3x + 2y - 11 = 0.$$

The graphs are shown in the figure.

69. Begin by finding the first and second derivatives of the function.

$$y = 2\sin x + 3\cos x$$

$$y' = 2\cos x - 3\sin x$$

$$y'' = -2\sin x - 3\cos x$$

$$y'' + y = (-2\sin x - 3\cos x) + (2\sin x + 3\cos x) = 0$$

75. We assume that the stone is thrown from an initial height of $s_0 = 0$. Thus the position equation is

$$s = -16t^2 + v_0 t.$$

The maximum value of s occurs when $ds/dt = 0$ and thus we have

$$\frac{ds}{dt} = -32t + v_0 = 0$$

$$-32t = -v_0$$

$$t = \frac{v_0}{32}.$$

This means that the maximum height is

$$s = -16\left(\frac{v_0}{32}\right)^2 + v_0\left(\frac{v_0}{32}\right) = \frac{v_0^2}{64}.$$

If s is to attain a value of 49, we must have

$$\frac{v_0^2}{64} = 49$$

$$v_0^2 = 3136$$

$$v_0 = 56 \text{ ft/sec}.$$

81. The figure is a cross section of the trough when the water is a depth of h meters.

$$\frac{s}{h} = \frac{1/2}{2}$$

$$s = \frac{1}{4}h$$

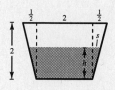

A = area of cross section of water at depth h

$$= 2h + 2\left(\frac{1}{2}sh\right)$$

$$= 2h + \left(\frac{1}{4}h\right)h = 2h + \frac{1}{4}h^2$$

V = volume of water in trough at depth h

$$= 5A = 5\left(2h + \frac{1}{4}h^2\right)$$

Differentiating with respect to t, yields

$$\frac{dV}{dt} = 5\left(2 + \frac{1}{2}h\right)\frac{dh}{dt} = \frac{5}{2}(4 + h)\frac{dh}{dt}$$

$$\frac{2(dV/dt)}{5(4 + h)} = \frac{dh}{dt}$$

Therefore, when $dV/dt = 1$ and $h = 1$, we have

$$\frac{dh}{dt} = \frac{2(1)}{5(4 + 1)} = \frac{2}{25} \text{ meters per minute}.$$

CHAPTER 3
Applications of Differentiation

C H A P T E R 3
Applications of Differentiation

Section 3.1   Extrema on an Interval
Solutions to Selected Odd-Numbered Exercises

7. Let f be defined at c. If $f'(c) = 0$ or if f' is undefined at c, then c is called a critical number of f. Begin by finding f'.

$$f(x) = x^2(x - 3) = x^3 - 3x^2$$

$$f'(x) = 3x^2 - 6x = 3x(x - 2)$$

Since $f'(0) = 0$ and $f'(2) = 0$, $x = 0$ and $x = 2$ are critical numbers. f' is defined for all real numbers x.

15. $f(x) = -x^2 + 3x$

$$f'(x) = -2x + 3 = 0$$

Therefore, $x = \frac{3}{2}$ is a critical number in $[0, 3]$. We determine the extrema of f by evaluating f at the critical number and at the endpoints of $[0, 3]$.

$$f(0) = -0^2 + 3(0) = 0 \qquad \text{Minimum}$$

$$f\left(\tfrac{3}{2}\right) = -\left(\tfrac{3}{2}\right)^2 + 3\left(\tfrac{3}{2}\right) = \tfrac{9}{4} \qquad \text{Maximum}$$

$$f(3) = -3^2 + 3(3) = 0 \qquad \text{Minimum}$$

19. $f(x) = 3x^{2/3} - 2x$

$$f(x) = 2x^{-1/3} - 2$$

$$= 2\left(\frac{1 - \sqrt[3]{x}}{\sqrt[3]{x}}\right)$$

Therefore, $x = 1$ and $x = 0$ are critical numbers in the interval $[-1, 1]$. $[f'(0)$ is undefined.$]$

We determine the extrema of f by evaluating f at the critical numbers and at the endpoints of the interval $[-1, 1]$.

$$f(-1) = 3(-1)^{2/3} - 2(-1) = 5 \qquad \text{Maximum}$$

$$f(0) = 3(0)^{2/3} - 2(0) = 0 \qquad \text{Minimum}$$

$$f(1) = 3(1)^{2/3} - 2(1) = 1$$

25. $f(x) = \cos \pi x$

$$f'(x) = -\sin \pi x \frac{d}{dx}[\pi x] = -\pi \sin \pi x$$

Therefore, $x = 0$ is a critical number in the interval $[0, 1/6]$, and the extrema are at endpoints of the interval.

$$f(0) = \cos 0 = 1 \qquad \text{Maximum}$$

$$f\left(\frac{1}{6}\right) = \cos \frac{\pi}{6} = \frac{\sqrt{3}}{2} \qquad \text{Minimum}$$

31. (a) Since f is decreasing in the interval (a, c) and increasing in the interval (c, b), the only possible minimum in the interval (a, b) would occur at $x = c$. However, $f(c)$ is greater than $f(x)$ for x near c. Thus, there is no minimum.

(b) Since $f(c) \le f(x)$ for all x in (a, b), $f(c)$ is a minimum.

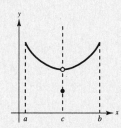

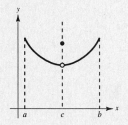

37. $f(x) = \begin{cases} 2x + 2, & 0 \le x \le 1 \\ 4x^2, & 1 < x \le 3 \end{cases}$

At $x = 1$, the derivative from the left is

$$\lim_{x \to 1^-} \frac{f(x) - f(1)}{x - 1} = \lim_{x \to 1^-} \frac{(2x + 2) - 4}{x - 1}$$

$$= \lim_{x \to 1^-} \frac{2(x - 1)}{x - 1}$$

$$= \lim_{x \to 1^-} 2 = 2.$$

At $x = 1$, the derivative from the right is

$$\lim_{x \to 1^+} \frac{f(x) - f(1)}{x - 1} = \lim_{x \to 1^+} \frac{4x^2 - 4}{x - 1}$$

$$= \lim_{x \to 1^+} \frac{4(x + 1)(x - 1)}{x - 1}$$

$$= \lim_{x \to 1^+} [4(x + 1)] = 8.$$

Since the one-sided limits are unequal, the derivative does not exist at $x = 1$, and $x = 1$ is a critical number. We determine the extrema of f by evaluating f at the critical number and at the endpoints of $[0, 3]$.

$f(0) = 2(0) + 2 = 2$ Minimum

$f(1) = 2(1) + 2 = 4$

$f(3) = 4(3^2) = 36$ Maximum

These results agree with the graph shown in the figure.

43. To find the maximum value of $|f''(x)|$ in $[0, 2]$, select the maximum of $|f''(0)|$, $|f''(2)|$, and $|f''(c)|$, where c is any critical number of $f''(x)$ [i.e., $f'''(c) = 0$ or $f'''(c)$ does not exist.] Using a symbolic differentiation utility yields the following derivatives.

$$f(x) = \sqrt{1 + x^3}$$

$$f'(x) = \frac{3x^2}{2\sqrt{1 + x^3}}$$

$$f''(x) = \frac{3x(x^3 + 4)}{4(1 + x^3)^{3/2}}$$

$$f'''(x) = \frac{-3(x^6 + 20x^3 - 8)}{8(1 + x^3)^{5/2}}$$

Therefore, $f'''(x) = 0$ when $x^6 + 20x^3 - 8 = 0$. Solving this equation using the symbolic differentiation utility yields the critical numbers

$$x = \begin{cases} \sqrt[3]{-10 + 6\sqrt{3}} & \approx 0.732 \\ \sqrt[3]{-10 - 6\sqrt{3}} & \approx -2.732 \end{cases}$$

The critical number in the interval $[0, 2]$ is $\sqrt[3]{-10 + 6\sqrt{3}}$. Evaluating the absolute value of the second derivative at the critical number and the endpoints of the interval, yields

$$\left| f''\left(\sqrt[3]{-10 + 6\sqrt{3}}\right) \right| \approx 1.468$$

$$|f''(0)| = 0$$

$$|f''(2)| = \frac{2}{3}.$$

The maximum value of $|f''(x)|$ in $[0, 2]$ is $\left| f''\left(\sqrt[3]{-10 + 6\sqrt{3}}\right) \right| \approx |f''(0.732)| \approx 1.468$.

47. Since $V = 12$ and $R = 0.5$, you have

$$P = 12I - \frac{1}{2}I^2$$

$$\frac{dP}{dI} = 12 - I.$$

Therefore, $I = 12$ is a critical number on the interval $[0, 15]$. We determine the maximum of P by evaluating P at the critical number and at the endpoints of the interval $[0, 15]$. Since

$$P(0) = 12(0) - \frac{1}{2}(0)^2 = 0$$

$$P(12) = 12(12) - \frac{1}{2}(12)^2 = 72$$

$$P(15) = 12(15) - \frac{1}{2}(15)^2 = 67.5,$$

it follows that the power P is maximum when $I = 12$.

Section 3.2   Rolle's Theorem and the Mean Value Theorem

5. Since $f(x) = (x - 1)(x - 2)(x - 3)$ is a polynomial, it is continuous and differentiable for all x. Also, the zeros of f are $x = 1$, $x = 2$, and $x = 3$. Thus, Rolle's Theorem can be applied on the intervals $[1, 2]$ and $[2, 3]$. Setting $f'(x) = 0$ yields

$$f'(x) = 3x^2 - 12x + 11 = 0$$

$$x = \frac{12 \pm \sqrt{144 - 132}}{6} = \frac{12 \pm 2\sqrt{3}}{6} = \frac{6 \pm \sqrt{3}}{3}.$$

Therefore, in the interval $[1, 2]$,

$$f'\left(\frac{6 - \sqrt{3}}{3}\right) = 0 \quad \text{where} \quad c = \frac{6 - \sqrt{3}}{3} \approx 1.423.$$

and in the interval $[2, 3]$,

$$f'\left(\frac{6 + \sqrt{3}}{3}\right) = 0 \quad \text{where} \quad c = \frac{6 + \sqrt{3}}{3} \approx 2.577.$$

7. We first observe that $f(x) = x^{2/3} - 1$ is continuous on the interval $[-8, 8]$. The zeros of f are found by solving the following equation.

$$f(x) = x^{2/3} - 1 = 0$$

$$x^{2/3} = 1$$

$$x^2 = 1 \quad \text{or} \quad x = \pm 1$$

Since

$$f'(x) = \frac{2}{3x^{1/3}},$$

we observe that $f'(0)$ is undefined and therefore, f is not differentiable at $x = 0$. Thus, Rolle's Theorem cannot be applied to this function on the specified interval.

19. f is continuous on $[-(1/4), (1/4)]$ and differentiable on $(-(1/4), (1/4))$. Therefore, Rolle's Theorem applies. Setting $f'(x) = 0$ yields the following.

$$f(x) = 4x - \tan \pi x$$

$$f'(x) = 4 - \pi \sec^2 \pi x$$

$$4 - \pi \sec^2 \pi x = 0$$

$$\sec^2 \pi x = \frac{4}{\pi}$$

$$\sec \pi x = \pm \frac{2}{\sqrt{\pi}}$$

$$x = \pm \frac{1}{\pi} \operatorname{arcsec} \frac{2}{\sqrt{\pi}}$$

$$= \pm \frac{1}{\pi} \arccos \frac{\sqrt{\pi}}{2} \approx \pm 0.1533 \text{ radian}$$

Therefore, $c \approx \pm 0.1433$ radian.

29. Since $f(x) = x^{2/3}$ is continuous on $[0, 1]$ and differentiable on $(0, 1)$, the Mean Value Theorem can be applied.

$$f(x) = x^{2/3}$$

(1) $\qquad f'(x) = \frac{2}{3}x^{-1/3} = \frac{2}{3\sqrt[3]{x}}$

(2) $\qquad \frac{f(1) - f(0)}{1 - 0} = \frac{1 - 0}{1 - 0} = 1$

Equating the right-hand members of the equations (1) and (2) yields

$$\frac{2}{3\sqrt[3]{x}} = 1$$

$$\frac{2}{3} = \sqrt[3]{x}$$

$$\frac{8}{27} = x.$$

Therefore, $c = 8/27$.

35. (a) The graph of $f(x) = x/(x + 1)$ on the interval $\left[-\frac{1}{2}, 2\right]$ is shown in the figure.

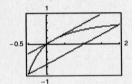

(b) The slope of the required secant line is

$$\frac{f(2) - f(-1/2)}{2 - (-1/2)} = \frac{(2/3) - (-1)}{5/2} = \frac{2}{3}.$$

The secant line passes through the point $\left(-\frac{1}{2}, -1\right)$. The equation of the secant line is

$$y - (-1) = \frac{2}{3}\left[x - \left(-\frac{1}{2}\right)\right]$$

$$y = \frac{2}{3}x - \frac{2}{3}.$$

The graph of the secant line is shown on the graph in part (a).

(c) To find the equation of the tangent line, we need to find the point of tangency where the slope of the tangent line equals $\frac{2}{3}$, the slope of the secant line. We begin by finding the derivative.

$$f(x) = \frac{x}{x + 1}$$

$$f'(x) = \frac{(x + 1)(1) - x(1)}{(x + 1)^2} = \frac{1}{(x + 1)^2}$$

$$\frac{1}{(x + 1)^2} = \frac{2}{3}$$

$$x + 1 = \pm\sqrt{\frac{3}{2}} \implies x = -1 \pm \frac{\sqrt{6}}{2}$$

Therefore, on the interval $\left[-\frac{1}{2}, 2\right]$, the value of c in the Mean Value Theorem is $c = -1 + \left(\sqrt{6}/2\right)$.

$$f\left(-1 + \frac{\sqrt{6}}{2}\right) = \frac{-1 + \left(\sqrt{6}/2\right)}{\left(-1 + \sqrt{6}/2\right) + 1} = 1 - \frac{\sqrt{6}}{3}$$

Therefore, the point of tangency is

$$\left(-1 + \frac{\sqrt{6}}{2}, 1 - \frac{\sqrt{6}}{3}\right)$$

and the equation of tangent line is

$$y - \left(1 - \frac{\sqrt{6}}{3}\right) = \frac{2}{3}\left[x - \left(-1 + \frac{\sqrt{6}}{2}\right)\right]$$

$$y = \frac{2}{3}x + \frac{1}{3}\left(5 - 2\sqrt{6}\right).$$

The graph of the tangent line is shown on the graph of part (a).

43. (a) The position function is $s(t) = -4.9t^2 + 500$. The average velocity during the first three seconds is

$$\frac{s(3) - s(0)}{3 - 0} = \frac{455.9 - 500}{3} = -\frac{44.1}{3} = -14.7.$$

(b) The instantaneous velocity of the object is given by $s'(t) = -9.8t$. We must find the time t in the interval $[0, 3]$ such that $s'(t) = -14.7$.

$$s'(t) = -14.7$$

$$-9.8t = -14.7 \implies t = 1.5$$

The instantaneous velocity equals the average velocity over the interval $[0, 3]$ when $t = 1.5$ seconds.

51. The polynomial $f(x) = x^{2n+1} + ax + b$ is continuous and differentiable for all x. Therefore, by Rolle's Theorem, if $f(x) = 0$ for two distinct values of x, there must be at least one value of x such that $f'(x) = 0$. However,

$$f(x) = x^{2n+1} + ax + b$$

$$f'(x) = (2n + 1)x^{2n} + a = 0$$

$$(x^n)^2 = -\frac{a}{2n + 1}$$

(positive number) = (negative number).

Therefore, $f'(x) = 0$ has no solution and consequently $f(x) = 0$ cannot have two real zeros.

Section 3.3   Increasing and Decreasing Functions and The First Derivative Test

11. $f(x) = 2x^3 + 3x^2 - 12x$

$f'(x) = 6x^2 + 6x - 12 = 6(x + 2)(x - 1)$

Therefore, $f'(x) = 0$ when $x = -2$ or $x = 1$. Since f is a polynomial, it is differentiable for all x and the only critical numbers are $x = -2$ and $x = 1$.

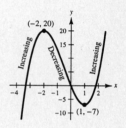

Interval	$-\infty < x < -2$	$-2 < x < 1$	$1 < x < \infty$
Test value	$x = -3$	$x = 0$	$x = 2$
Sign of $f'(x)$	$f'(-3) > 0$	$f'(0) < 0$	$f'(2) > 0$
Conclusion	f is increasing	f is decreasing	f is increasing

When $x = -2$, we have $f(-2) = 2(-2)^3 + 3(-2)^2 - 12(-2) = 20$. When $x = 1$, we have $f(1) = 2 + 3 - 12 = -7$. Therefore, it follows that $(-2, 20)$ is a relative maximum and $(1, -7)$ is a relative minimum.

15. $f(x) = x^{1/3} + 1$

$f'(x) = \left(\frac{1}{3}\right)x^{-2/3} = \frac{1}{3x^{2/3}}$

Since f is continuous for all x and differentiable for all x other than $x = 0$, the only critical number is $x = 0$. ($f'(0)$ is undefined.) We also observe that $f'(x) > 0$ for all x not equal to zero. Therefore, it follows that f is increasing for all x and there are no relative extrema.

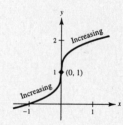

27. The function f is continuous for all x other than $x = -1$. The graph of f has a vertical asymptote at $x = -1$.

$$f(x) = \frac{x^2 - 2x + 1}{x + 1}$$

$$f'(x) = \frac{(x + 1)(2x - 2) - (x^2 - 2x + 1)(1)}{(x + 1)^2} = \frac{(x + 1)(2)(x - 1) - (x - 1)^2}{(x + 1)^2}$$

$$= \frac{(x - 1)(2x + 2 - x + 1)}{(x + 1)^2} = \frac{(x - 1)(x + 3)}{(x + 1)^2}$$

Since f is differentiable for all x other than $x = -1$, the only critical numbers are $x = 1$ and $x = -3$.

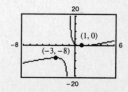

Interval	$-\infty < x < -3$	$-3 < x < -1$	$-1 < x < 1$	$1 < x \infty$
Test value	$x = -4$	$x = -2$	$x = 0$	$x = 2$
Sign of $f'(x)$	$f'(-4) > 0$	$f'(-2) < 0$	$f'(0) < 0$	$f'(2) > 0$
Conclusion	f is increasing	f is decreasing	f is decreasing	f is increasing

When $x = -3$, we have $f(-3) = (9 + 6 + 1)/(-2) = -8$, and when $x = 1$, we have $f(1) = (1 - 2 + 1)/2 = 0$. Therefore, $(-3, -8)$ is a relative maximum and $(1, 0)$ is a relative minimum. The figure is the graph of f produced by a graphing utility.

29. $f(x) = \dfrac{x}{2} + \cos x$

$f'(x) = \dfrac{1}{2} - \sin x$

Hence, $f'(x) = 0$ when $x = \pi/6$ or $x = 5\pi/6$. Since f is continuous and differentiable for all x, the only critical numbers are $x = \pi/6$ and $5\pi/6$.

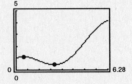

Interval	$0 < x < \pi/6$	$\pi/6 < x < 5\pi/6$	$5\pi/6 < x < 2\pi$
Test value	$x = \pi/12$	$x = \pi/2$	$x = \pi$
Sign of $f'(x)$	$f'(\pi/12) > 0$	$f'(\pi/2) < 0$	$f'(\pi) > 0$
Conclusion	f is increasing	f is decreasing	f is increasing

We conclude that a relative maximum occurs at $\left(\dfrac{\pi}{6}, \dfrac{\pi + 6\sqrt{3}}{12}\right)$ and a relative minimum at $\left(\dfrac{5\pi}{6}, \dfrac{5\pi - 6\sqrt{3}}{12}\right)$. The figure is the graph of f produced by a graphing utility.

35. (a) $f(t) = t^2 \sin t$

$f'(t) = t^2 \cos t + 2t \sin t = t(t \cos t + 2 \sin t)$

(b) The graphs of f and f' are shown in the figure.

(c) To find the critical numbers, solve the equation $f'(t) = 0$ on the interval $[0, 2\pi]$.

$$t(t \cos t + 2 \sin t) = 0$$

$$t = 0 \text{ or } t \cos t + 2 \sin t = 0$$

To solve the second equation, use the root finding capabilities of the symbolic differentiation utility to obtain $t \approx 2.2889$ and $t \approx 5.0870$ in the interval $[0, 2\pi]$.

Interval	$0 < t < 2.2889$	$2.2889 < t < 5.0870$	$5.0870 < t < 2\pi$
Test value	$t = 1$	$t = 3$	$t = 6$
Sign of $f'(x)$	$f'(1) > 0$	$f'(3) < 0$	$f'(6) > 0$
Conclusion	f is increasing	f is decreasing	f is increasing

f is increasing when f' is positive and decreasing when f' is negative.

45. The graph of f is decreasing in the interval $(-\infty, 2)$, and therefore, f' is negative in this interval. The graph of f is increasing in the interval $(2, \infty)$, and therefore, f' is positive in this interval. The graph of f has a horizontal tangent at $x = 2$, and therefore $f'(2) = 0$. Using this information, we can sketch a possible graph of f' as shown in the figure.

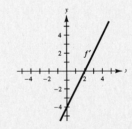

55. $v = k(R - r)r^2 = k(Rr^2 - r^3)$

$\dfrac{dv}{dr} = k(2Rr - 3r^2) = kr(2R - 3r)$

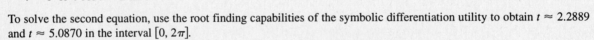

Interval	$-\infty < r < 0$	$0 < r < \frac{2}{3}R$	$\frac{2}{3}R < r < \infty$
Test value	$r = -R$	$r = \frac{R}{3}$	$r = \frac{4R}{3}$
Sign of $v'(r)$	$v'(-R) < 0$	$v'\left(\frac{R}{3}\right) > 0$	$v'\left(\frac{4R}{3}\right) < 0$
Conclusion	v is decreasing	v is increasing	v is decreasing

Therefore, $dv/dr = 0$ when $r = 0$ or $r = \frac{2}{3}R$. Since v is continuous and differentiable for all r, the only critical numbers are $r = 0$ and $r = 2R/3$.

We conclude that the velocity v is maximum when $r = (2R)/3$.

61. (a) Since there are two relative extrema, there are two critical numbers. This implies that the minimum degree of f' is 2. Therefore, the minimum degree of f is 3 and $f(x) = a_3x^3 + a_2x^2 + a_1x + a_0$.

(c) Using a graphing utility to solve the system in part (b) yields $a_3 = -\frac{1}{2}$ and $a_2 = \frac{3}{2}$. The required polynomial function is

$$f(x) = -\tfrac{1}{2}x^3 + \tfrac{3}{2}x^2.$$

(d) The graph of f is shown in the figure.

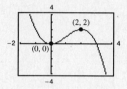

(b) Since $(0, 0)$ and $(2, 2)$ are solution points, we have the following.

$$f(0) = 0 = a_3(0^3) + a_2(0^2) + a_1(0) + a_0$$

$$a_0 = 0$$

$$f(2) = 2 = a_3(2^3) + a_2(2^2) + a_1(2) + a_0$$

$$2 = 8a_3 + 4a_2 + 2a_1 + a_0$$

(1) $1 = 4a_3 + 2a_2 + a_1$

Since $(0, 0)$ and $(2, 2)$ are extrema, $x = 0$ and $x = 2$ are critical numbers. The derivative of f is

$$f'(x) = 3a_3x^2 + 2a_2x + a_1.$$

$$f'(0) = 0 = 3a_3(0^2) + 2a_2(0) + a_1$$

$$0 = a_1$$

$$f'(2) = 0 = 3a_3(2^2) + 2a_2(2) + a_1$$

$$0 = 12a_3 + 4a_2 + 0$$

(2) $0 = 3a_3 + a_2$

Hence, $a_0 = a_1 = 0$. To find a_2 and a_3 solve the linear system comprised of equations (1) and (2).

$$4a_3 + 2a_2 = 1$$

$$3a_3 + a_2 = 0$$

Section 3.4   Concavity and the Second Derivative Test

11. $f(x) = x^3 - 3x^2 + 3$

$f'(x) = 3x^2 - 6x = 3x(x - 2)$

$f'(x) = 0$ when $x = 0, 2$

$f''(x) = 6x - 6$

At $x = 0$, we have $f(0) = 3, f'(0) = 0$, and $f''(0) = -6$. Therefore, by the Second Derivative Test, $(0, 3)$ is a relative maximum.

At $x = 2$, we have $f(2) = -1, f'(2) = 0$, and $f''(2) = 6$. Therefore, by the Second Derivative Test, $(2, -1)$ is a relative minimum.

17. $f(x) = x + \dfrac{4}{x}$

$f'(x) = 1 - \dfrac{4}{x^2} = \dfrac{x^2 - 4}{x^2}$

$f'(x) = 0$ when $x = \pm 2$

(Note that f' is undefined when $x = 0$. It is **not** a critical number since it is not in the domain of f.)

$f''(x) = (-4)(-2)x^{-3} = \dfrac{8}{x^3}$

At $x = -2$, we have $f(-2) = -4, f'(-2) = 0$, and $f''(-2) = -1$. Therefore, by the Second Derivative Test, $(-2, -4)$ is a relative maximum.

Since the graph of f is symmetric to the origin, $(2, 4)$ is a relative minimum.

19. $f(x) = \cos x - x$

$f'(x) = -\sin x - 1$

Therefore, $f'(x) = 0$ when $\sin x = -1$. The critical numbers in the interval $[0, 4\pi]$ are

$$x = \frac{3\pi}{2} \quad \text{and} \quad x = \frac{7\pi}{2}.$$

However, there are no relative extrema since $f'(x) \le 0$ for all x.

27. $f(x) = x(x - 4)^3$

$f'(x) = x[3(x - 4)^2(1)] + (x - 4)^3(1)$

$\quad = (x - 4)^2[3x + (x - 4)] = 4(x - 4)^2(x - 1)$

$f'(x) = 0$ when $x = 1, 4$

$f''(x) = 4[(x - 4)^2(1) + (x - 1)(2)(x - 4)(1)]$

$\quad = 4(x - 4)[(x - 4) + 2(x - 1)] = 12(x - 4)(x - 2)$

Possible points of inflection occur at $x = 2$ and $x = 4$. By testing the intervals determined by these x-values, we can conclude that they both yield points of inflection.

Interval	$-\infty < x < 2$	$2 < x < 4$	$4 < x < \infty$
Test value	$x = 0$	$x = 3$	$x = 5$
Sign of $f''(x)$	$f''(0) > 0$	$f''(3) < 0$	$f''(5) > 0$
Conclusion	Concave upward	Concave downward	Concave upward

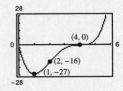

Therefore, $(1, -27)$ is a relative minimum, and $(2, -16)$ and $(4, 0)$ are points of inflection. Note that there is **not** a relative extrema at $(4, 0)$ even though $x = 4$ is a critical number. The first derivative remains nonnegative as x increases through the critical number $x = 4$.

29. $f(x) = x\sqrt{x + 3}$

$f'(x) = x\left(\dfrac{1}{2}\right)(x + 3)^{-1/2}(1) + (x + 3)^{1/2}(1)$

$\quad = \dfrac{x}{2\sqrt{x + 3}} + \sqrt{x + 3}$

$\quad = \dfrac{x + 2(x + 3)}{2\sqrt{x + 3}} = \dfrac{3(x + 2)}{2\sqrt{x + 3}}$

$f'(x) = 0$ when $x = -2$

$f''(x) = \dfrac{3}{2}\left[\dfrac{\sqrt{x + 3}(1) - (x + 2)(1/2)(x + 3)^{-1/2}(1)}{x + 3}\right]$

$\quad = \dfrac{3}{2}\left[\dfrac{\sqrt{x + 3} - (x + 2)/(2\sqrt{x + 3})}{x + 3}\right]$

$\quad = \dfrac{3(x + 4)}{4(x + 3)^{3/2}} > 0$ for all x in $(-3, \infty)$

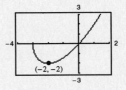

Domain: $(-3, \infty)$

We conclude that the graph is concave upward for each x in the domain of f and therefore that $(-2, -2)$ is a relative minimum.

35. $f(x) = 2 \sin x + \sin 2x$

$f'(x) = 2 \cos x + 2 \cos 2x$

$\quad = 2[\cos x + 2 \cos^2 x - 1] = 2(2 \cos x - 1)(\cos x + 1)$

$f'(x) = 0$ when $x = \dfrac{\pi}{3}, \pi, \dfrac{5\pi}{3}$

$f''(x) = -2 \sin x - 4 \sin 2x$

$\quad = -2(\sin x + 4 \sin x \cos x) = -2 \sin x(1 + 4 \cos x)$

$f''(x) = 0$ when $x = 0, \pi, \arccos\left(-\dfrac{1}{4}\right), 2\pi - \arccos\dfrac{1}{4}$

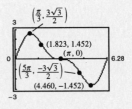

Since $f''(\pi/3) < 0$, there is a relative maximum at $(\pi/3, 3\sqrt{3}/2)$, and since $f''(5\pi/3) > 0$, there is a relative minimum at $(5\pi/3, -3\sqrt{3}/2)$. The inflection points are $(\pi, 0)$, $(1.823, 1.452)$, and $(4.460, -1.452)$.

45. Since f' is an increasing function, its derivative f'' is a positive function. Since $f'' > 0$, f is concave upward.

(a) Since $f' < 0$, the function is decreasing. The graph shows a function that is decreasing and concave upward.

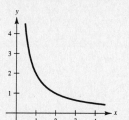

(b) Since $f' > 0$, the function is increasing. The graph shows a function that is increasing and concave upward.

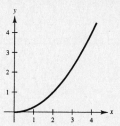

51. $f(x) = ax^3 + bx^2 + cx + d, f'(x) = 3ax^2 + 2bx + c,\ f''(x) = 6ax + 2b = 2(3ax + b)$

$\quad f(3) = 3 \implies \quad 27a + 9b + 3c + d = 3 \quad (1)$

$\quad f(4) = 2 \implies \quad 64a + 16b + 4c + d = 2 \quad (2)$

$\quad f(5) = 1 \implies 125a + 25b + 5c + d = 1 \quad (3)$

$\quad f'(3) = 0 \implies \quad\quad 27a + 6b + c = 0 \quad (4)$

$\quad f'(5) = 0 \implies \quad\quad 75a + 10b + c = 0 \quad (5)$

$\quad f''(4) = 0 \implies \quad\quad\quad 2(12a + b) = 0 \quad (6)$

From (6) we get $b = -12a$. Substituting this into (5), we obtain

$$75a + 10(-12a) + c = 0$$

$$-45a + c = 0$$

$$c = 45a$$

Substitution into (1) and (2) yields

$$27a + 9(-12a) + 3(45a) + d = 3$$

$$54a + d = 3 \quad (7)$$

$$64a + 16(-12a) + 4(45a) + d = 2$$

$$52a + d = 2. \quad (8)$$

To solve simultaneously, subtract (8) from (7) and obtain $2a = 1 \quad a = \frac{1}{2}$

From (7) $\quad d = 3 - 54\left(\frac{1}{2}\right) = -24$

$$c = 45\left(\frac{1}{2}\right) = \frac{45}{2}$$

$$b = -12\left(\frac{1}{2}\right) = -6.$$

Therefore, $f(x) = \frac{1}{2}x^3 - 6x^2 + \frac{45}{2}x - 24 = \frac{1}{2}(x^3 - 12x^2 + 45x - 48).$

61. The domain of D is $[0, L]$.

$$D(x) = 2x^4 - 5Lx^3 + 3L^2x^2$$

$$D'(x) = 8x^3 - 15Lx^2 + 6L^2x = x(8x^2 - 15Lx + 6L^2)$$

$$D''(x) = 24x^2 - 30Lx + 6L^2 = 6(4x^2 - 5Lx + L^2)$$

To find the critical numbers, solve the equation $D'(x) = 0$. At the critical number $x = 0$, there is no deflection since $D(0) = 0$. To find additional critical numbers use the Quadratic Formula to solve the equation

$$8x^2 - 15Lx + 6L^2 = 0$$

$$x = \frac{15L \pm \sqrt{(-15L)^2 - 4(8)(6L^2)}}{2(8)}$$

$$= \left(\frac{15 \pm \sqrt{33}}{16}\right)L.$$

—CONTINUED—

61. **—CONTINUED—**

The critical number in the domain of the function is

$$x = \left(\frac{15 - \sqrt{33}}{16}\right)L \approx 0.578L.$$

$$D''\left(\frac{15 - \sqrt{33}}{16}\right)L = \frac{3}{16}(11 - 5\sqrt{33})L^2 \approx -3.323L^2 < 0.$$

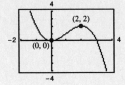

By the Second Derivative Test, the deflection is maximum when

$$x = \left(\frac{15 - \sqrt{33}}{16}\right)L \approx 0.578L.$$

69. We begin by evaluating the function f and its first and second derivatives at $x = \pi/4$.

$$f(x) = 2(\sin x + \cos x) \qquad f\left(\frac{\pi}{4}\right) = 2\sqrt{2}$$

$$f'(x) = 2(\cos x - \sin x) \qquad f'\left(\frac{\pi}{4}\right) = 0$$

$$f''(x) = -2(\sin x + \cos x) \qquad f''\left(\frac{\pi}{4}\right) = -2\sqrt{2}$$

$$P_1(x) = 2\sqrt{2} + 0\left(x - \frac{\pi}{4}\right) = 2\sqrt{2}$$

$$P_2(x) = 2\sqrt{2} + 0\left(x - \frac{\pi}{4}\right) + \frac{1}{2}(-2\sqrt{2})\left(x - \frac{\pi}{4}\right)^2 = 2\sqrt{2} - \sqrt{2}\left(x - \frac{\pi}{4}\right)^2.$$

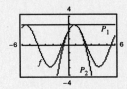

The graphs of f, P_1, and P_2 produced by a graphing utility are shown in the figure.

73. We assume the three zeros of the cubic are r_1, r_2, and r_3. Then,

$$f(x) = a(x - r_1)(x - r_2)(x - r_3)$$

$$f'(x) = a[(x - r_1)(x - r_2) + (x - r_1)(x - r_3) + (x - r_2)(x - r_3)]$$

$$f''(x) = a[(x - r_1) + (x - r_2) + (x - r_1) + (x - r_3) + (x - r_2) + (x - r_3)]$$

$$= a[6x - 2(r_1 + r_2 + r_3)].$$

Consequently, $f''(x) = 0$ if

$$x = \frac{2(r_1 + r_2 + r_3)}{6} = \frac{r_1 + r_2 + r_3}{3} = \text{ average of } r_1, r_2, \text{ and } r_3.$$

Section 3.5   Limits at Infinity

13. Divide both the numerator and denominator by x^2 (the highest power of x in the denominator).

$$\lim_{x \to \infty} \frac{x}{x^2 - 1} = \lim_{x \to \infty} \frac{x/x^2}{(x^2/x^2) - (1/x^2)}$$

$$= \lim_{x \to \infty} \frac{1/x}{1 - (1/x^2)}$$

$$= \frac{0}{1 - 0} = 0$$

(Note that the degree of the numerator is less than the degree of the denominator. See Exercise 75 in this section.)

15. Divide both the numerator and denominator by x (the highest power of x in the denominator). Since

$$\lim_{x \to -\infty} \frac{5x^2}{x + 3} = \lim_{x \to -\infty} \frac{5x^2/x}{(x/x) + (3/x)}$$

$$= \lim_{x \to -\infty} \frac{5x}{1 + (3/x)} = -\infty,$$

the limit does not exist. (Note that the degree of the numerator is greater than the degree of the denominator. See Exercise 75 in this section.)

17. Divide the numerator and denominator of the function by x and note that $x = -\sqrt{x^2}$ when $x < 0$.

$$\lim_{x \to -\infty} \frac{x}{\sqrt{x^2 - x}} = \lim_{x \to -\infty} \frac{x/x}{\sqrt{x^2 - x}/\left(-\sqrt{x^2}\right)}$$

$$= \lim_{x \to -\infty} \frac{1}{-\sqrt{1 - (1/x)}}$$

$$= -\frac{1}{\sqrt{1 + 0}} = -1$$

25. If we make the substitution $t = 1/x$, then finding the limit as $t \to 0^+$ in the resulting function is equivalent to finding the limit as $x \to \infty$ in the given function.

$$\lim_{x \to \infty} x \sin \frac{1}{x} = \lim_{t \to 0^+} \frac{1}{t} \sin t$$

$$= \lim_{t \to 0^+} \frac{\sin t}{t} = 1.$$

29. $\lim_{x \to \infty} \left(x - \sqrt{x^2 + x} \right) = \lim_{x \to \infty} \left(x - \sqrt{x^2 + x} \right) \dfrac{x + \sqrt{x^2 + x}}{x + \sqrt{x^2 + x}}$

$$= \lim_{x \to \infty} \frac{x^2 - (x^2 + x)}{x + \sqrt{x^2 + x}}$$

$$= \lim_{x \to \infty} \frac{-x}{x + \sqrt{x^2 + x}}$$

$$= \lim_{x \to \infty} \frac{-x/x}{(x/x) + \sqrt{x^2 + x}/\sqrt{x^2}}$$

$$= \lim_{x \to \infty} \frac{-1}{1 + \sqrt{1 + (1/x)}}$$

$$= \frac{-1}{1 + \sqrt{1 + 0}} = -\frac{1}{2}$$

(Note: For $x > 0$, $x = \sqrt{x^2}$.)

35. If $x = 0$, then $y = 2$ and the y-intercept occurs at $(0, 2)$. If $y = 0$, then $2 + x = 0$, $x = -2$, and the x-intercept is $(-2, 0)$. There is no symmetry with respect to either axis or to the origin. There is a vertical asymptote at $x = 1$. Furthermore,

$$\lim_{x \to 1^-} \frac{2 + x}{1 - x} = \infty \quad \text{and} \quad \lim_{x \to 1^+} \frac{2 + x}{1 - x} = -\infty$$

$$\lim_{x \to \pm\infty} \frac{2 + x}{1 - x} = \lim_{x \to \pm\infty} \frac{2/x + x/x}{1/x - x/x} = \lim_{x \to \pm\infty} \frac{(2/x) + 1}{(1/x) - 1} = \frac{0 + 1}{0 - 1} = -1.$$

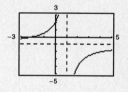

Therefore, there is a horizontal asymptote (to the right and left) at $y = -1$. Start the graph by plotting the following solution points.

x	-3	-1	0.5	2	3	4
y	-0.25	0.5	5	-4	-2.5	-2

51. Since $x^2 - 4$ must be positive, the domain is $(-\infty, -2)$ and $(2, \infty)$. There is no symmetry with respect to either axis. However, there is symmetry with respect to the origin since

$$(-y) = \frac{(-x)^3}{\sqrt{(-x)^2 - 4}}$$

$$-y = \frac{-x^3}{\sqrt{x^2 - 4}}$$

$$y = \frac{x^3}{\sqrt{x^2 - 4}}$$

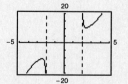

which is equivalent to the original equation. Since the denominator of $x^3/\sqrt{x^2 - 4}$ is zero when $x = 2$ or $x = -2$, there are vertical asymptotes at $x = 2$ and $x = -2$.

x	2.25	2.50	2.75	3.00	4.00
y	11.05	10.45	11.02	12.07	18.48

63. (a) The figure produced by a graphing utility shows graphically that *f* and *g* appear to represent the same function.
Note that $x = 0$ and $x = 3$ are vertical asymptotes.

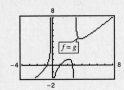

(b) $f(x) = \dfrac{x^3 - 3x^2 + 2}{x(x - 3)}$

$\qquad = \dfrac{x(x^2 - 3x) + 2}{x^2 - 3x}$

$\qquad = \dfrac{x(x^2 - 3x)}{x^2 - 3x} + \dfrac{2}{x^2 - 3x}$

$\qquad = x + \dfrac{2}{x(x - 3)} = g(x)$

(c) When we zoom out sufficiently far on a graphing utility the graph of the function appears as a line. This line is the slant
asymptote $y = x$ and is shown in the figure.

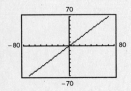

67. $\quad C = 0.5x + 500$

$\overline{C} = \dfrac{C}{x} = \dfrac{0.5x + 500}{x} = 0.5 + \dfrac{500}{x}$

$\displaystyle\lim_{x \to \infty} \overline{C} = \lim_{x \to C} \left(0.5 + \dfrac{500}{x} \right) = 0.5 + 0 = 0.5$

Section 3.6   A Summary of Curve Sketching

13. $y = 3x^4 + 4x^3 = x^3(3x + 4)$

$y' = 12x^3 + 12x^2 = 12x^2(x + 1)$

$y'' = 36x^2 + 24x = 12x(3x + 2)$

Intercepts: $(0, 0), \left(-\tfrac{4}{3}, 0\right)$

Critical numbers: $x = 0, x = -1$

Possible inflection points: $\left(-\tfrac{2}{3}, -\tfrac{16}{27}\right), (0, 0)$

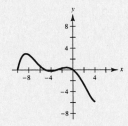

x	y	y'	y''	Shape of graph
$-\infty < x < -1$		$-$	$+$	decreasing, concave up
$x = -1$	-1	0	$+$	relative minimum
$-1 < x < -\tfrac{2}{3}$		$+$	$+$	increasing, concave up
$x = -\tfrac{2}{3}$	$-\tfrac{16}{27}$	$+$	0	point of inflection
$-\tfrac{2}{3} < x < 0$		$+$	$-$	increasing, concave down
$x = 0$	0	0	0	point of inflection
$0 < x < \infty$		$+$	$+$	increasing, concave up

21. $y = x\sqrt{4 - x}$

$$y' = x\left(\frac{1}{2}\right)(-1)(4 - x)^{-1/2} + (4 - x)^{1/2}$$

$$= \frac{-x}{2\sqrt{4 - x}} + \sqrt{4 - x}$$

$$= \frac{-x + 8 - 2x}{2\sqrt{4 - x}}$$

$$= \frac{-3x + 8}{2\sqrt{4 - x}}$$

$$y'' = \frac{2(4 - x)^{1/2}(-3) - (-3x + 8)(2)(1/2)(-1)(4 - x)^{-1/2}}{4(4 - x)}$$

$$= \frac{-6(4 - x) + (-3x + 8)}{4(4 - x)^{3/2}}$$

$$= \frac{-24 + 6x - 3x + 8}{4(4 - x)^{3/2}} = \frac{3x - 16}{4(4 - x)^{3/2}}$$

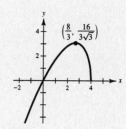

Domain: $x \le 4$

Intercepts: $(0, 0), (4, 0)$

Critical numbers: $\frac{8}{3}, 4$

x	y	y'	y''	Shape of graph
$-\infty < x < \frac{8}{3}$		$+$	$-$	increasing, concave down
$x = \frac{8}{3}$	$\frac{16\sqrt{3}}{9}$	0	$-$	relative maximum
$\frac{8}{3} < x < 4$		$-$	$-$	decreasing, concave down
$x = 4$	0	undefined	undefined	

(**Note:** $x = \frac{16}{3}$ is not in the domain of the function.)

23. $y = 3x^{2/3} - 2x = x^{2/3}(3 - 2x^{1/3})$

$$y' = 2x^{-1/3} - 2 = \frac{2(1 - x^{1/3})}{x^{1/3}}$$

$$y'' = \left(\frac{-1}{3}\right)(2)x^{-4/3} = \frac{-2}{3x^{4/3}}$$

Intercepts: $(0, 0), \left(\frac{27}{8}, 0\right)$

Critical numbers: $x = 0, x = 1$

x	y	y'	y''	Shape of graph
$-\infty < x < 0$		$-$	$-$	decreasing, concave down
$x = 0$	0	undefined	undefined	relative minimum
$0 < x < 1$		$+$	$-$	increasing, concave down
$x = 1$	1	0	$-$	relative maximum
$1 < x < \infty$		$-$	$-$	decreasing, concave down

25. $y = \sin x - \dfrac{1}{18} \sin 3x$

$y' = \cos x - \dfrac{1}{6} \cos 3x$

$ = \cos x - \dfrac{1}{6}(4 \cos^3 x - 3 \cos x)$

$ = -\dfrac{1}{6} \cos x (4 \cos^2 x - 9)$

$y'' = -\dfrac{1}{6}[\cos x(8 \cos x)(-\sin x) + (4 \cos^2 x - 9)(-\sin x)]$

$ = \dfrac{1}{6} \sin x(8 \cos^2 x + 4 \cos^2 x - 9)$

$ = \dfrac{1}{2} \sin x(4 \cos^2 x - 3)$

$y'' = 0$ when $x = 0, \dfrac{\pi}{6}, \dfrac{5\pi}{6}, \pi, \dfrac{7\pi}{6}, \dfrac{11\pi}{6}$

Critical numbers: $x = \dfrac{\pi}{2}, \dfrac{3\pi}{2}$

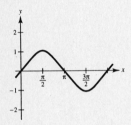

x	y	y'	y''	Shape of graph
$0 < x < \dfrac{\pi}{6}$		+	+	increasing, concave up
$x = \dfrac{\pi}{6}$	$\dfrac{4}{9}$	+	0	point of inflection
$\dfrac{\pi}{6} < x < \dfrac{\pi}{2}$		+	−	increasing, concave down
$x = \dfrac{\pi}{2}$	$\dfrac{19}{18}$	0	−	relative maximum
$\dfrac{\pi}{2} < x < \dfrac{5\pi}{6}$		−	−	decreasing, concave down
$x = \dfrac{5\pi}{6}$	$\dfrac{4}{9}$	−	0	point of inflection
$\dfrac{5\pi}{6} < x < \pi$		−	+	decreasing, concave up
$x = \pi$	0	−	0	point of inflection
$\pi < x < \dfrac{7\pi}{6}$		−	−	decreasing, concave down
$x = \dfrac{7\pi}{6}$	$-\dfrac{4}{9}$	−	0	point of inflection
$\dfrac{7\pi}{6} < x < \dfrac{3\pi}{2}$		−	+	decreasing, concave up
$x = \dfrac{3\pi}{2}$	$-\dfrac{19}{18}$	0	+	relative minimum
$\dfrac{3\pi}{2} < x < \dfrac{11\pi}{6}$		+	+	increasing, concave up
$x = \dfrac{11\pi}{6}$	$-\dfrac{4}{9}$	+	0	point of inflection
$\dfrac{11\pi}{6} < x < 2\pi$		+	−	increasing, concave down

31. $y = \dfrac{1}{x - 2} - 3$

$ = \dfrac{1 - 3(x - 2)}{x - 2}$

$ = \dfrac{7 - 3x}{x - 2}$

$y' = \dfrac{-1}{(x - 2)^2}$

$y'' = \dfrac{2}{(x - 2)^3}$

Domain: all $x \neq 2$

Intercepts: $\left(\dfrac{7}{3}, 0\right), \left(0, -\dfrac{7}{2}\right)$

The graph has a vertical asymptote at $x = 2$. Since

$$\lim_{x \to \pm\infty} \left(\dfrac{1}{x - 2} - 3\right) = -3,$$

there is a horizontal asymptote at $y = -3$.

x	y'	y''	Shape of graph
$-\infty < x < 2$	−	−	decreasing, concave down
$2 < x < \infty$	−	+	decreasing, concave up

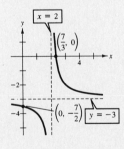

43. Begin by using a symbolic differentiation utility to find f' and f''.

$$f(x) = \frac{x}{\sqrt{x^2 + 7}}$$

$$f'(x) = \frac{7}{(x^2 + 7)^{3/2}}$$

$$f''(x) = \frac{-21x}{(x^2 + 7)^{5/2}}$$

x	$f(x)$	$f'(x)$	$f''(x)$	Shape of graph
$-\infty < x < 0$		+	+	increasing, concave up
$x = 0$	0	+	0	point of inflection
$0 < x < \infty$		+	−	increasing, concave down

Note that $f'(x) > 0$ for all x and there are no critical numbers. Since $f''(0) = 0$ there is a possible point of inflection at $(0, 0)$.

There are no vertical asymptotes since the function is continuous for all x. To determine if there are horizontal asymptotes, use a symbolic differentiation utility to find the following limits.

$$\lim_{x \to \infty} \frac{x}{\sqrt{x^2 + 7}} = 1$$

$$\lim_{x \to -\infty} \frac{x}{\sqrt{x^2 + 7}} = -1$$

Therefore, $y = 1$ and $y = -1$ are horizontal asymptotes. The graph of f is shown in the figure.

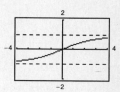

53. Since $x = 5$ is an asymptote, f is a rational function such that the numerator is not zero and the denominator is zero at $x = 5$. Since $y = 3x + 2$ is a slant asymptote,

$$\lim_{x \to \infty} f(x) = 3x + 2.$$

Therefore, $f(x) = 3x + 2 + \dfrac{1}{x - 5} = \dfrac{3x^2 - 13x - 9}{x - 5}$, is one function that satisfies the requirements.

65. Since $f(x) = ax^3 + bx^2 + cx + d$ and

$$\lim_{x \to \infty} f(x) = -\infty, \, a < 0.$$

Also, $f(x)$ is a decreasing function, and therefore $f'(x) = 3ax^2 + 2bx + c < 0$ for all x. Hence, the discriminant must be negative and we have

$$(2b)^2 - 4(3a)(c) < 0 \implies 4(b^2 - 3ac) < 0$$

$$\implies b^2 < 3ac.$$

Section 3.7   Optimization Problems

5. Let x = first number, y = second number, and S = the sum to be minimized. To minimize S, use the *primary* equation

$$S = x + y.$$

Since the second number is the reciprocal of the first, the *secondary* equation is $y = \dfrac{1}{x}$, and therefore,

$$S = x + \frac{1}{x}.$$

Differentiation yields

$$\frac{dS}{dx} = 1 - \frac{1}{x^2}$$

$$\frac{dS}{dx} = 0 \text{ when } x = \pm 1.$$

$$\frac{d^2S}{dx^2} = \frac{2}{x^3}$$

Finally, since the second derivative is positive when $x = 1$, it follows that S is minimum when $x = 1$ and $y = 1$.

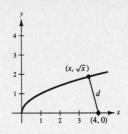

11. Let (x, y) to be a point on the graph of $y = \sqrt{x}$. The distance between (x, y) and the point $(4, 0)$ is given by the *primary* equation

$$d(x) = \sqrt{(x - 4)^2 + (y - 0)^2} = [(x - 4)^2 + y^2]^{1/2}.$$

Since $y = \sqrt{x}$, we have $d(x) = [(x - 4)^2 + x]^{1/2}$.

To minimize $d(x)$, solve $d'(x) = 0$ as follows:

$$d'(x) = \frac{1}{2}[(x - 4)^2 + x]^{-1/2}[2(x - 4) + 1]$$

$$= \frac{2(x - 4) + 1}{2\sqrt{(x - 4)^2 + x}} = \frac{2x - 7}{2\sqrt{(x - 4)^2 + x}}$$

We observe that $d'(x) = 0$ when

$$2x - 7 = 0$$

$$x = \frac{7}{2} \quad \text{and} \quad y = \sqrt{\frac{7}{2}}.$$

Therefore, the required point is $\left(\frac{7}{2}, \sqrt{\frac{7}{2}}\right)$.

21. The area A of the window is given by the *primary* equation

$$A = xy + \frac{1}{2}\pi r^2 = xy + \frac{1}{2}\pi\left(\frac{x}{2}\right)^2.$$

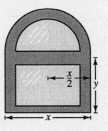

The *secondary* equation is formulated from the known perimeter of the window.

$$2y + x + \pi r = 16$$

$$2y + x + \pi\left(\frac{x}{2}\right) = 16$$

$$4y + 2x + \pi x = 32 \implies y = \frac{32 - 2x - \pi x}{4}$$

Substituting this expression for y into the primary equation and differentiating yields

$$A = x\left(\frac{32 - 2x - \pi x}{4}\right) + \frac{1}{2}\pi\left(\frac{x}{2}\right)^2 = 8x - \frac{1}{2}x^2 - \frac{\pi}{4}x^2 + \frac{\pi}{8}x^2$$

$$\frac{dA}{dx} = 8 - x - \frac{\pi}{2}x + \frac{\pi}{4}x = 8 - \left(1 + \frac{\pi}{4}\right)x$$

$$\frac{d^2A}{dx^2} = -\left(1 + \frac{\pi}{4}\right) < 0.$$

To find any critical numbers solve the equation $dA/dx = 0$.

$$8 - \left(1 + \frac{\pi}{4}\right)x = 0 \implies x = \frac{8}{1 + (\pi/4)} = \frac{32}{4 + \pi}$$

Substituting this value of x into the secondary equation yields

$$y = \frac{32 - 2x - \pi x}{4}$$

$$= \frac{32 - 2\left(\dfrac{32}{4 + \pi}\right) - \pi\left(\dfrac{32}{4 + \pi}\right)}{4} = \frac{16}{4 + \pi}$$

Since $d^2A/dx^2 < 0$, the area is maximum when $x = \dfrac{32}{4 + \pi}$ and $y = \dfrac{16}{4 + \pi}$.

29. (a) See the first six rows of the table in part (b).

(b)

Radius r	Height	Surface Area
0.2	$\dfrac{22}{\pi(0.2)^2}$	$2\pi(0.2)\left[0.2 + \dfrac{22}{\pi(0.2)^2}\right] \approx 220.3$
0.4	$\dfrac{22}{\pi(0.4)^2}$	$2\pi(0.4)\left[0.4 + \dfrac{22}{\pi(0.4)^2}\right] \approx 111.1$
0.6	$\dfrac{22}{\pi(0.6)^2}$	$2\pi(0.6)\left[0.6 + \dfrac{22}{\pi(0.6)^2}\right] \approx 75.6$
0.8	$\dfrac{22}{\pi(0.8)^2}$	$2\pi(0.8)\left[0.8 + \dfrac{22}{\pi(0.8)^2}\right] \approx 59.0$
1.0	$\dfrac{22}{\pi(1.0)^2}$	$2\pi(1.0)\left[1.0 + \dfrac{22}{\pi(1.0)^2}\right] \approx 50.3$
1.2	$\dfrac{22}{\pi(1.2)^2}$	$2\pi(1.2)\left[1.2 + \dfrac{22}{\pi(1.2)^2}\right] \approx 45.7$
1.4	$\dfrac{22}{\pi(1.4)^2}$	$2\pi(1.4)\left[1.4 + \dfrac{22}{\pi(1.4)^2}\right] \approx 43.7$
1.6	$\dfrac{22}{\pi(1.6)^2}$	$2\pi(1.6)\left[1.6 + \dfrac{22}{\pi(1.6)^2}\right] \approx 43.6$
1.8	$\dfrac{22}{\pi(1.8)^2}$	$2\pi(1.8)\left[1.8 + \dfrac{22}{\pi(1.8)^2}\right] \approx 44.8$
2.0	$\dfrac{22}{\pi(2.0)^2}$	$2\pi(2.0)\left[2.0 + \dfrac{22}{\pi(2.0)^2}\right] \approx 47.1$

The estimate of the minimum surface area is 43.6 square inches.

(c) The volume of the right circular cylinder (see figure) is $V = \pi r^2 h$ and its surface area is

$S = 2(\text{area of base}) + (\text{lateral surface})$

$\quad = 2\pi r^2 + 2\pi rh = 2\pi r(r + h).$

From the formula for the volume it follows that

$h = V/(\pi r^2)$. Since $V = 22$ cubic inches, we have

$S = 2\pi r\left(r + \dfrac{22}{\pi r^2}\right) = 2\pi r^2 + \dfrac{44}{r}.$

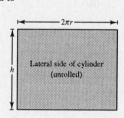

(d) The graph of area function is given in the figure. From the graph, the estimate of the minimum surface area is 43.46 square inches.

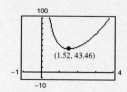

(e) $\dfrac{dS}{dr} = 4\pi r - \dfrac{44}{r^2} = 0$

$$2r = \dfrac{22}{\pi r^2}$$

$$r^3 = \dfrac{11}{\pi} \text{ or } r = \sqrt[3]{\dfrac{11}{\pi}}$$

Since

$$\dfrac{d^2 S}{dr^2} = 4\pi + \dfrac{88}{r^3}$$

is positive at the critical number $r = \sqrt[3]{11/\pi}$, this radius yields the minimum surface area of the cylinder. The corresponding height is

$$h = \dfrac{22}{\pi r^2} = \dfrac{22r}{\pi r^3} = \dfrac{22r}{\pi(11/\pi)} = 2r.$$

Note that the surface area is minimum when the height of the cylinder equals the diameter.

39. Letting S be the strength and k the constant of proportionality, we have $S = kwh^2$. Since $w^2 + h^2 = 24^2$, we have $h^2 = 24^2 - w^2$.

$S = kw(24^2 - w^2) = k(576w = w^3)$

Differentiating and solving $dS/dw = 0$, yields

$\dfrac{dS}{dw} = k(576 - 3w^2) = 0$

$$3w^2 = 576$$

$$w^2 = 192$$

$$w = \pm 8\sqrt{3}.$$

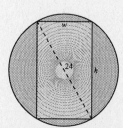

Since w is positive, it follows that $w = 8\sqrt{3}$ inches and $h = \sqrt{24^2 - \left(8\sqrt{3}\right)^2} = 8\sqrt{6}$ inches will produce the strongest beam.

45. Let F be the illumination at point P which is x units from Source 1 (see figure)

$$F = \frac{kI_1}{x^2} + \frac{kI_2}{(d-x)^2}$$

$$\frac{dF}{dx} = \frac{-2kI_1}{x^3} + \frac{2kI_2}{(d-x)^3}$$

Find any critical numbers by solving the equation $dF/dx = 0$.

$$\frac{-2kI_1}{x^3} + \frac{2kI_2}{(d-x)^3} = 0$$

$$\frac{2kI_1}{x^3} = \frac{2kI_2}{(d-x)^3}$$

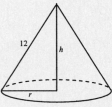

$$\frac{\sqrt[3]{I_1}}{\sqrt[3]{I_2}} = \frac{x}{d-x}$$

$$(d-x)\sqrt[3]{I_1} = x\sqrt[3]{I_2}$$

$$d\sqrt[3]{I_1} = \left(\sqrt[3]{I_1} + \sqrt[3]{I_2}\right)x$$

$$x = \frac{d\sqrt[3]{I_1}}{\sqrt[3]{I_1} + \sqrt[3]{I_2}}$$

$$\frac{d^2F}{dx^2} = \frac{6kI_1}{x^4} + \frac{6kI_2}{(d-x)^4} > 0 \text{ for all } x$$

Therefore, the illumination is minimum when $x = \dfrac{d\sqrt[3]{I_1}}{\sqrt[3]{I_1} + \sqrt[3]{I_2}}$.

51. We begin by determining the radius of the cone. The circumference of the given circle is $c = 2\pi r = 24\pi$ and the length of the arc of the sector (see figure) removed from the circle is $s = r\theta = 12\theta$. Hence, the circumference of the cone is $C = 24\pi - 12\theta$ and the radius R of the cone is

$$R = \frac{C}{2\pi} = \frac{24\pi - 12\theta}{2\pi} = \frac{6}{\pi}(2\pi - \theta).$$

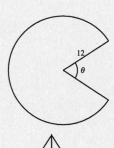

We next find the height of h of the cone. From the figure we observe that the slant height of the cone is the radius of the given circle. Since the radius of the cone has been determined, we find h by the Pythagorean Theorem.

$$h = \sqrt{12^2 - \left[\frac{6}{\pi}(2\pi - \theta)\right]^2} = \frac{6}{\pi}\sqrt{4\pi^2 - (2\pi - \theta)^2}$$

Therefore, we can determine the volume V of the cone and the value of θ for which V is maximum by solving the equation $dV/d\theta = 0$.

$$V = \frac{1}{3}\pi r^2 h = \frac{1}{3}\pi\left(\frac{6}{\pi}\right)^3 (2\pi - \theta)^2 \sqrt{4\pi^2 - (2\pi - \theta)^2}$$

$$\frac{dV}{d\theta} = \frac{1}{3}\pi\left(\frac{6}{\pi}\right)^3\left[(2\pi - \theta)^2 \frac{2\pi - \theta}{\sqrt{4\pi^2 - (2\pi - \theta)^2}} + 2(2\pi - \theta)(-1)\sqrt{4\pi^2 - (2\pi - \theta)^2}\right]$$

$$= \frac{1}{3}\pi\left(\frac{6}{\pi}\right)^3 (2\pi - \theta) \cdot \frac{(2\pi - \theta)^2 - 2[4\pi^2 - (2\pi - \theta)^2]}{\sqrt{4\pi^2 - (2\pi - \theta)^2}}$$

$$= \frac{1}{3}\pi\left(\frac{6}{\pi}\right)^3 \frac{2\pi - \theta}{\sqrt{4\pi^2 - (2\pi - \theta)^2}}[3(2\pi - \theta)^2 - 8\pi^2]$$

Therefore, $dV/d\theta = 0$ when

$$3(2\pi - \theta)^2 = 8\pi^2$$

$$2\pi - \theta = \pm\frac{2\sqrt{2}\pi}{\sqrt{3}}$$

$$\theta = 2\pi \pm \frac{2\sqrt{2}\pi}{\sqrt{3}} = \frac{2\pi}{3}\left(3 \pm \sqrt{6}\right).$$

The maximum volume of the cone will occur when $\theta = \dfrac{2\pi}{3}\left(3 - \sqrt{6}\right) \approx 66°$.

Section 3.8   Newton's Method

5. $f(x) = x^3 + x - 1$

$f'(x) = 3x^2 + 1$

The approximate root is $x = 0.682$.

n	x_n	$f(x_n)$	$f'(x_n)$	$\dfrac{f(x_n)}{f'(x_n)}$	$x_n - \dfrac{f(x_n)}{f'(x_n)}$
1	0.5000	−0.3750	1.7500	−0.2143	0.7143
2	0.7143	0.0787	2.5306	0.0311	0.6832
3	0.6832	0.0021	2.4002	0.0009	0.6823
4	0.6823	0.0000	2.3967	0.0000	0.6823

7. $f(x) = 3\sqrt{x-1} - x$

$f'(x) = \dfrac{3}{2\sqrt{x-1}} - 1$

The approximate zero is $x = 1.146$.

n	x_n	$f(x_n)$	$f'(x_n)$	$\dfrac{f(x_n)}{f'(x_n)}$	$x_n - \dfrac{f(x_n)}{f'(x_n)}$
1	1.2000	0.1416	2.3541	0.0602	1.1398
2	1.1398	−0.0180	3.0113	−0.0060	1.1458
3	1.1458	−0.0003	2.9283	−0.0001	1.1459

15. To approximate the x-value of the point of intersection of the graphs of $f(x) = x$ and $g(x) = \tan x$, let $x = \tan x$. Since this implies that $x - \tan x = 0$, it is necessary to find the zeros of the function $h(x) = x - \tan x$ where $h'(x) = 1 - \sec^2 x$. The iterative formula for Newton's Method has the form

$$x_{n+1} = x_n - \frac{x_n - \tan x_n}{1 - \sec^2 x_n}.$$

n	x_n	$h(x_n)$	$h'(x_n)$	$\dfrac{h(x_n)}{h'(x_n)}$	$x_n - \dfrac{f(x_n)}{f'(x_n)}$
1	4.5000	−0.1373	−21.5048	0.0064	4.4936
2	4.4936	−0.0039	−20.2271	0.0002	4.4934

The calculations are shown in the table, beginning with an initial guess of $x_n = 4.5$. Therefore, the point of intersection of the graphs of f and g occurs when $x \approx 4.493$.

25. Let $f(x) = x^2 - a$. Then $f'(x) = 2x$. Since $\sqrt{a}$ is a zero of $f(x) = 0$, we can use Newton's Method to approximate $\sqrt{a}$ as follows:

$$x_{i+1} = x_i - \frac{x_i^2 - a}{2x_i} = \frac{x_i^2 + a}{2x_i}$$

For example, if $a = 2$, and $x_1 = 1$, then we approximate $\sqrt{2}$ as follows:

$x_1 = 1$

$x_2 = \dfrac{1^2 + 2}{2(1)} = \dfrac{1 + 2}{2} = 1.5000$

$x_3 = \dfrac{(1.5)^2 + 2}{2(1.5)} = \dfrac{4.25}{3} = 1.41667$

$x_4 = \dfrac{(1.41667)^2 + 2}{2(1.41667)} = \dfrac{4.00697}{2.8333} = 1.41421$

(Note: To five decimal places, $\sqrt{2} = 1.41421$.)

35. $f(x) = x \cos x$

$f'(x) = x(-\sin x) + \cos x = 0$

$x - \cot x = 0$

Thus the iterative formula for Newton's Method takes the form

$$x_{n+1} = x_n - \frac{x_n - \cot x_n}{1 + \csc^2 x_n}$$

Interval	$f(x)$	$f'(x)$	$f''(x)$	Shape of graph
$0 < x < 0.860$		$+$	$-$	increasing, concave down
$x = 0.860$	0.561	0	$-$	relative maximum
$0.860 < x < 2.289$		$-$	$-$	decreasing, concave down
$x = 2.289$	-1.506	$-$	0	point of inflection
$2.289 < x < \pi$		$-$	$+$	decreasing, concave up

Since the critical number occurs at the x-value, where $\cot x = x$ in the interval $[0, \pi]$, select $x_1 = 1$ as the initial guess and obtain the critical number $x \approx 0.860$.

To find possible points of inflection solve the equation

$$f''(x) = -(2 \sin x + x \cos x) = 0.$$

Newton's Method yields the iterative formula

$$x_{n+1} = x_n - \frac{-(2 \sin x + x \cos x)}{x \sin x - 3 \cos x}.$$

Beginning with an initial guess $3\pi/4$, you obtain a possible point of inflection at $(2.289, -1.506)$.

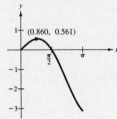

Section 3.9   Differentials

3. Given the function $f(x) = x^5, f'(x) = 5x^4$, and $f'(2) = 80$. Therefore, the equation of the tangent line is

$$y - 32 = 80(x - 2)$$

$$y = 80x - 160 + 32$$

$$T(x) = y = 80x - 128.$$

x	1.9	1.99	2	2.01	2.1
$f(x)$	24.761	31.208	32	32.808	40.841
$T(x)$	24.000	31.200	32	32.800	40.000

Notice that the closer x is to 2 the better the linear approximation.

9. $y = f(x) = x^4 + 1$

$\Delta y = f(x + \Delta x) - f(x)$

$\quad = f(-1 + 0.01) - f(1)$

$\quad = f(-0.99) - f(1)$

$\quad = [(-0.99)^4 + 1] - [(-1)^4 + 1] \approx -0.0394$

Since $f(x) = x^4 + 1, f'(x) = 4x^3$. Therefore,

$dy = f'(x)\, dx$

$\quad = f'(-1)(0.01) = (-4)(0.01) = -0.04.$

The difference between Δy and dy is

$\Delta y - dy \approx -0.0394 - (-0.04) = 0.0006.$

15. $y = x\sqrt{1 - x^2}$

$$dy = \left[x\left(\frac{1}{2}\right)(1 - x^2)^{-1/2}(-2x) + (1 - x^2)^{1/2}(1) \right] dx$$

$$= \left[\frac{-x^2}{\sqrt{1 - x^2}} + \frac{1 - x^2}{\sqrt{1 - x^2}} \right] dx = \frac{1 - 2x^2}{\sqrt{1 - x^2}}\, dx$$

27. The radius of the sphere is $r = 6 \pm 0.02$.

(a) The formula for the volume of the sphere is

$$V = \frac{4}{3}\pi r^3.$$

The approximate ΔV by dV, let $r = 6$ and $dr = \pm 0.02$.

$$dV = \frac{4}{3}\pi (3r^2)\, dr$$

$$dV = 4\pi (36)(\pm 0.02) = \pm 2.88\pi \text{ in}^3$$

(b) The formula for the surface area of the sphere is $S = 4\pi r^2$. To approximate ΔS by dS, let $r = 6$ and $dr = \pm 0.02$.

$$dS = 8\pi r\, dr$$

$$dS = 8\pi (6)(\pm 0.02) = \pm 0.96\pi \text{ in}^2$$

(c) For part (a) the relative error is approximately

$$\frac{dV}{V} = \frac{2.88\pi}{(4/3)\pi(6^3)} = \frac{2.88}{288} = 0.01 = 1\%.$$

For (b) the relative error is approximately

$$\frac{dS}{S} = \frac{0.96\pi}{4\pi(6^2)} = \frac{0.96}{144} = 0.0067 = \frac{2}{3}\%.$$

31. (a) $T = 2\pi \sqrt{\dfrac{L}{g}}$

$$dT = 2\pi \left(\frac{1}{2}\right)\left(\frac{L}{g}\right)^{-1/2}\left(\frac{1}{g}\right) dL = \frac{\pi}{g\sqrt{L/g}}\, dL$$

$$\frac{dT}{T}(100) = \text{percentage error}$$

$$= \frac{(\pi/g\sqrt{L/g})\, dL}{2\pi\sqrt{L/g}}(100) = \frac{1}{2}\left(\frac{dL}{L}100\right)$$

$$= \frac{1}{2}(\text{percentage change in } L)$$

$$= \frac{1}{2}\left(\frac{1}{2}\right) = \frac{1}{4}\%$$

(b) approximate error $= \left(\frac{1}{4}\%\right)(\text{number of seconds per day})$

$$= (0.0025)(60)(60)(24)$$

$$= 216 \text{ seconds} = 3.6 \text{ minutes}$$

41. Let $f(x) = \sqrt[4]{x}$, $x = 625$, and $dx = \Delta x = -1$.

$$f(x + \Delta x) = f(x) + \Delta y$$

$$\approx f(x) + dy$$

$$= f(x) + f'(x)\, dx$$

$$= \sqrt[4]{x} + \frac{1}{4\sqrt[4]{x^3}}\, dx$$

$$f(x + \Delta x) = \sqrt[4]{624}$$

$$\approx \sqrt[4]{625} + \frac{1}{4(\sqrt[4]{625})^3}(-1)$$

$$= 5 - \frac{1}{500} = 4.998$$

Using a calculator: $\sqrt[4]{624} \approx 4.9980$.

Section 3.10   Business and Economics Applications

7. $C = 0.125x^2 + 20x + 5000$

$$\overline{C} = \frac{C}{x} = 0.125x + 20 + \frac{5000}{x}$$

$$\frac{d\overline{C}}{dx} = 0.125 - \frac{5000}{x^2}$$

$$\frac{d^2\overline{C}}{dx^2} = \frac{10,000}{x^3}$$

To find the critical numbers, solve the following equation.

$$\frac{d\overline{C}}{dx} = 0$$

$$0.125 - \frac{5000}{x^2} = 0$$

$$0.125x^2 - 5000 = 0$$

$$x^2 = 40,000 \implies x = 200$$

Since the second derivative is positive for when $x > 0$, the average cost function is concave upward. By the Second Derivative Test, $x = 200$ yields the minimum average cost.

11. The profit is given by

$$P = \text{(price per unit)(number of units)} - \text{(cost)}$$

$$= px - C$$

$$= (90 - x)(x) - (100 + 30x) = -x^2 + 60x - 100$$

To maximize P, solve $dP/dx = 0$ as follows:

$$\frac{dP}{dx} = -2x + 60 = 0$$

$$2x = 60 \text{ and } x = 30$$

Therefore, the profit is maximum when the price is $p = 90 - 30 = 60$.

21. Since the speed for the 110-mile is v miles per hour, the total time is $t = 110/v$ hours. Therefore, the total cost is

$$\text{(Total cost)} = \text{(Fuel cost)} + \text{(Wages)}$$

$$C = \frac{v^2}{600}\left(\frac{110}{v}\right) + 5\left(\frac{110}{v}\right) \cdot = \frac{11}{60}v + 5(110)v^{-1}.$$

To minimize C, solve $dC/dv = 0$ as follows:

$$\frac{dC}{dv} = \frac{11}{60} - 5(110)v^{-2} = 0$$

$$\frac{11}{60} = \frac{5(110)}{v^2}$$

$$v^2 = \frac{5(110)(60)}{11}$$

$$v^2 = 3000$$

$$v = 10\sqrt{30} \approx 54.8 \text{ mi/hr}$$

Thus, a speed 54.8 mi/hr will yield the minimum cost.

29. Let

$d = $ amount in the bank

$i = $ interest rate paid by the bank

$p = $ profit

The bank can take the deposited money d and reinvest to obtain 12% or $(0.12)d$. Since the bank pays out interest to its depositors, its profit is

$$P = (0.12)d - id.$$

Finally, since d is proportional to the square of i, we have

$$d = ki^2.$$

Thus

$$P = (0.12)(ki^2) - i(ki^2) = k[(0.12)i^2 - i^3]$$

To maximize P, solve $dP/di = 0$ as follows:

$$\frac{dP}{di} = k(0.24i - 3i^2) = 0$$

$$ki(0.24 - 3i) = 0$$

(We disregard the critical number $i = 0$.)

$$i = \frac{0.24}{3} = 0.08$$

Thus the bank can maximize its profit by setting $i = 8\%$.

31. $C = 100\left(\dfrac{200}{x^2} + \dfrac{x}{x + 30}\right),\ 1 \le x$

$\dfrac{dC}{dx} = 100\left(-\dfrac{400}{x^3} + \dfrac{30}{(x + 30)^2}\right)$

$\qquad = 1000\left[\dfrac{-40(x + 30)^2 + 3x^3}{x^3(x + 30)^2}\right]$

To find the critical number of C use Newton's Method to find the zero of the function $f(x) = -40(x + 30)^2 + 3x^2$ using $x_1 = 30$ as our first estimate.

$f(x) = 3x^3 - 40x^2 - 2400x - 36{,}000 \quad \text{and} \quad f'(x) = 9x^2 - 80x - 2400$

n	x_n	$f(x_n)$	$f'(x_n)$	$\dfrac{f(x_n)}{f'(x_n)}$	$x_n - \dfrac{f(x_n)}{f'(x_n)}$
1	30	$-63{,}000$	3300	-19.091	49.091
2	49.091	104,702	15,362	6.816	42.275
3	42.275	17,712	10,303	1.719	40.556
4	40.556	992.239	9158.622	0.108	40.448

Therefore, it follows that the critical number is $x \approx 40.4$ and the minimum cost occurs when 40 units are ordered.

41. Since $p = 400 - 3x,\ \dfrac{dp}{dx} = -3.$

$\eta = \dfrac{p/x}{dp/dx}$

$\quad = \dfrac{(400 - 3x)/x}{-3} = 1 - \dfrac{400}{3x}$

At $x = 20,\ \eta = 1 - \dfrac{400}{3(20)} = 1 - \dfrac{20}{3} = -\dfrac{17}{3}.$ Therefore,

$|\eta| = \dfrac{17}{3} > 1,$

and demand is elastic.

Review Exercises for Chapter 3

9. Since $f(x) = x - \cos x$ is continuous on $\left[-\frac{\pi}{2}, \frac{\pi}{2}\right]$ and differentiable on $\left(-\frac{\pi}{2}, \frac{\pi}{2}\right)$, the Mean Value Theorem can be applied.

$\qquad f(x) = x - \cos x$

(1) $\qquad f'(x) = 1 + \sin x$

(2) $\qquad \dfrac{f(b) - f(a)}{b - a} = \dfrac{(\pi/2) - (-\pi/2)}{(\pi/2) - (-\pi/2)} = 1$

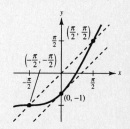

Equating the right-hand members of the equations (1) and (2) yields

$\quad 1 + \sin x = 1$

$\qquad \sin x = 0 \implies x = 0.$

Therefore, $c = 0$ and the point on the graph where the instantaneous rate of change of f equals the average rate of change over the specified interval is $(0, -1)$. This is shown in the figure.

11. Since f is continuous and differentiable for all real x, the Mean Value Theorem can be applied over the specified interval. It is necessary to find all c in $[x_1, x_2]$ such that

$$f'(c) = \frac{f(x_2) - f(x_1)}{x_2 - x_1}.$$

$$f'(x) = 2Ax + B$$

$$f'(c) = 2Ac + B = \frac{f(x_2) - f(x_1)}{x_2 - x_1}$$

$$= \frac{Ax_2{}^2 + Bx_2 + C - Ax_1{}^2 - Bx_1 - C}{x_2 - x_1}$$

$$= \frac{A(x_2{}^2 - x_1{}^2) + B(x_2 - x_1)}{x_2 - x_1}$$

$$= A(x_2 + x_1) + B$$

$$2c = x_2 + x_1$$

$$c = \frac{x_1 + x_2}{2}$$

For a quadratic function, the required value of c is the average of x_1 and x_2.

17. $h(t) = \frac{1}{4}t^4 - 8t$

   $h'(t) = t^3 - 8 = (t - 2)(t^2 + 2t + 4)$

Therefore, $h'(t) = 0$ when $t = 2$. Since h is a polynomial, it is differentiable for all t and the only critical number is $t = 2$.

Interval	$-\infty < t < 2$	$2 < t < \infty$
Test Value	$t = 0$	$t = 2$
Sign of $h'(t)$	$h'(0) < 0$	$h'(3) > 0$
Conclusion	h is decreasing	h is increasing

Hence, h has a minimum at $(2, -12)$. The graph of h is shown in the figure.

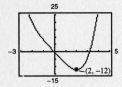

35. The domain of $f(x) = x\sqrt{16 - x^2}$ is all real numbers in the interval $[-4, 4]$ and the graph of f is symmetric to the origin since

$$f(-x) = (-x)\sqrt{16 - (-x)^2} = -x\sqrt{16 - x^2} = -f(x).$$

$$f'(x) = x\left(\frac{1}{2}\right)(16 - x^2)^{-1/2}(-2x) + (16 - x^2)^{1/2} = \frac{16 - 2x^2}{\sqrt{16 - x^2}}$$

$$f''(x) = \frac{\sqrt{16 - x^2}(-4x) - (16 - 2x^2)(1/2)(16 - x^2)^{-1/2}(-2x)}{16 - x^2} = \frac{2x(x^2 - 24)}{(16 - x^2)^{3/2}}$$

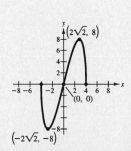

Thus, $f'(x) = 0$ when $x = \pm 2\sqrt{2}$ and undefined when $x = \pm 4$. Since $f''(2\sqrt{2}) < 0$, the graph is concave downward and $(2\sqrt{2}, 8)$ is a maximum. By symmetry, $(-2\sqrt{2}, -8)$ is a minimum. There is a point of inflection at $(0, 0)$.

39. $f(x) = x^{1/3}(x + 3)^{2/3}$

$$f'(x) = x^{1/3}\left(\frac{2}{3}\right)(x + 3)^{-1/3} + (x + 3)^{2/3}\left(\frac{1}{3}\right)x^{-2/3}$$

$$= \frac{2x^{1/3}}{3(x + 3)^{1/3}} + \frac{(x + 3)^{2/3}}{3x^{2/3}}$$

$$= \frac{2x + x + 3}{3x^{2/3}(x + 3)^{1/3}}$$

$$= \frac{x + 1}{x^{2/3}(x + 3)^{1/3}} = \frac{x + 1}{(x^3 + 3x^2)^{1/3}}$$

$$f''(x) = \frac{(x^3 + 3x^2)^{1/3}(1) - (x + 1)(1/3)(x^3 + 3x^2)^{-2/3}(3x^2 + 6x)}{(x^3 + 3x^2)^{2/3}}$$

$$= \frac{(x^3 + 3x^2) - (x + 1)(x^2 + 2x)}{(x^3 + 3x^2)^{4/3}}$$

$$= \frac{x^3 + 3x^2 - x^3 - 2x^2 - x^2 - 2x}{(x + 3x^4)^{2/3}}$$

$$= \frac{-2x}{(x^3 + 3x^2)^{4/3}} = \frac{-2}{x^{5/3}(x + 3)^{4/3}}$$

Intercepts: $(0, 0), (-3, 0)$

Critical numbers: $x = -1, x = 0, x = -3$

Possible point of inflection: $(0, 0)$

x	$f(x)$	$f'(x)$	$f''(x)$	Shape of graph
$-\infty < x < -3$		$+$	$+$	increasing, concave up
$x = -3$	0	undefined	undefined	relative maximum
$-3 < x < -1$		$-$	$+$	decreasing, concave up
$x = -1$	$-\sqrt[3]{4}$	0	$+$	relative minimum
$-1 < x < 0$		$+$	$+$	increasing, concave up
$x = 0$	0	undefined	undefined	point of inflection
$0 < x < \infty$		$+$	$-$	increasing, concave down

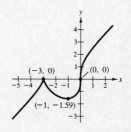

49. Domain: $[0, 2\pi]$

Range: $[1, 1 + 2\pi]$

Symmetry: None

Asymptotes: None

Intercepts: $(0, 1)$

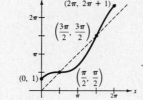

$$f(x) = x + \cos x$$

$$f'(x) = 1 - \sin x$$

$$f''(x) = -\cos x$$

Since $f'(x) \geq 0$, f is increasing. $f''(x) = 0$ when $x = \pi/2$ and $x = 3\pi/2$. The graph is shown in the figure.

—CONTINUED—

49. —CONTINUED—

x	$f(x)$	$f'(x)$	$f''(x)$	Shape of graph
$0 < x < \dfrac{\pi}{2}$		$+$	$-$	increasing, concave down
$x = \dfrac{\pi}{2}$	$\dfrac{\pi}{2}$	$+$	0	point of inflection
$\dfrac{\pi}{2} < x < \dfrac{3\pi}{2}$		$+$	$+$	increasing, concave up
$x = \dfrac{3\pi}{2}$	$\dfrac{3\pi}{2}$	$+$	0	point of inflection
$\dfrac{3\pi}{2} < x < 2\pi$		$+$	$-$	increasing, concave down

57. The longest pipe that will go around the corner will have a length equal to the minimum length of the hypotenuse [through the point $(4, 6)$] of the triangle whose vertices are $(0, 0)$, $(x, 0)$, and $(0, y)$. Begin by relating x and y as follows:

$$m = \frac{y - 6}{0 - 4} = \frac{6 - 0}{4 - x}$$

$$y - 6 = \frac{-24}{4 - x}$$

$$y = \frac{24}{x - 4} + 6 = \frac{6x}{x - 4}$$

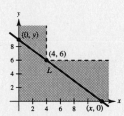

[Note that $dy/dx = -24/(x - 4)^2$.] Now the length of the hypotenuse is given by

$$L = \sqrt{x^2 + y^2}.$$

To minimize L, solve $dL/dx = 0$ as follows.

$$\frac{dL}{dx} = \frac{(1/2)[2x + (2y)(dy/dx)]}{\sqrt{x^2 + y^2}} = 0$$

$$x = -y\frac{dy}{dx} = -\left(\frac{6x}{x - 4}\right)\left[\frac{-24}{(x - 4)^2}\right]$$

$$x(x - 4)^3 = 144x$$

$$(x - 4)^3 = 144$$

$$x - 4 = \sqrt[3]{144}$$

$$x = \sqrt[3]{144} + 4$$

Therefore, the minimum length of L and the maximum length of pipe are given by

$$L = \sqrt{x^2 + y^2} = \sqrt{x^2 + \frac{36x^2}{(x - 4)^2}} = \frac{x}{x - 4}\sqrt{(x - 4)^2 + 36}$$

$$= \frac{\sqrt[3]{144} + 4}{\sqrt[3]{144}}\sqrt{144^{2/3} + 36} \approx 14.05 \text{ ft}$$

59. From the figure observe that:

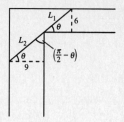

$$\csc \theta = \frac{L_1}{6} \quad \text{or} \quad L_1 = 6 \csc \theta$$

$$\csc\left(\frac{\pi}{2} - \theta\right) = \frac{L_2}{9} \quad \text{or} \quad L_2 = 9 \csc\left(\frac{\pi}{2} - \theta\right)$$

Therefore, the length of the pipe is given by

$$L = L_1 + L_2 = 6 \csc \theta + 9 \csc\left(\frac{\pi}{2} - \theta\right) = 6 \csc \theta + 9 \sec \theta.$$

Note that $\csc[(\pi/2) - \theta] = \sec \theta$. To maximize L, solve $dL/d\theta = 0$ as follows.

$$\frac{dL}{d\theta} = -6 \csc \theta \cot \theta + 9 \sec \theta \tan \theta = 0$$

$$9 \sec \theta \tan \theta = 6 \csc \theta \cot \theta$$

$$\frac{\sec \theta \tan \theta}{\csc \theta \cot \theta} = \frac{6}{9}$$

$$\tan^3 \theta = \frac{2}{3}$$

$$\tan \theta = \frac{2^{1/3}}{3^{1/3}}$$

From the figure observe that

$$\csc \theta = \frac{\sqrt{2^{2/3} + 3^{2/3}}}{2^{1/3}} \quad \text{and} \quad \sec \theta = \frac{\sqrt{2^{2/3} + 3^{2/3}}}{3^{1/3}}.$$

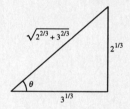

$$L = (6)\left(\frac{\sqrt{2^{2/3} + 3^{2/3}}}{2^{1/3}}\right) + (9)\left(\frac{\sqrt{2^{2/3} + 3^{2/3}}}{2^{1/3}}\right)$$

$$= 3\sqrt{2^{2/3} + 3^{2/3}}(2^{2/3} + 3^{2/3}) = 3(2^{2/3} + 3^{2/3})^{3/2}$$

61. $y = \frac{1}{3}\cos(12t) - \frac{1}{4}\sin(12t)$

$v = y' = -4\sin(12t) - 3\cos(12t)$

(a) When $t = \pi/8$, we have

$$y = \frac{1}{3}\cos\frac{3\pi}{2} - \frac{1}{4}\sin\frac{3\pi}{2} = \frac{1}{4} \text{ inch} \quad \text{and} \quad v = -4\sin\frac{3\pi}{2} - 3\cos\frac{3\pi}{2} = 4 \text{ inches per second.}$$

(b) The maximum displacement occurs when $y' = v = 0$.

$$-4\sin(12t) - 3\cos(12t) = 0$$

$$-4\sin(12t) = 3\cos(12t)$$

$$\frac{\sin(12t)}{\cos(12t)} = -\frac{3}{4} \implies \tan(12t) = -\frac{3}{4}$$

From the figure observe that when $\tan \theta = \frac{3}{4}$, $\sin \theta = \frac{3}{5}$, and $\cos \theta = \frac{4}{5}$. Since $\tan(12t)$ is negative, $\sin(12t)$ and $\cos(12t)$ must have opposite signs. If $\sin(12t) = -\frac{3}{5}$ and $\cos(12t) = \frac{4}{5}$, then

$$y = \frac{1}{3}\left(\frac{4}{5}\right) - \frac{1}{4}\left(-\frac{3}{5}\right) = \frac{25}{60} = \frac{5}{12} \text{ inches.}$$

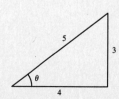

If $\sin(12t) = \frac{3}{5}$ and $\cos(12t) = -\frac{4}{5}$, then

$$y = \frac{1}{3}\left(-\frac{4}{5}\right) - \frac{1}{4}\left(\frac{3}{5}\right) = -\frac{25}{60} = -\frac{5}{12} \text{ inches.}$$

In either case the maximum displacement from equilibrium is $\frac{5}{12}$ inches.

61. **—CONTINUED—**

(c) The period of the function is

$$\frac{2\pi}{12} = \frac{\pi}{6}.$$

Since the frequency is the reciprocal of the period, the frequency is

$$\frac{6}{\pi}.$$

73. The profit is given by

$$P = \text{(price per unit)(number of units)} - \text{(cost)} = px - C$$

$$= (36 - 4x)(x) - (2x^2 + 6) = 36x - 4x^2 - 2x^2 - 6$$

$$= -6x^2 + 36x - 6.$$

To maximize P, solve $dP/dx = 0$ as follows:

$$\frac{dP}{dx} = -12x + 36 = 0$$

$$12x = 36$$

$$x = 3 \text{ units}$$

Thus the maximum profit is

$$P = -6(3)^2 + 36(3) - 6 = -54 + 108 - 6 = \$48.$$

C H A P T E R   4
Integration

CHAPTER 4
Integration

Section 4.1   Antiderivatives and Indefinite Integration

Solutions to Selected Odd-Numbered Exercises

7.

Given	*Rewrite*	*Integrate*	*Simplify*
$\displaystyle\int \frac{1}{x\sqrt{x}}\,dx$	$\displaystyle\int x^{-3/2}\,dx$	$\dfrac{x^{-1/2}}{-1/2} + C$	$\dfrac{-2}{\sqrt{x}} + C$

13. $\displaystyle\int (x^{3/2} + 2x + 1)\,dx = \frac{x^{5/2}}{5/2} + 2\left(\frac{x^2}{2}\right) + x + C$

$$= \frac{2x^{5/2}}{5} + x^2 + x + C$$

Check

If $y = \dfrac{2x^{5/2}}{5} + x^2 + x + C$, then

$$\frac{dy}{dx} = \left(\frac{2}{5}\right)\left(\frac{5}{2}\right)x^{3/2} + 2x + 1 + 0 = x^{3/2} + 2x + 1.$$

25. $\displaystyle\int (x + 1)(3x - 2)\,dx = \int (3x^2 + x - 2)\,dx$

$$= x^3 + \frac{1}{2}x^2 - 2x + C$$

Check

If $y = x^3 + \dfrac{1}{2}x^2 - 2x + C$, then

$$\frac{dy}{dx} = 3x^2 + \frac{1}{2}(2x) - 2 + 0$$

$$= 3x^2 + x - 2 = (x + 1)(3x - 2).$$

31. $\displaystyle\int (2 \sin x + 3 \cos x)\,dx = 2\int \sin x\,dx + 3\int \cos x\,dx$

$$= -2 \cos x + 3 \sin x + C$$

Check

If $y = -2 \cos x + 3 \sin x + C$, then

$$\frac{dy}{dx} = -2(-\sin x) + 3 \cos x + 0 = 2 \sin x + 3 \cos x.$$

37. $\displaystyle\int (\tan^2 y + 1)\,dy = \int \sec^2 y\,dy = \tan y + C$

Check

If $f(y) = \tan y + C$, then $f'(y) = \sec^2 y = \tan^2 y + 1$.

45. $\dfrac{dy}{dx} = 2x - 1$

$$y = \int (2x - 1)\,dx = x^2 - x + C$$

Substituting the coordinates of the solution point into the antiderivative yields

$$y = x^2 - x + C$$

$$1 = (1)^2 - (1) + C \quad \text{or} \quad C = 1.$$

Therefore, the required equation is $y = x^2 - x + 1$.

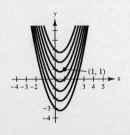

53. $f''(x) = x^{-3/2}$

$$f'(x) = \int x^{-3/2}\,dx$$

$$= \frac{x^{-1/2}}{-1/2} + C_1 = \frac{-2}{\sqrt{x}} + C_1$$

$$f'(4) = \frac{-2}{\sqrt{4}} + C_1 = 2 \implies C_1 = 3$$

$$f'(x) = -2x^{-1/2} + 3$$

$$f(x) = \int (-2x^{-1/2} + 3)\,dx$$

$$= \frac{(-2)x^{1/2}}{1/2} + 3x + C_2 = -4x^{1/2} + 3x + C_2$$

$$f(0) = -4(0)^{1/2} + 3(0) + C_2 = 0 \implies C_2 = 0$$

Therefore, $f(x) = -4x^{1/2} + 3x = -4\sqrt{x} + 3x.$

59. If $s = f(t)$ is the position of the object at any time t, then $f'(t)$ is its velocity and $f''(t)$ its acceleration. Therefore, $f''(t) = -32$ since -32 ft/sec^2 is the acceleration due to gravity.

$$f'(t) = \int -32 \, dt = -32t + C_1 = -32t + v_0$$

where v_0 is the initial velocity. Furthermore,

$$f(t) = \int (-32t + v_0) \, dt$$
$$= -16t^2 + v_0 t + C_2 = -16t^2 + v_0 t + s_0$$

where $s_0 = 0$ is the initial height. Thus,

$$s = f(t) = -16t^2 + v_0 t.$$

Now since s is a maximum when $f'(t) = 0$, we have

$$-32t + v_0 = 0 \quad \text{or} \quad t = \frac{v_0}{32}.$$

Finally, in order for s to attain a height of 550 feet, we have

$$s = -16\left(\frac{v_0}{32}\right)^2 + v_0\left(\frac{v_0}{32}\right) = 550$$

$$\frac{v_0^2}{64} = 550$$

$$v_0^2 = 35,200$$

$$v_0 = \sqrt{35,200}$$

$$= 40\sqrt{22}$$

$$\approx 187.617 \text{ ft/sec}$$

65. $a(t) = -1.6$

$$v(t) = \int (-1.6) \, dt$$
$$= -1.6t + v_0$$
$$= -1.6t$$

Since the stone was dropped, $v_0 = 0$.

$$s(t) = \int (-1.6t) \, dt = -0.8t^2 + s_0$$

Since the stone hit the surface of the moon $t = 20$, we have

$$s(20) = -0.8(20)^2 + s_0$$
$$= -320 + s_0 = 0 \implies s_0 = 320.$$

Therefore, $s(t) = -0.8t^2 + 320$, and $v(t) = -1.6t$. The stone was dropped from the height $s(0) = 320$ meters, and its velocity at the time of impact was $v(20) = -1.6(20) = -32$ meters per second.

73. Let $T(t)$ and $A(t)$ represent the position functions of the truck and auto. It follows that

$$T'(t) = 30, \quad T(0) = 0$$
$$A''(t) = 6, \quad A'(0) = 0, \quad \text{and} \quad A(0) = 0.$$

For the truck,

$$T(t) = \int 30 \, dt = 30t + C_1$$

$$T(0) = 30(0) + C_1 = 0 \implies C_1 = 0.$$

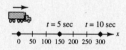

For the auto,

$$A'(t) = \int 6 \, dt = 6t + C_2$$

$$A'(0) = 6(0) + C_2 = 0 \implies C_2 = 0$$

$$A'(t) = 6t$$

$$A(t) = \int 6t \, dt = 3t^2 + C_3$$

$$A(0) = 3(0)^2 + C_3 = 0 \implies C_3 = 0$$

$$A(t) = 3t^2.$$

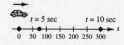

—CONTINUED—

73. —CONTINUED—

Therefore, when the auto catches up with the truck, we have

$$A(t) = T(t)$$

$$3t^2 = 30t$$

$$3t^2 - 30t = 0$$

$$3t(t - 10) = 0$$

$$t = 10 \text{ sec.} \quad \text{We disregard } t = 0.$$

(a) When $t = 10$ sec, the auto will have traveled

$$A(10) = 3(10^2) = 300 \text{ ft.}$$

(b) It will be traveling

$$A'(10) = 60 \text{ ft/sec} = \frac{60(3600)}{5280} \approx 41 \text{ mi/hr.}$$

79. $\dfrac{dR}{dx} = 100 - 5x$

$$R(x) = \int (100 - 5x)\, dx = 100x - \frac{5}{2}x^2 + C$$

Since there is no revenue generated if no units are sold, $R(0) = 0$. Therefore, $C = 0$ and the revenue function is

$$R(x) = 100x - \frac{5}{2}x^2.$$

The revenue generated is the number of units sold x times the price per unit p.

$$R(x) = xp = 100x - \frac{5}{2}x^2 = x\left(100 - \frac{5}{2}x\right)$$

Thus, the demand function is $p = 100 - \dfrac{5}{2}x.$

Section 4.2   Area

3. $\displaystyle\sum_{k=0}^{4} \frac{1}{k^2 + 1} = \frac{1}{0^2 + 1} + \frac{1}{1^2 + 1} + \frac{1}{2^2 + 1} + \frac{1}{3^2 + 1} + \frac{1}{4^2 + 1}$

$$= \frac{1}{1} + \frac{1}{2} + \frac{1}{5} + \frac{1}{10} + \frac{1}{17}$$

$$= \frac{170 + 85 + 34 + 17 + 10}{170}$$

$$= \frac{316}{170} = \frac{158}{85}$$

11. We begin by noting that the n terms in this sum are each of the form

$$f(i) = \left[\left(\frac{2i}{n}\right)^3 - \frac{2i}{n}\right]\left(\frac{2}{n}\right).$$

Furthermore, observe that in the first term $i = 1$, in the second term $i = 2$, and so on until we reach the nth term. Thus our index i runs from 1 to n, and the sigma notation for the given sum is

$$\sum_{i=1}^{n} f(i) = \sum_{i=1}^{n} \left[\left(\frac{2i}{n}\right)^3 - \frac{2i}{n}\right]\left(\frac{2}{n}\right)$$

$$= \frac{2}{n}\sum_{i=1}^{n} \left[\left(\frac{2i}{n}\right)^3 - \frac{2i}{n}\right].$$

19. $\displaystyle\sum_{i=1}^{15} i(i - 1)^2 = \sum_{i=1}^{15} (i^3 - 2i^2 + i)$

$$= \left[\sum_{i=1}^{15} i^3 - 2\sum_{i=1}^{15} i^2 + \sum_{i=1}^{15} i\right]$$

$$= \left[\frac{15^2(16)^2}{4} - 2\frac{15(16)(31)}{6} + \frac{15(16)}{2}\right]$$

$$= 14{,}400 - 2480 + 120 = 12{,}040$$

25. $\displaystyle\lim_{n\to\infty} s(n) = \lim_{n\to\infty} \frac{81}{n^4}\left[\frac{n^2(n+1)^2}{4}\right]$

$\displaystyle = \lim_{n\to\infty} \frac{81}{4}\left(\frac{n^4 + 2n^3 + n^2}{n^4}\right)$

$\displaystyle = \lim_{n\to\infty} \frac{81}{4}\left(1 + \frac{2}{n} + \frac{1}{n^2}\right) = \frac{81}{4}$

31. $\displaystyle\sum_{i=1}^{n} \frac{1}{n^3}(i-1)^2 = \frac{1}{n^3}\sum_{i=1}^{n}(i^2 - 2i + 1)$

$\displaystyle = \frac{1}{n^3}\left[\sum_{i=1}^{n} i^2 - 2\sum_{i=1}^{n} i + \sum_{i=1}^{n} 1\right]$

$\displaystyle = \frac{1}{n^3}\left[\frac{n(n+1)(2n+1)}{6} - 2\frac{n(n+1)}{2} + n\right]$

$\displaystyle = \frac{1}{n^3}\left(\frac{n^3}{3} - \frac{n^2}{2} + \frac{n}{6}\right)$

$\displaystyle = \frac{1}{3} - \frac{1}{2n} + \frac{1}{6n^2}$

Therefore,

$$\lim_{n\to\infty}\sum_{i=1}^{n} \frac{1}{n^3}(i-1)^2 = \lim_{n\to\infty}\left(\frac{1}{3} - \frac{1}{2n} + \frac{1}{6n^2}\right) = \frac{1}{3}.$$

37. Dividing the interval into five subintervals of equal width yields

$$x_0 = 1, \quad x_1 = 1.2, \quad x_2 = 1.4, \quad x_3 = 1.6, \quad x_4 = 1.8, \quad x_5 = 2.$$

Since y is decreasing from 1 to 2, the lower sum is obtained by using the *right* endpoints of the five subintervals.

$s = \dfrac{1}{1.2}\left(\dfrac{1}{5}\right) + \dfrac{1}{1.4}\left(\dfrac{1}{5}\right) + \dfrac{1}{1.6}\left(\dfrac{1}{5}\right) + \dfrac{1}{1.8}\left(\dfrac{1}{5}\right) + \dfrac{1}{2}\left(\dfrac{1}{5}\right)$

$= \dfrac{1}{6} + \dfrac{1}{7} + \dfrac{1}{8} + \dfrac{1}{9} + \dfrac{1}{10} \approx 0.646$

Similarly, the upper sum is obtained by using the *left* endpoints of the five subintervals.

$S = 1\left(\dfrac{1}{5}\right) + \dfrac{1}{1.2}\left(\dfrac{1}{5}\right) + \dfrac{1}{1.4}\left(\dfrac{1}{5}\right) + \dfrac{1}{1.6}\left(\dfrac{1}{5}\right) + \dfrac{1}{1.8}\left(\dfrac{1}{5}\right)$

$= \dfrac{1}{5} + \dfrac{1}{6} + \dfrac{1}{7} + \dfrac{1}{8} + \dfrac{1}{9} \approx 0.746$

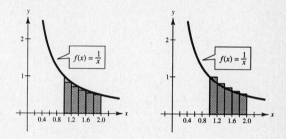

47. Let $\Delta x = [1 - (-1)]/n = 2/n$. Choosing right endpoints, we have

$$c_i = -1 + i\left(\frac{2}{n}\right) = -1 + \frac{2i}{n}.$$

Therefore, for $f(x) = x^2 - x^3$, we have

$$f\left(-1 + \frac{2i}{n}\right)\left(\frac{2}{n}\right).$$

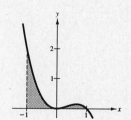

$\text{Area} = \displaystyle\lim_{n\to\infty}\sum_{i=1}^{n} f\left(-1 + \frac{2i}{n}\right)\left(\frac{2}{n}\right)$

$\displaystyle = \lim_{n\to\infty}\frac{2}{n}\sum_{i=1}^{n}\left[\left(-1 + \frac{2i}{n}\right)^2 - \left(-1 + \frac{2i}{n}\right)^3\right]$

$\displaystyle = \lim_{n\to\infty}\frac{2}{n}\sum_{i=1}^{n}\left[2 - \frac{10i}{n} + \frac{16i^2}{n^2} - \frac{8i^3}{n^3}\right]$

$\displaystyle = \lim_{n\to\infty}\left[\frac{2}{n}\sum_{i=1}^{n} 2 - \frac{20}{n^2}\sum_{i=1}^{n} i + \frac{32}{n^3}\sum_{i=1}^{n} i^2 - \frac{16}{n^4}\sum_{i=1}^{n} i^3\right]$

$\displaystyle = \lim_{n\to\infty}\left[\frac{2}{n}(2n) - \left(\frac{20}{n^2}\right)\frac{n(n+1)}{2} + \left(\frac{32}{n^3}\right)\frac{n(n+1)(2n+1)}{6} - \left(\frac{16}{n^4}\right)\frac{n^2(n+1)^2}{4}\right]$

$\displaystyle = \lim_{n\to\infty}\left(4 - 10 - \frac{10}{n} + \frac{32}{3} + \frac{16}{n} + \frac{16}{3n^2} - 4 - \frac{8}{n} - \frac{4}{n^2}\right) = 4 - 10 + \frac{32}{3} - 4 = \frac{2}{3}.$

53. Consider $f(x) = \tan x$ on the interval $[0, \pi/4]$. Let

$$\Delta x = \frac{\pi/4 - 0}{4} = \frac{\pi}{16}.$$

Dividing the interval into four parts of equal lengths yields endpoints

$$x_0 = 0, \; x_1 = \frac{\pi}{16}, \; x_2 = \frac{2\pi}{16}, \; x_3 = \frac{3\pi}{16}, \text{ and } x_4 = \frac{4\pi}{16}.$$

The midpoints of these subintervals are

$$\frac{x_0 + x_1}{2} = \frac{\pi}{32}, \; \frac{x_1 + x_2}{2} = \frac{3\pi}{32}, \; \frac{x_2 + x_3}{2} = \frac{5\pi}{32}, \text{ and}$$

$$\frac{x_3 + x_4}{2} = \frac{7\pi}{32}.$$

$$\text{Area} \approx \sum_{i=1}^{4} f\left(\frac{x_i + x_{i-1}}{2}\right) \Delta x$$

$$= \frac{\pi}{16}\left(\tan\frac{\pi}{32} + \tan\frac{3\pi}{32} + \tan\frac{5\pi}{32} + \tan\frac{7\pi}{32}\right)$$

$$\approx 0.345$$

61. A sketch of the region is given in the figure. Since area is positive, -2 cannot be correct. The area of a rectangle of height 4 over the interval $[0, 2]$ is 8. Thus, the area of the specified region is less than 8 which eliminates 8 and 10 as possible answers. A rectangle of height 3 over the interval $[0, 1]$ has area 3 which appears significantly less than the area of the specified region. It follows that the best approximation of the area is 6 square units given in part (b).

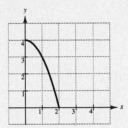

Section 4.3 Riemann Sums and Definite Integrals

15. The region whose area is given by $\int_0^2 (2x + 5)\, dx$ is shown by the accompanying figure to be a trapezoid. Since the height of the trapezoid is $h = 2$ and the lengths of the two bases are $b_1 = 5$ and $b_2 = 9$, the area of the trapezoid is

$$A = h\left[\frac{b_1 + b_2}{2}\right] = 2\left[\frac{5 + 9}{2}\right] = 14.$$

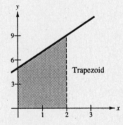

23. (a) $\displaystyle\int_2^6 [f(x) + g(x)]\, dx = \int_2^6 f(x)\, dx + \int_2^6 g(x)\, dx$

$$= 10 + (-2) = 8$$

(b) $\displaystyle\int_2^6 [g(x) - f(x)]\, dx = \int_2^6 g(x)\, dx - \int_2^6 f(x)\, dx$

$$= -2 - 10 = -12$$

(c) $\displaystyle\int_2^6 2g(x)\, dx = 2\int_2^6 g(x)\, dx = 2(-2) = -4$

(d) $\displaystyle\int_2^6 3f(x)\, dx = 3\int_2^6 f(x)\, dx = 3(10) = 30$

27. Let $\Delta x = [1 - (-1)]/n = 2/n$. Using right-hand endpoints, we have $c_i = -1 + (2i/n)$, and the definite integral is given by the limit

$$\int_{-1}^{1} x^3\, dx = \lim_{n\to\infty} \sum_{i=1}^{n}\left(-1 + \frac{2i}{n}\right)^3\left(\frac{2}{n}\right)$$

$$= \lim_{n\to\infty} \sum_{i=1}^{n}\left(\frac{2}{n}\right)\left(-1 + \frac{6i}{n} - \frac{12i^2}{n^2} + \frac{8i^3}{n^3}\right)$$

$$= \lim_{n\to\infty} \left(\frac{2}{n}\right)\left[-\sum_{i=1}^{n} 1 + \frac{6}{n}\sum_{i=1}^{n} i - \frac{12}{n^2}\sum_{i=1}^{n} i^2 + \frac{8}{n^3}\sum_{i=1}^{n} i^3\right]$$

$$= \lim_{n\to\infty} \left(\frac{2}{n}\right)\left[-n + \frac{6n(n+1)}{2n} - \frac{12n(n+1)(2n+1)}{6n^2} + \frac{8n^2(n+1)^2}{4n^3}\right]$$

$$= \lim_{n\to\infty} \left[\frac{-2n}{n} + \frac{6n(n+1)}{n^2} - \frac{4n(n+1)(2n+1)}{n^3} + \frac{4n^2(n+1)^2}{n^4}\right]$$

$$= -2 + 6 - 8 + 4 = 0.$$

39. Since the function is decreasing and x_i is the left endpoint of the ith subinterval, $f(x_i)\,\Delta x$ represents the area of a circumscribed rectangle extending outside the ith subregion. The sum of the areas of the circumscribed rectangles is greater than the area of the shaded region shown in the figure. Since

$$\int_1^5 f(x)\,dx$$

represents the area of the shaded region, it follows that

$$\sum_{i=1}^n f(x_i)\,\Delta x > \int_1^5 f(x)\,dx.$$

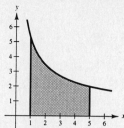

43. (a) Since f is negative over the interval $[0, 2]$, the definite integral will be negative. The definite integral equals the negative of the area of one-fourth a circle of radius 2. Therefore,

$$\int_0^2 f(x)\,dx = -\frac{1}{4}\pi(2^2) = -\pi.$$

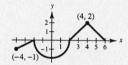

(b) The definite integral over the interval $[2, 6]$ represents the area of a triangle with base 4 and height 2. Therefore,

$$\int_2^6 f(x)\,dx = \frac{1}{2}(4)(2) = 4.$$

(c) Since f is negative over the interval $[-4, 2]$, the definite integral will be negative. The definite integral equals the negative of the combined area of a triangle with base 2 and height 1 and a semicircle of radius 2. Therefore,

$$\int_{-4}^2 f(x)\,dx = -\left(\frac{1}{2}(2)(1) + \frac{1}{2}\pi(2^2)\right) = -(1 + 2\pi).$$

(d) Since

$$\int_{-4}^6 f(x)\,dx = \int_{-4}^2 f(x)\,dx + \int_2^6 f(x)\,dx,$$

it follows from parts (b) and (c) that

$$\int_{-4}^6 f(x)\,dx = -(1 + 2\pi) + 4 = 3 - 2\pi.$$

(e) Since

$$\int_{-4}^6 |f(x)|\,dx = \int_{-4}^2 |f(x)|\,dx + \int_2^6 f(x)\,dx,$$

it follows from parts (b) and (c) that

$$\int_{-4}^6 |f(x)|\,dx = |-(1 + 2\pi)| + 4 = 5 + 2\pi.$$

(f) $\displaystyle\int_{-4}^6 (f(x) + 2)\,dx = \int_{-4}^6 f(x)\,dx + \int_{-4}^6 2\,dx$

$\int_{-4}^6 2\,dx$ represents the area of a rectangle with base 10 and height 2. Therefore, from part (d) it follows that

$$\int_{-4}^6 (f(x) + 2)\,dx = (3 - 2\pi) + 20 = 23 - 2\pi.$$

57. The function is not integrable on the specified interval since it has a nonremovable discontinuity at $x = 4$ (see figure).

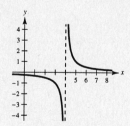

Section 4.4   The Fundamental Theorem of Calculus

11. $\displaystyle\int_0^1 (2t - 1)^2 \, dt = \int_0^1 (4t^2 - 4t + 1) \, dt$

$$= \left[\frac{4t^3}{3} - \frac{4t^2}{2} + t\right]_0^1$$

$$= \left(\frac{4}{3} - \frac{4}{2} + 1\right) - (0 - 0 + 0)$$

$$= \frac{4}{3} - \frac{6}{3} + \frac{3}{3} = \frac{1}{3}$$

15. $\displaystyle\int_1^4 \frac{u - 2}{\sqrt{u}} \, du = \int_1^4 (u^{1/2} - 2u^{-1/2}) \, du$

$$= \left[\frac{2}{3}u^{3/2} - 4u^{1/2}\right]_1^4$$

$$= \left[\frac{2}{3}(\sqrt{4})^3 - 4\sqrt{4}\right] - \left(\frac{2}{3} - 4\right) = \frac{2}{3}$$

23. Since $2x - 3 < 0$ if $x < \frac{3}{2}$, it follows that

$$|2x - 3| = \begin{cases} -(2x - 3), & x < \frac{3}{2} \\ 2x - 3, & x \geq \frac{3}{2} \end{cases}. \qquad \text{(See figure)}$$

$$\int_0^3 |2x - 3| \, dx = \int_0^{3/2} (3 - 2x) \, dx + \int_{3/2}^3 (2x - 3) \, dx$$

$$= 2\int_0^{3/2} (3 - 2x) \, dx \qquad \text{(By Symmetry)}$$

$$= 2\left[3x - x^2\right]_0^{3/2} = 2\left(\frac{9}{2} - \frac{9}{4}\right) = \frac{9}{2}$$

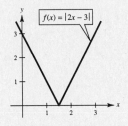

29. $\displaystyle\int_{-\pi/3}^{\pi/3} 4 \sec \theta \tan \theta \, d\theta = \left[4 \sec \theta\right]_{-\pi/3}^{\pi/3} = 4\left[\sec\frac{\pi}{3} - \sec\left(-\frac{\pi}{3}\right)\right] = 4[2 - 2] = 0$

35. $\displaystyle A = \int_0^3 (3 - x)\sqrt{x} \, dx = \int_0^3 (3x^{1/2} - x^{3/2}) \, dx$

$$= \left[2x^{3/2} - \frac{2}{5}x^{5/2}\right]_0^3 = \left[\frac{2x\sqrt{x}}{5}(5 - x)\right]_0^3$$

$$= \frac{6\sqrt{3}}{5}(2) - 0 = \frac{12\sqrt{3}}{5}$$

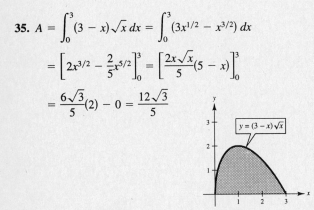

41. Using the figure we can see that the area of the region is

$$\text{Area} = \int_0^2 (x^3 + x) \, dx$$

$$= \left[\frac{1}{4}x^4 + \frac{1}{2}x^2\right]_0^2$$

$$= (4 + 2) - (0) = 6.$$

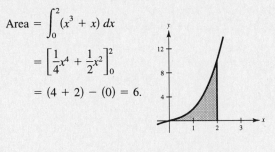

47. The average value is given by

$$\frac{1}{b - a}\int_a^b f(x) \, dx = \frac{1}{2 - (-2)}\int_{-2}^2 (4 - x^2) \, dx = \frac{1}{4}\int_{-2}^2 (4 - x^2) \, dx$$

$$= 2\left(\frac{1}{4}\right)\int_0^2 (4 - x^2) \, dx \qquad \text{(By Symmetry)}$$

$$= \frac{1}{2}\left[4x - \frac{1}{3}x^3\right]_0^2$$

$$= \frac{1}{2}\left(8 - \frac{8}{3}\right) = \frac{8}{3}.$$

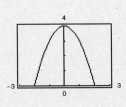

—CONTINUED—

47. —CONTINUED—

To find the values of x for which $f(x) = 8/3$ in the interval $[-2, 2]$, solve the equation

$$f(x) = 4 - x^2 = \frac{8}{3}$$

$$x^2 = \frac{4}{3}$$

$$x = \pm\frac{2}{\sqrt{3}} = \pm\frac{2\sqrt{3}}{3} \approx \pm 1.155.$$

57. (a) Begin by rewriting the given integral as a sum of integrals as follows.

$$\int_1^7 f(x)\, dx = \int_1^2 f(x)\, dx + \int_2^3 f(x)\, dx + \int_3^4 f(x)\, dx + \int_4^7 f(x)\, dx$$

The first three integrals each yield the area of a trapezoid and the fourth yields the area of a rectangle. Using the formulas for the area of trapezoids and rectangles, we have the following.

$$\int_1^7 f(x)\, dx = \frac{1}{2}(3 + 1) + \frac{1}{2}(1 + 2) + \frac{1}{2}(2 + 1) + 3(1) = 8.$$

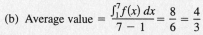

(b) Average value $= \dfrac{\int_1^7 f(x)\, dx}{7 - 1} = \dfrac{8}{6} = \dfrac{4}{3}$

(c) Since the graph is translated two units upward, it is given by the function $g(x) = f(x) + 2$. Therefore,

$$\int_1^7 g(x)\, dx = \int_1^7 [f(x) + 2]\, dx = \int_1^7 f(x)\, dx + \int_1^7 2\, dx = 8 + 6(2) = 20.$$

The average value of the function is translated 2 units upward. Therefore,

$$\text{Average value} = \frac{4}{3} + 2 = \frac{10}{3}.$$

61. (a) $F(x) = k \sec^2 x$

$F(0) = k \sec^2 0 = k = 500$

$F(x) = 500 \sec^2 x, \quad 0 \le x \le \dfrac{\pi}{3}$

(b) The average force over the interval $[0, \pi/3]$ is

$$\frac{1}{(\pi/3) - 0}\int_0^{\pi/3} 500 \sec^2 x\, dx = \frac{1500}{\pi}\int_0^{\pi/3} \sec^2 x\, dx$$

$$= \frac{1500}{\pi}\Big[\tan x\Big]_0^{\pi/3}$$

$$= \frac{1500}{\pi}\big(\sqrt{3}\big)$$

$$\approx 827 \text{ newtons.}$$

71. (a) $F(x) = \displaystyle\int_{x/4}^x \sec^2 t\, dt = \Big[\tan t\Big]_{\pi/4}^x = \tan x - 1$

(b) $F'(x) = \dfrac{d}{dx}[\tan x - 1] = \sec^2 x$

81. Since the function is easily integrated, we can find the derivative as follows.

$$F(x) = \int_0^{\sin x} \sqrt{t}\, dt = \Big[\frac{2}{3}t^{3/2}\Big]_0^{\sin x} = \frac{2}{3}(\sin x)^{3/2}$$

$$F'(x) = \frac{2}{3}\Big(\frac{3}{2}\Big)(\sin x)^{1/2}(\cos x) = \cos x\sqrt{\sin x}$$

An alternate method is to use the Second Fundamental Theorem of Calculus and the Chain Rule where $u = \sin x$.

$$F'(x) = \frac{dF}{du}\frac{du}{dx} = \frac{d}{du}\Big[\int_0^u \sqrt{t}\, dt\Big]\frac{du}{dx}$$

$$= \sqrt{u}(\cos x) = \cos x\sqrt{\sin x}$$

87. (a) $C(x) = 5000\Big(25 + 3\displaystyle\int_0^x t^{1/4} dt\Big)$

$$= 5000\Big(25 + 3\Big[\frac{4}{5}t^{5/4}\Big]_0^x\Big)$$

$$= 5000\Big(25 + \frac{12}{5}x^{5/4}\Big) = 1000(125 + 12x^{5/4})$$

(b) $C(1) = 1000[125 + 12(1)^{5/4}] = \$137,000$

$C(5) = 1000[125 + 12(5)^{5/4}] \approx \$214,721$

$C(10) = 1000[125 + 12(10)^{5/4}] \approx \$338,394$

Section 4.5   Integration by Substitution

11. To evaluate $\int x^2(x^3 - 1)^4 \, dx$, use the method of pattern recognition by letting $g(x) = x^3 - 1$, and $g'(x) = 3x^2$. Thus, by Theorem 4.13, we have

$$\int x^2(x^3 - 1)^4 \, dx = \int (x^3 - 1)^4 \left(\frac{1}{3}\right)(3x^2) \, dx$$

$$= \frac{1}{3} \int \overbrace{(x^3 - 1)^4}^{[g(x)]^4} \overbrace{(3x^2)}^{g'(x)} \, dx$$

$$= \frac{1}{3} \frac{[g(x)]^5}{5} + C$$

$$= \left(\frac{1}{3}\right)\left[\frac{(x^3 - 1)^5}{5}\right] + C$$

$$= \frac{1}{15}(x^3 - 1)^5 + C.$$

Check

If $y = \frac{1}{15}(x^3 - 1)^5 + C$, then

$$\frac{dy}{dx} = \frac{1}{15}(5)(x^3 - 1)^4(3x^2) + 0 = x^2(x^3 - 1)^4.$$

13. To evaluate $\int 5x\sqrt[3]{1 - x^2} \, dx$, use the method of pattern recognition by letting $g(x) = 1 - x^2$, and $g'(x) = -2x$.

$$\int 5x\sqrt[3]{1 - x^2} \, dx = -\frac{5}{2} \int \overbrace{(1 - x^2)^{1/3}}^{[g(x)]^{1/3}} \overbrace{(-2x)}^{g'(x)} \, dx$$

$$= -\frac{5}{2} \frac{[g(x)]^{4/3}}{4/3} + C$$

$$= -\left(\frac{5}{2}\right)\left[\frac{(1 - x^2)^{4/3}}{4/3}\right] + C$$

$$= -\frac{15}{8}(1 - x^2)^{4/3} + C$$

Check

If $y = -\frac{15}{8}(1 - x^2)^{4/3} + C$, then

$$\frac{dy}{dx} = -\left(\frac{15}{8}\right)\left(\frac{4}{3}\right)(1 - x^2)^{1/3}(-2x) + 0$$

$$= 5x(1 - x^2)^{1/3}.$$

17. We evaluate this integral by changing the variable by letting $u = 1 + (1/t)$. Then,

$$du = -\frac{1}{t^2} \, dt.$$

Therefore,

$$\int \left(1 + \frac{1}{t}\right)^3 \left(\frac{1}{t^2}\right) dt = -\int \left(1 + \frac{1}{t}\right)^3 \left(\frac{-1}{t^2}\right) dt$$

$$= -\int u^3 \, du$$

$$= -\frac{u^4}{4} + C$$

$$= -\frac{1}{4}\left(1 + \frac{1}{t}\right)^4 + C$$

Check

If $y = -\frac{1}{4}\left(1 + \frac{1}{t}\right)^4 + C$, then

$$\frac{dy}{dt} = \left(-\frac{1}{4}\right)(4)\left(1 + \frac{1}{t}\right)^3\left(-\frac{1}{t^2}\right) + 0$$

$$= \left(1 + \frac{1}{t}\right)^3\left(\frac{1}{t^2}\right).$$

19.
$$\int \frac{1}{\sqrt{2x}} \, dx = \int \frac{1}{\sqrt{2}\sqrt{x}} \, dx$$

$$= \frac{1}{\sqrt{2}} \int x^{-1/2} \, dx$$

$$= \frac{1}{\sqrt{2}}\left(\frac{x^{1/2}}{1/2}\right) + C = \sqrt{2x} + C$$

Check

If $y = \sqrt{2x} + C = (2x)^{1/2} + C$, then

$$\frac{dy}{dx} = \frac{1}{2}(2x)^{-1/2}(2) + 0 = \frac{1}{\sqrt{2x}}.$$

27. $\dfrac{dy}{dx} = 4x + \dfrac{4x}{\sqrt{16 - x^2}}$

$$y = \int \left(4x + \frac{4x}{\sqrt{16 - x^2}} \right) dx$$

$$= 4\int x\,dx + (-2)\int \overbrace{(16 - x^2)^{-1/2}}^{u^{-1/2}}\,\overbrace{(-2x)\,dx}^{du}$$

$$= 4\left(\frac{x^2}{2}\right) - 2\left[\frac{(16 - x^2)^{1/2}}{1/2} \right] + C$$

$$= 2x^2 - 4\sqrt{16 - x^2} + C$$

35. If we let $u = 1/\theta$, then

$$du = -\frac{1}{\theta^2}\,d\theta \ \text{ and } \ -du = \frac{1}{\theta^2}\,d\theta.$$

Therefore,

$$\int \frac{1}{\theta^2} \cos \frac{1}{\theta}\,d\theta = \int \cos \frac{1}{\theta}\left(\frac{1}{\theta^2}\,d\theta \right)$$

$$= \int \cos u\,(-du)$$

$$= -\int \cos u\,du$$

$$= -\sin u + C = -\sin \frac{1}{\theta} + C.$$

37. Using the double angle identity we obtain $2 \sin 2x \cos 2x = \sin 4x$. If we let $u = 4x$, then $du = 4\,dx$ and $dx = \frac{1}{4}\,du$. Therefore,

$$\int \sin 2x \cos 2x\,dx = \frac{1}{2}\int \sin 4x\,dx$$

$$= \frac{1}{2}\int \sin u\left(\frac{1}{4}\right) du$$

$$= \frac{1}{8}\int \sin u\,du$$

$$= -\frac{1}{8}\cos u + C_1 = -\frac{1}{8}\cos 4x + C_1.$$

We could also evaluate this integral by letting $u = \sin 2x$. Then $du = 2\cos 2x\,dx$ and $\cos 2x\,dx = \frac{1}{2}\,du$. Therefore,

$$\int \sin 2x \cos 2x\,dx = \int u\left(\frac{1}{2}\right) du$$

$$= \frac{1}{2}\int u\,du$$

$$= \frac{1}{4}u^2 + C_2$$

$$= \frac{1}{4}\sin^2 2x + C_2.$$

Through the use of trigonometric identities it can be proved that the results of the two methods of integration are equivalent. This exercise shows that the method of evaluating an integral may not be unique and the results, even though equivalent, may appear unrelated.

43. $\displaystyle\int \cot^2 x\,dx = \int (\csc^2 x - 1)\,dx = -\cot x - x + C$

49. Let $u = \sqrt{1-x}$. Then $u^2 = 1 - x$, $x = 1 - u^2$, and $dx = -2u\,du$. Thus,

$$\int x^2 \sqrt{1-x}\,dx = (1 - u^2)^2\, u(-2u)\,du$$

$$= -\int (2u^2 - 4u^4 + 2u^6)\,du$$

$$= -\left(\frac{2u^3}{3} - \frac{4u^5}{5} + \frac{2u^7}{7}\right) + C$$

$$= \frac{-2u^3}{105}(35 - 42u^2 + 15u^4) + C$$

$$= \frac{-2}{105}(1-x)^{3/2}[35 - 42(1-x) + 15(1-x)^2] + C$$

$$= \frac{-2}{105}(1-x)^{3/2}(15x^2 + 12x + 8) + C.$$

51. Let $u = \sqrt{2x-1}$. Then $u^2 = 2x - 1$, $x = \dfrac{u^2+1}{2}$, and $dx = u\,du$. Thus,

$$\int \frac{x^2 - 1}{\sqrt{2x-1}}\,dx = \int \frac{[(u^2+1)/2]^2 - 1}{u}\,(u\,du)$$

$$= \frac{1}{4}\int (u^4 + 2u^2 - 3)\,du$$

$$= \frac{1}{4}\left(\frac{u^5}{5} + \frac{2u^3}{3} - 3u\right) + C$$

$$= \frac{u}{60}(3u^4 + 10u^2 - 45) + C$$

$$= \frac{1}{60}\sqrt{2x-1}[3(2x-1)^2 + 10(2x-1) - 45] + C$$

$$= \frac{1}{60}\sqrt{2x-1}(12x^2 + 8x - 52) + C$$

$$= \frac{1}{15}\sqrt{2x-1}(3x^2 + 2x - 13) + C.$$

55. Let $u = x^2 + 1$. Then $du = 2x\,dx$, and $dx = du/2$. Furthermore, when $x = -1$, $u = 2$ and when $x = 1$, $u = 2$. Since the upper and lower limits are equal, we have

$$\int_{-1}^{1} x(x^2 + 1)^3\,dx = \int_{2}^{2} u^3\left(\frac{1}{2}\right) du = 0.$$

59. Let $u = 1 + \sqrt{x}$. Then $du = 1/(2\sqrt{x})\,dx$. Furthermore, if $x = 1$, then $u = 2$, and if $x = 9$, then $u = 4$. Hence,

$$\int_{1}^{9} \frac{1}{\sqrt{x}\left(1 + \sqrt{x}\right)^2}\,dx = 2\int_{1}^{9} \frac{1}{\left(1 + \sqrt{x}\right)^2}\left(\frac{1}{2\sqrt{x}}\right) dx$$

$$= 2\int_{2}^{4} \frac{1}{u^2}\,du$$

$$= 2\int_{2}^{4} u^{-2}\,du$$

$$= \left[\frac{-2}{u}\right]_{2}^{4}$$

$$= -\frac{1}{2} - (-1) = \frac{1}{2}.$$

65. Let $u = \sqrt[3]{x + 1}$. Then $u^3 = x + 1$, $x = u^3 - 1$, and $dx = 3u^2\,du$. Furthermore, if $x = 0$, then $u = 1$, and if $x = 7$, then $u = 2$. Thus,

$$\text{Area} = \int_0^7 x\sqrt[3]{x + 1}\,dx = \int_1^2 (u^3 - 1)(u)(3u^2\,du)$$

$$= 3\int_1^2 (u^6 - u^3)\,du$$

$$= 3\left[\frac{u^7}{7} - \frac{u^4}{4}\right]_1^2$$

$$= 3\left(\frac{128}{7} - \frac{16}{4} - \frac{1}{7} + \frac{1}{4}\right)$$

$$= 3\left(\frac{127}{7} - \frac{15}{4}\right) = 3\left(\frac{508 - 105}{28}\right) = \frac{1209}{28}.$$

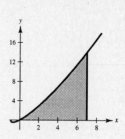

67. $\text{Area} = \displaystyle\int_0^\pi (2\sin x + \sin 2x)\,dx$

$$= 2\int_0^\pi \sin x\,dx + \frac{1}{2}\int_0^\pi \sin 2x(2)\,dx$$

$$= \left[-2\cos x - \frac{1}{2}\cos 2x\right]_0^\pi = 4$$

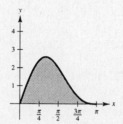

69. Let $u = x/2$. Then $du = (1/2)\,dx$ and $dx = 2\,du$. Also, if $x = \pi/2$, then $u = \pi/4$, and if $x = 2\pi/3$, then $u = \pi/3$. Thus,

$$\text{Area} = \int_{\pi/2}^{2\pi/3} \sec^2\!\left(\frac{x}{2}\right)dx = \int_{\pi/4}^{\pi/3} \sec^2 u\,(2)\,du$$

$$= 2\Big[\tan u\Big]_{\pi/4}^{\pi/3}$$

$$= 2\left[\tan\frac{\pi}{3} - \tan\frac{\pi}{4}\right] = 2\left(\sqrt{3} - 1\right).$$

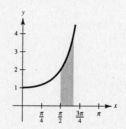

77. Method 1: $\displaystyle\int (2x - 1)^2\,dx = \frac{1}{2}\int (2x - 1)^2(2)\,dx$

$$= \frac{1}{6}(2x - 1)^3 + C_1$$

$$= \frac{4}{3}x^3 - 2x^2 + x - \frac{1}{6} + C_1$$

Method 2: $\displaystyle\int (2x - 1)^2\,dx = \int (4x^2 - 4x + 1)\,dx$

$$= \frac{4}{3}x^3 - 2x^2 + x + C_2$$

The two differ by a constant: $C_2 = C_1 - \dfrac{1}{6}$

79. The function $y = x^2$ is even (see figure).

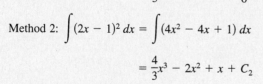

(a) $\displaystyle\int_{-2}^0 x^2\,dx = \int_0^2 x^2\,dx = \frac{8}{3}$

(b) $\displaystyle\int_{-2}^2 x^2\,dx = 2\int_0^2 x^2\,dx = \frac{16}{3}$

(c) $\displaystyle\int_0^2 (-x^2)\,dx = -\int_0^2 x^2\,dx = -\frac{8}{3}$

(d) $\displaystyle\int_{-2}^0 3x^2\,dx = 3\int_0^2 x^2\,dx = 8$

83. Since the rate of depreciation, dV/dt, is inversely proportional to the square of $t + 1$, it follows that

$$\frac{dV}{dt} = \frac{k}{(t + 1)^2}$$

$$V = \int \frac{k}{(t + 1)^2}\, dt$$

$$= k \int (t + 1)^{-2}\, dt$$

$$= k \int u^{-2}\, du \qquad (u = t + 1, du = dt)$$

$$= k \frac{u^{-1}}{-1} + C$$

$$= \frac{-k}{t + 1} + C.$$

Since the initial value of the machine was $500,000, we have

$$V(0) = \frac{-k}{0 + 1} + C = -k + C = 500,000.$$

During the first year the value of the machine decreased $100,000. Therefore,

$$V(1) = \frac{-k}{1 + 1} + C = -\frac{1}{2}k + C = 400,000.$$

Solving the two equations simultaneously yields the solution

$$V(t) = \frac{200,000}{t + 1} + 300,000.$$

The approximate value of the machine after 4 years is

$$V(4) = \frac{200,000}{4 + 1} + 300,000 = \$340,000.$$

Note that according to this model the value of the machine will always be greater than $300,000 since

$$\lim_{t \to \infty} V(t) = 300,000.$$

87. The average sales over the interval of time $a \le t \le b$ is

$$\frac{1}{b - a}\int_a^b \left(74.50 + 43.75 \sin \frac{\pi t}{6}\right) dt = \frac{1}{b - a}\left[74.50t - \frac{262.5}{\pi} \cos \frac{\pi t}{6}\right]_a^b.$$

(a) $\dfrac{1}{3}\left[74.50t - \dfrac{262.5}{\pi} \cos \dfrac{\pi t}{6}\right]_0^3 = \dfrac{1}{3}\left(223.5 + \dfrac{262.5}{\pi}\right) \approx 102.352$ thousand units

(b) $\dfrac{1}{3}\left[74.50t - \dfrac{262.5}{\pi} \cos \dfrac{\pi t}{6}\right]_3^6 = \dfrac{1}{3}\left(447 + \dfrac{262.5}{\pi} - 223.5\right) \approx 102.352$ thousand units

(c) $\dfrac{1}{12}\left[74.50t - \dfrac{262.5}{\pi} \cos \dfrac{\pi t}{6}\right]_0^{12} = \dfrac{1}{12}\left(894 - \dfrac{262.5}{\pi} + \dfrac{262.5}{\pi}\right) \approx 74.5$ thousand units

Section 4.6   Numerical Integration

5. Trapezoidal Rule ($n = 8$)

$$\int_0^2 x^3\, dx \approx \frac{2}{2(8)}\left[0 + 2\left(\frac{1}{4}\right)^3 + 2\left(\frac{2}{4}\right)^3 + 2\left(\frac{3}{4}\right)^3 + 2\left(\frac{4}{4}\right)^3 + 2\left(\frac{5}{4}\right)^3 + 2\left(\frac{6}{4}\right)^3 + 2\left(\frac{7}{4}\right)^3 + 2^3\right]$$

$$= \frac{1}{8}\left[\frac{2(1^3 + 2^3 + 3^3 + 4^3 + 5^3 + 6^3 + 7^3)}{4^3} + 8\right]$$

$$= \frac{1}{8}\left[\frac{2(784)}{164} + 8\right] = \frac{65}{16} = 4.0625$$

Simpson's Rule ($n = 8$)

$$\int_0^2 x^3\, dx \approx \frac{2}{3(8)}\left[0 + 4\left(\frac{1}{4}\right)^3 + 2\left(\frac{2}{4}\right)^3 + 4\left(\frac{3}{4}\right)^3 + 2\left(\frac{4}{4}\right)^3 + 4\left(\frac{5}{4}\right)^3 + 2\left(\frac{6}{4}\right)^3 + 4\left(\frac{7}{4}\right)^3 + 2^3\right]$$

$$= \frac{1}{12}\left[\frac{4(1^3 + 3^3 + 5^3 + 7^3) + 2(2^3 + 4^3 + 6^3)}{4^3} + 8\right]$$

$$= \frac{1}{12}\left[\frac{4(496) + 2(288)}{4^3} + 8\right]$$

$$= \frac{1}{12}\left(\frac{2560}{64} + 8\right) = \frac{1}{12}(48) = 4$$

In this particular case, Simpson's Rule is exact since

$$\int_0^2 x^3\, dx = \left[\frac{x^4}{4}\right]_0^2 = \frac{16}{4} = 4.$$

13. Trapezoidal Rule ($n = 4$)

$$\int_0^1 \sqrt{x}\sqrt{1-x}\, dx \approx \frac{1}{2(4)}\left[0 + 2\sqrt{\frac{1}{4}}\sqrt{\frac{3}{4}} + 2\sqrt{\frac{2}{4}}\sqrt{\frac{2}{4}} + 2\sqrt{\frac{3}{4}}\sqrt{\frac{1}{4}} + 0\right]$$

$$= \frac{1}{8}\left[\frac{2\sqrt{3}}{4} + \frac{2(2)}{4} + \frac{2\sqrt{3}}{4}\right] = \frac{1}{8}\left(1 + \sqrt{3}\right) \approx 0.342$$

Simpson's Rule ($n = 4$)

$$\int_0^1 \sqrt{x}\sqrt{1-x}\, dx \approx \frac{1}{3(4)}\left[0 + 4\sqrt{\frac{1}{4}}\sqrt{\frac{3}{4}} + 2\sqrt{\frac{2}{4}}\sqrt{\frac{2}{4}} + 4\sqrt{\frac{3}{4}}\sqrt{\frac{1}{4}} + 0\right]$$

$$= \frac{1}{2}\left[\frac{4\sqrt{3}}{4} + \frac{2(2)}{4} + \frac{4\sqrt{3}}{4}\right] = \frac{1}{12}\left(2\sqrt{3} + 1\right) \approx 0.372$$

15. Using the numerical integration capabilities of a graphing utility yields

$$\int_0^1 \sqrt{x}\sqrt{1-x}\, dx \approx 0.393.$$

Trapezoidal Rule with $n = 4$

$$\int_0^{\sqrt{\pi/2}} \cos x^2\, dx \approx \frac{\sqrt{\pi/2}}{2(4)}\left[\cos(0) + 2\cos\left(\frac{\sqrt{\pi/2}}{4}\right)^2 + 2\cos\left(\frac{2\sqrt{\pi/2}}{4}\right)^2 + 2\cos\left(\frac{3\sqrt{\pi/2}}{4}\right)^2 + \cos\left(\frac{4\sqrt{\pi/2}}{4}\right)^2\right]$$

$$\approx 0.957$$

Simpson's Rule with $n = 4$

$$\int_0^{\sqrt{\pi/2}} \cos x^2\, dx \approx \frac{\sqrt{\pi/2}}{3(4)}\left[\cos(0) + 4\cos\left(\frac{\sqrt{\pi/2}}{4}\right)^2 + 2\cos\left(\frac{2\sqrt{\pi/2}}{4}\right)^2 + 4\cos\left(\frac{3\sqrt{\pi/2}}{4}\right)^2 + \cos\left(\frac{4\sqrt{\pi/2}}{4}\right)^2\right]$$

$$\approx 0.978$$

Using the numerical integration capabilities of a graphing utility yields

$$\int_0^{\sqrt{\pi/2}} \cos x^2\, dx \approx 0.977.$$

27. (a) Begin by letting $f(x) = \tan x^2$ and finding the second derivative of f using the symbolic differentiation utility. (Note: The simplification of the derivatives will differ depending on which differentiation utility is used.)

$$f'(x) = 2x \sec^2 x^2$$

$$f''(x) = \frac{2(\cos x^2 + 4x^2 \sin x^2)}{\cos^3 x^2}$$

The graph of the second derivative (produced by the differentiation utility) is shown in the figure. The maximum value of $|f''(x)|$ on the interval $[0, 1]$ is $|f''(1)| \approx 50$ Thus, by Theorem 4.19, we can write

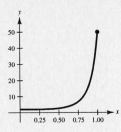

$$E \le \frac{(b - a)^3}{12n^2}|f''(1)| \le \frac{50}{12n^2}.$$

To obtain an error E that is less than 0.00001, we must choose n so that $50/(12n^2) \le 0.00001$. Thus,

$$50(100{,}000) \le 12n^2 \implies 645.5 \approx \sqrt{\frac{50(100{,}000)}{12}} \le n.$$

Therefore, choose $n = 646$.

(b) Begin by letting $f(x) = \tan x^2$ and finding the fourth derivative of f using the symbolic differentiation utility. The first and second derivative are given in part (a).

$$f'''(x) = -\frac{8x(4x^2 \cos^2 x^2 - 3 \sin x^2 \cos x^2 - 6x^2)}{\cos^4 x^2}$$

$$f^{(4)}(x) = -\frac{8}{\cos^5 x^2}[24x^2 \cos^3 x^2 + (16x^4 - 3) \sin x^2 \cos^2 x^2 - 36x^2 \cos x^2 - 48x^4 \sin x^2]$$

The graph of the fourth derivative (produced by the differentiation utility) is shown in the figure. The maximum value of $|f^{(4)}(x)|$ on the interval $[0, 1]$ is $|f^{(4)}(1)| \approx 9185$. Thus, by Theorem 4.19, we can write

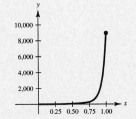

$$E \le \frac{(b - a)^5}{180n^4}|f^{(4)}(1)| \le \frac{9185}{180n^4}.$$

To obtain an error E that is less than 0.00001, we must choose n so that $9185/(180n^4) \le 0.00001$. Thus,

$$9185(100{,}000) \le 180n^4 \implies 47.5 \approx \sqrt[4]{\frac{9185(100{,}000)}{180}} \le n.$$

Therefore, choose $n = 48$.

37. Simpson's Rule with $n = 12$:

$$\int_0^5 100x\sqrt{125 - x^3}\, dx \approx \frac{5}{3(12)}\Big[0 + 400\sqrt{125 - (5/12)^3} + 200\sqrt{125 - (10/12)^3}$$

$$+ 400\sqrt{125 - (15/12)^3} + 200\sqrt{125 - (20/12)^3}$$

$$+ 400\sqrt{125 - (25/12)^3} + 200\sqrt{125 - (30/12)^3}$$

$$+ 400\sqrt{125 - (35/12)^3} + 200\sqrt{125 - (40/12)^3}$$

$$+ 400\sqrt{125 - (45/12)^3} + 200\sqrt{125 - (50/12)^3}$$

$$+ 400\sqrt{125 - (55/12)^3} + 0\Big] \approx 10{,}233.58 \text{ ft-lb}$$

Review Exercises for Chapter 4

5. $\displaystyle\int \frac{x^3 + 1}{x^2}\, dx = \int \left(\frac{x^3}{x^2} + \frac{1}{x^2}\right) dx$

$$= \int (x + x^{-2})\, dx = \frac{1}{2}x^2 + \frac{x^{-1}}{-1} + C = \frac{1}{2}x^2 - \frac{1}{x} + C$$

11. Let the position function of the plane be given by $s(t)$ where s is measured in feet and t is time in seconds. If $t = 0$ is the time the plane starts its take off roll, then $s(0) = 0$, $s(30) = 3600$, $s'(0) = 0$, and $s''(t) = a$ where a is constant.

$$s'(t) = \int a\, dt = at + C_1$$

$$s'(0) = 0 + C_1 = 0 \implies C_1 = 0$$

$$s(t) = \int at\, dt = \frac{a}{2}t^2 + C_2$$

$$s(0) = 0 + C_2 = 0 \implies C_2 = 0 \implies s(t) = \frac{a}{2}t^2$$

$$s(30) = \frac{a}{2}(30)^2 = 3600 \ \ or \ \ a = \frac{3600(2)}{30^2} = 8 \text{ ft/sec}^2$$

Therefore,

$$s(t) = 4t^2,\ s'(t) = v(t) = 8t \quad \text{and}$$

$$v(30) = 8(30) = 240 \text{ ft/sec.}$$

19. (a) $\displaystyle\int_2^6 [f(x) + g(x)]\, dx = \int_2^6 f(x)\, dx + \int_2^6 g(x)\, dx$

$$= 10 + 3 = 13$$

(b) $\displaystyle\int_2^6 [f(x) - g(x)]\, dx = \int_2^6 f(x)\, dx - \int_2^6 g(x)\, dx$

$$= 10 - 3 = 7$$

(c) $\displaystyle\int_2^6 [2f(x) - 3g(x)]\, dx = 2\int_2^6 f(x)\, dx - 3\int_2^6 g(x)\, dx$

$$= 2(10) - 3(3) = 11$$

(d) $\displaystyle\int_2^6 5f(x)\, dx = 5\int_2^6 f(x)\, dx = 5(10) = 50$

23. To find $\int x^2/\sqrt{x^3 + 3}\, dx$, let $u = x^3 + 3$. Then $du = 3x^2\, dx$.

$$\int \frac{x^2}{\sqrt{x^3 + 3}}\, dx = \frac{1}{3}\int (x^3 + 3)^{-1/2}(3x^2)\, dx$$

$$= \frac{1}{3}\int u^{-1/2}\, du$$

$$= \frac{2}{3}u^{1/2} + C$$

$$= \frac{2}{3}(x^3 + 3)^{1/2} + C$$

$$= \frac{2}{3}\sqrt{x^3 + 3} + C$$

31. To find $\int \tan^n x \sec^2 x\, dx$, let

$$u = \tan x \text{ and } du = \sec^2 x\, dx.$$

$$\int \tan^n x \sec^2 x\, dx = \int u^n\, du$$

$$= \frac{u^{n+1}}{n + 1} + C$$

$$= \frac{\tan^{n+1}}{n + 1} + C,\ n \neq -1$$

39. If we let $u = 1 + x$, then $du = dx$. Also, when $x = 0$, $u = 1$, and when $x = 3$, $u = 4$. Therefore,

$$\int_0^3 \frac{1}{\sqrt{1 + x}}\, dx = \int_1^4 u^{-1/2}\, du$$

$$= \left[2u^{1/2}\right]_1^4 = 2(2 - 1) = 2$$

43. If we let $u = \sqrt{1 - y}$. Then

$y = 1 - u^2$ and $dy = -2u\, du$.

Furthermore, when $y = 0$, $u = 1$ and when $y = 1$, $u = 0$.

$$2\pi \int_0^1 (y + 1)\sqrt{1 - y}\, dy = 2\pi \int_1^0 (2 - u^2)(u)(-2u)\, du$$

$$= -4\pi \int_1^0 (2u^2 - u^4)\, du$$

$$= -4\pi \left[\frac{2}{3}u^3 - \frac{1}{5}u^5 \right]_1^0$$

$$= \frac{28\pi}{15}$$

53. The graph of the region is shown in the figure.

$$A = \int_0^8 \frac{4}{\sqrt{x + 1}}\, dx = 4\int_0^8 (x + 1)^{-1/2}\, dx$$

$$= \left[8\sqrt{x + 1} \right]_0^8$$

$$= 8\sqrt{9} - 8(1) = 16$$

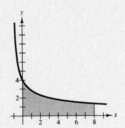

57. The average value is given by

$$\frac{1}{10 - 5}\int_5^{10} \frac{1}{\sqrt{x - 1}}\, dx = \frac{1}{5}\int_5^{10} (x - 1)^{-1/2}(1)\, dx$$

$$= \left[\frac{2}{5}(x - 1)^{1/2} \right]_5^{10} = \frac{2}{5}.$$

To find the value of x where the function assumes its mean value on $[5, 10]$, solve

$$\frac{1}{\sqrt{x - 1}} = \frac{2}{5}$$

$$\sqrt{x - 1} = \frac{5}{2}$$

$$x - 1 = \frac{25}{4}$$

$$x = \frac{29}{4}.$$

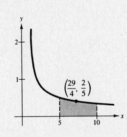

61. Trapezoidal Rule: $n = 4$

$$\int_1^2 \frac{1}{1 + x^3}\, dx \approx \frac{2 - 1}{2(4)}\left(\frac{1}{1 + 1^3} + \frac{2}{1 + (1.25)^3} + \frac{2}{1 + (1.5)^3} + \frac{2}{1 + (1.75)^3} + \frac{1}{1 + 2^3} \right)$$

$$\approx 0.257$$

Simpson's Rule: $n = 4$

$$\int_1^2 \frac{1}{1 + x^3}\, dx \approx \frac{2 - 1}{3(4)}\left(\frac{1}{1 + 1^3} + \frac{4}{1 + (1.25)^3} + \frac{2}{1 + (1.5)^3} + \frac{4}{1 + (1.75)^3} + \frac{1}{1 + 2^3} \right)$$

$$\approx 0.254$$

Using the numerical integration capabilities of a graphing utility, we have

$$\int_1^2 \frac{1}{1 + x^3}\, dx \approx 0.254.$$

67. Since $p = 1.20 + 0.04t$, the annual cost is given by

$$C = \frac{15,000}{M}\int_t^{t+1} (1.20 + 0.04s)\, ds = \frac{15,000}{M}\left[1.20s + 0.02s^2 \right]_t^{t+1}.$$

(a) For the year 2005, $t = 10$.

$$C = \frac{15,000}{M}\left[1.20s + 0.02s^2 \right]_{10}^{11} = \frac{24,300}{M}$$

(b) For the year 2005, $t = 15$.

$$C = \frac{15,000}{M}\left[1.20s + 0.02s^2 \right]_{15}^{16} = \frac{27,300}{M}$$

C H A P T E R   5
Logarithmic, Exponential, and
Other Transcendental Functions

CHAPTER 5
Logarithmic, Exponential, and Other Transcendental Functions

Section 5.1   The Natural Logarithmic Function and Differentiation
Solutions to Selected Odd-Exercises

13. (a) $\ln 6 = \ln(2 \cdot 3)$

$\qquad = \ln 2 + \ln 3 \approx 0.6931 + 1.0986 = 1.7917$

(b) $\ln \frac{2}{3} = \ln 2 - \ln 3$

$\qquad \approx 0.6931 - 1.0986 = -0.4055$

(c) $\ln 81 = \ln(3^4) = 4 \ln 3 \approx 4(1.0986) = 4.3944$

(d) $\ln \sqrt{3} = \ln(3^{1/2}) = \frac{1}{2} \ln 3 \approx \frac{1}{2}(1.0986) = 0.5493$

21. $\ln\left(\dfrac{x^2 - 1}{x^3}\right)^3 = 3[\ln(x^2 - 1) - \ln x^3]$

$\qquad = 3\{\ln[(x + 1)(x - 1)] - 3 \ln x\}$

$\qquad = 3[\ln(x + 1) + \ln(x - 1) - 3 \ln x]$

27. $\frac{1}{3}[2 \ln(x + 3) + \ln x - \ln(x^2 - 1)] = \frac{1}{3}[\ln(x + 3)^2 + \ln x - \ln(x^2 - 1)]$

$\qquad\qquad\qquad = \frac{1}{3}\{\ln[x(x + 3)^2] - \ln(x^2 - 1)\}$

$\qquad\qquad\qquad = \frac{1}{3} \ln \dfrac{x(x + 3)^2}{x^2 - 1}$

$\qquad\qquad\qquad = \ln\left(\dfrac{x(x + 3)^2}{x^2 - 1}\right)^{1/3} = \ln \sqrt[3]{\dfrac{x(x + 3)^2}{x^2 - 1}}$

37. $y = \ln x^3 = 3 \ln x$

$\dfrac{dy}{dx} = 3\left(\dfrac{1}{x}\right) = \dfrac{3}{x}$

At the point $(1, 0)$,

$\dfrac{dy}{dx} = \dfrac{3}{1} = 3.$

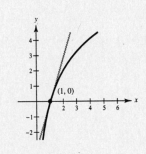

45. $y = \ln\left(x\sqrt{x^2 - 1}\right) = \ln x + \ln\sqrt{x^2 - 1}$

$\qquad = \ln x + \frac{1}{2} \ln(x^2 - 1)$

$\dfrac{dy}{dx} = \dfrac{1}{x} + \dfrac{1}{2}\left(\dfrac{1}{x^2 - 1}\right)(2x)$

$\qquad = \dfrac{1}{x} + \dfrac{x}{x^2 - 1} = \dfrac{2x^2 - 1}{x(x^2 - 1)}$

49. $g(t) = \dfrac{\ln t}{t^2}$

$g'(x) = \dfrac{(t^2)(1/t) - (\ln t)(2t)}{t^4} = \dfrac{t - 2t \ln t}{t^4} = \dfrac{1 - 2 \ln t}{t^3}$

57. $y = \dfrac{-\sqrt{x^2 + 1}}{x} + \ln\left(x + \sqrt{x^2 + 1}\right)$

$\dfrac{dy}{dx} = -\dfrac{x(1/2)(x^2 + 1)^{-1/2}(2x) - (x^2 + 1)^{1/2}(1)}{x^2} + \dfrac{1}{x + \sqrt{x^2 + 1}}\left[1 + \left(\dfrac{1}{2}\right)(x^2 + 1)^{-1/2}(2x)\right]$

$\qquad = -\dfrac{x^2(x^2 + 1)^{-1/2} - (x^2 + 1)^{1/2}}{x^2} + \left(\dfrac{1}{x + \sqrt{x^2 + 1}}\right)\left(1 + \dfrac{x}{\sqrt{x^2 + 1}}\right)$

$\qquad = \dfrac{-x^2 + (x^2 + 1)}{x^2\sqrt{x^2 + 1}} + \dfrac{1}{\sqrt{x^2 + 1}} = \dfrac{1 + x^2}{x^2\sqrt{x^2 + 1}} = \dfrac{\sqrt{x^2 + 1}}{x^2}$

61. $y = \ln\left|\dfrac{\cos x}{\cos x - 1}\right| = \ln|\cos x| - \ln|\cos x - 1|$

$$\frac{dy}{dx} = \frac{1}{\cos x}(-\sin x) - \frac{1}{\cos x - 1}(-\sin x)$$

$$= -\tan x + \frac{\sin x}{\cos x - 1}$$

71. $y = 2(\ln x) + 3$

$$y' = 2\left(\frac{1}{x}\right)$$

$$y'' = 2\left(\frac{-1}{x^2}\right) = \frac{-2}{x^2}$$

$$x(y'') + y' = x\left(\frac{-2}{x^2}\right) + \left(\frac{2}{x}\right) = \left(\frac{-2}{x}\right) + \left(\frac{2}{x}\right) = 0$$

73. $y = \dfrac{x^2}{2} - \ln x$

$$y' = x - \frac{1}{x}$$

$$y'' = 1 + \frac{1}{x^2} > 0$$

Domain: $0 < x$

We first observe that $y' = 0$ when $x = \pm 1$. However, $x = -1$ is not in the domain of the function. Since y'' is positive for all x in the domain, the graph is concave up and $\left(1, \frac{1}{2}\right)$ is a relative minimum point. Furthermore, since y'' is never zero, there are no points of inflection. By plotting a few points, we sketch the graph shown in the figure.

x	0.25	0.5	1	1.5	2	3
y	1.418	0.818	0.5	0.720	1.307	3.401

81. Approximating the x-coordinate of the point of intersection of the graphs of $y = \ln x$ and $y = -x$ is equivalent to approximating the zero of the function $f(x) = \ln x + x$. The iterative formula for Newton's Method is

$$x_{n+1} = x_n - \frac{f(x_n)}{f'(x_n)} = x_n - \frac{\ln x_n + x_n}{(1/x_n) + 1}.$$

Using the figure, it appears appropriate to choose $x = 0.5$ as the initial estimate.

n	x_n	$f(x_n)$	$f'(x_n)$	$\dfrac{f(x_n)}{f'(x_n)}$	$x_n - \dfrac{f(x_n)}{f'(x_n)}$
1	0.5000	-0.1931	3	-0.0644	0.5644
2	0.5644	-0.0076	2.7718	-0.0027	0.5671
3	0.5671	0.0000	2.7632	0.000	0.5671

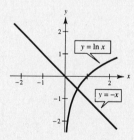

Therefore, the approximate root is $x = 0.567$.

85. Begin by taking the natural logarithm of each member of the equation.

$$y = \frac{x^2\sqrt{3x - 2}}{(x - 1)^2}$$

$$\ln y = \ln \frac{x^2\sqrt{3x - 2}}{(x - 1)^2} = 2\ln x + \frac{1}{2}\ln(3x - 2) - 2\ln(x - 1)$$

$$\frac{1}{y}\frac{dy}{dx} = \frac{2}{x} + \left(\frac{1}{2}\right)\frac{3}{3x - 2} - 2\frac{1}{x - 1}$$

$$\frac{dy}{dx} = y\left[\frac{3x^2 - 15x + 8}{2x(3x - 2)(x - 1)}\right]$$

The derivative can be written in terms of x by replacing y with its equivalent in the original equation.

$$y' = \frac{3x^3 - 15x^2 + 8x}{2(x - 1)^3\sqrt{3x - 2}}$$

91. (a) The plot of the data and the graph of the model are given in the figure.

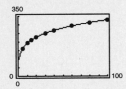

(c) The graph of T' is given in the figure. As the altitude increases, the pressure decreases at a slower rate.

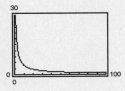

(b)
$$T = 87.97 + 34.96 \ln p + 7.91\sqrt{p}$$
$$= 87.97 + 34.96 \ln p + 7.91 p^{1/2}$$

$$T'(p) = 34.96\left(\frac{1}{p}\right) + 7.91\left(\frac{1}{2}\right)p^{-1/2} = \frac{34.96}{p} + \frac{3.955}{\sqrt{p}}$$

$$T'(10) = \frac{34.96}{10} + \frac{3.955}{\sqrt{10}} \approx 4.75°F/\text{lb/sq in.}$$

$$T'(70) = \frac{34.96}{70} + \frac{3.955}{\sqrt{70}} \approx 0.97°F/\text{lb/sq in.}$$

Section 5.2   The Natural Logarithmic Function and Integration

5. Letting $u = x^2 + 1$, you have $du = 2x\, dx$. By multiplying and dividing the integral by 2 yields

$$\int \frac{x}{x^2 + 1}\, dx = \frac{1}{2}\int \frac{2x}{x^2 + 1}\, dx$$

$$= \frac{1}{2}\int \frac{1}{u}\, du$$

$$= \frac{1}{2} \ln u + C$$

$$= \frac{1}{2} \ln(x^2 + 1) + C = \ln\sqrt{x^2 + 1} + C.$$

[Note that the absolute value signs around $(x^2 + 1)$ are unnecessary since $x^2 + 1 > 0$ for all x.]

7.
$$\int \frac{x^2 - 4}{x}\, dx = \int \left(\frac{x^2}{x} - \frac{4}{x}\right) dx$$

$$= \int \left(x - \frac{4}{x}\right) dx = \frac{x^2}{2} - 4 \ln|x| + C$$

13.
$$\int \frac{1}{\sqrt{x + 1}}\, dx = \int (x + 1)^{-1/2}(1)\, dx$$

$$= \int u^{-1/2}\, du$$

$$= \frac{u^{1/2}}{1/2}$$

$$= 2(x + 1)^{1/2} + C = 2\sqrt{x + 1} + C$$

15. If $u = \sqrt{x} - 3$, then
$$x = (u + 3)^2 \text{ and } dx = 2(u + 3)\, du.$$
Therefore,

$$\int \frac{\sqrt{x}}{\sqrt{x} - 3}\, dx = \int \frac{u + 3}{u}[2(u + 3)]\, du$$

$$= 2\int \frac{u^2 + 6u + 9}{u}\, du$$

$$= 2\int \left(u + 6 + \frac{9}{u}\right) du$$

$$= 2\left(\frac{u^2}{2} + 6u + 9 \ln|u|\right) + C_1$$

$$= \left(\sqrt{x} - 3\right)^2 + 12\left(\sqrt{x} - 3\right)$$

$$\qquad\qquad + 18 \ln|\sqrt{x} - 3| + C_1$$

$$= x + 6\sqrt{x} + 18 \ln|\sqrt{x} - 3| + C.$$

17. $\displaystyle\int \frac{2x}{(x-1)^2}\,dx = \int \frac{2x-2+2}{(x-1)^2}\,dx$

$\displaystyle\qquad = \int \frac{2(x-1)}{(x-1)^2}\,dx + 2\int \frac{1}{(x-1)^2}\,dx$

$\displaystyle\qquad = 2\int \frac{1}{x-1}\,dx + 2\int (x-1)^{-2}\,dx$

$\displaystyle\qquad = 2\int \frac{1}{u}\,du + 2\int u^{-2}\,du$

$\displaystyle\qquad = 2\ln|u| + 2\,\frac{u^{-1}}{-1} + C$

$\displaystyle\qquad = 2\ln|x-1| - \frac{2}{x-1} + C$

25. Letting $u = \sec x - 1$, we have $du = \sec x \tan x\,dx$. Therefore,

$$\int \frac{\sec x \tan x}{\sec x - 1}\,dx = \int \frac{1}{u}\,du$$

$$= \ln|u| + C$$

$$= \ln|\sec x - 1| + C.$$

29. $\dfrac{ds}{d\theta} = \tan 2\theta$

$\displaystyle s = \int \tan 2\theta\,d\theta$

$\displaystyle\quad = -\frac{1}{2}\int \overset{1/u}{\overbrace{\frac{1}{\cos 2\theta}}}\,\overset{du}{\overbrace{(-2\sin 2\theta)}}\,d\theta$

$\displaystyle\quad = -\frac{1}{2}\ln|\cos 2\theta| + C$

The graphs of the solution for $C = -2$, $C = 0$, and $C = 2$ are given in the figure. The solution passes through the point $(0, 2)$ when $C = 2$.

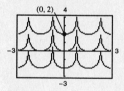

35. Letting $u = 1 + \ln x$, we have $du = (1/x)\,dx$. When $x = 1$, $u = 1$ and when $x = e$, $u = 2$. Therefore,

$$\int_1^e \frac{(1 + \ln x)^2}{x}\,dx = \int_1^e (1 + \ln x)^2\left(\frac{1}{x}\right) dx$$

$$= \int_1^2 u^2\,du = \left[\frac{u^3}{3}\right]_1^2 = \frac{7}{3}.$$

47. $-\ln|\cos x| + C = \ln|(\cos x)^{-1}| + C$

$\displaystyle\qquad = \ln\left|\frac{1}{\cos x}\right| + C = \ln|\sec x| + C$

57. Area $\displaystyle = \int_1^4 \frac{x^2 + 4}{x}\,dx = \int_1^4 \left(x + \frac{4}{x}\right) dx$

$\displaystyle\qquad = \left[\frac{1}{2}x^2 + 4\ln x\right]_1^4$

$\displaystyle\qquad = 8 + 4\ln 4 - \frac{1}{2}$

$\displaystyle\qquad = \frac{1}{2}(15 + 16\ln 2)$

$\displaystyle\qquad \approx 13.045 \text{ square units}$

63. The average price is

$$\frac{1}{50-40}\int_{40}^{50} \frac{90{,}000}{400 + 3x}\,dx = \frac{90{,}000}{10}\left(\frac{1}{3}\right)\int \frac{1}{400 + 3x}(3)\,dx$$

$$= 3000\Big[\ln|400 + 3x|\Big]_{40}^{50} \approx \$168.27.$$

Section 5.3   Inverse Functions

5. (a) $f(x) = \sqrt{x - 4}$ and $g(x) = x^2 + 4$ $(x \ge 0)$

The composite of f with g is given by

$$f(g(x)) = f(x^2 + 4)$$
$$= \sqrt{(x^2 + 4) - 4} = \sqrt{x^2} = x.$$

The composite of g with f is given by

$$g(f(x)) = g\left(\sqrt{x - 4}\right)$$
$$= \left(\sqrt{x - 4}\right)^2 + 4 = x - 4 + 4 = x.$$

Since $f(g(x)) = g(f(x)) = x$, we can conclude that f and g are inverses of each other.

(b) The graphs of f and g are shown in the figure. Note that the graph of g is a reflection of the graph of f in the line $y = x$.

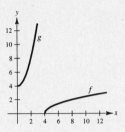

13. The function has an inverse because it is increasing on its entire domain. To find an equation for the inverse, let $y = f(x)$ and solve for x in terms of y.

$$2x - 3 = y$$
$$x = \frac{y + 3}{2}$$
$$f^{-1}(y) = \frac{y + 3}{2}$$

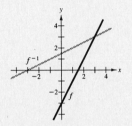

Interchanging x and y yields $f^{-1}(x) = \dfrac{x + 3}{2}$. Remember that any variable can be used to represent the independent variable. Thus,

$$f^{-1}(y) = \frac{y + 3}{2}, \quad f^{-1}(x) = \frac{x + 3}{2}, \quad \text{and} \quad f^{-1}(t) = \frac{t + 3}{2}$$

represent the same function. The graph of f^{-1} is a reflection of the graph of f in the line $y = x$.

23. The function has an inverse because it is increasing $[0, \infty)$. To find an equation for the inverse, let $y = f(x)$ and solve for x in terms of y.

$$x^{2/3} = y \qquad x \ge 0$$
$$x = y^{3/2} \qquad y \ge 0$$
$$f^{-1}(y) = y^{3/2} \qquad y \ge 0$$

Finally, using x as the independent variable yields

$$f^{-1}(x) = x^{3/2} \qquad x \ge 0.$$

The graph of f^{-1} is a reflection of the graph of f in the line $y = x$.

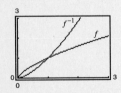

33. The function f passes through the points $(0, 1)$, $(1, 2)$, $(2, 3)$, and $(4, 4)$. From the reflective property of inverses, we know the graph of f contains the point (a, b) if and only if the graph f^{-1} contains the point (b, a). Therefore, we obtain the required table by interchanging the x and y coordinates of the preceding points.

x	1	2	3	4
$f^{-1}(x)$	0	1	2	4

The graph of f^{-1} is shown in the figure.

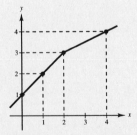
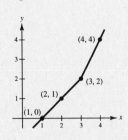

41. The graph of h is shown in the figure. Since h is strictly monotonic (decreasing) on its entire domain, it is one-to-one and, hence, possesses an inverse.

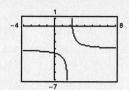

55. Begin by differentiating the function $f(x) = 4/x^2$ to obtain

$$f'(x) = \frac{-8}{x^3} < 0$$

on the interval $(0, \infty)$. Therefore, f is monotonically decreasing on $(0, \infty)$ and has an inverse.

65. One solution is to delete the part of the graph when $x < 3$. The remaining part of the graph is strictly increasing and has an inverse.

$$f(x) = (x - 3)^2, \quad x \geq 3$$
$$y = (x - 3)^2$$
$$\pm\sqrt{y} = x - 3$$
$$x = 3 \pm \sqrt{y}$$
$$y = 3 \pm \sqrt{x}, \quad \text{Interchange } x \text{ and } y$$

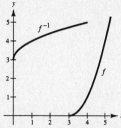

Because of the restriction that $x \geq 3$, the inverse must have the positive square root in the equation above.

$$f^{-1}(x) = 3 + \sqrt{x}, \quad x \geq 0.$$

The graph of f with the restricted domain and f^{-1} are shown in the figure. (*Note:* If the part of the graph is deleted where $x > 3$, then the inverse function would be

$$f^{-1}(x) = 3 - \sqrt{x}, \quad x \geq 0.)$$

73. Since $f(1) = 2$, it follows that $f^{-1}(2) = 1$.

$$f(x) = x^3 + 2x - 1$$
$$f'(x) = 3x^2 + 2$$
$$(f^{-1})'(2) = \frac{1}{f'(f^{-1}(2))} \quad \text{Theorem 5.9}$$
$$= \frac{1}{f'(1)} = \frac{1}{3(1)^2 + 2} = \frac{1}{5}$$

81. $f(x) = \sqrt{x - 4}$

$$f^{-1}(x) = x^2 + 4$$

(a) Domain of f: $[4, \infty)$
Domain of f^{-1}: $[0, \infty)$

(c) The graphs of f and f^{-1} are shown in the figure.

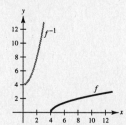

(b) Range of f: $[0, \infty)$
Range of f^{-1}: $[4, \infty)$

(d) For the function $f(x) = \sqrt{x - 4}$, we have

$$f'(x) = \frac{1}{2\sqrt{x - 4}} \quad \text{and} \quad f'(5) = \frac{1}{2\sqrt{5 - 4}} = \frac{1}{2}.$$

For the function $f^{-1}(x) = x^2 + 4$, we have

$$(f^{-1})'(x) = 2x \quad \text{and} \quad (f^{-1})'(1) = 2.$$

Therefore,

$$(f^{-1})'(1) = \frac{1}{f'(5)}.$$

89.
$$f(x) = x + 4 \implies f^{-1}(x) = x - 4$$
$$g(x) = 2x - 5 \implies g^{-1}(x) = \frac{x + 5}{2}$$
$$(g^{-1} \circ f^{-1}) = g^{-1}(f^{-1}(x)) = f^{-1}(x - 4) = \frac{(x - 4) + 5}{2} = \frac{x + 1}{2}$$

Section 5.4 Exponential Functions: Differentiation and Integration

3. By definition,

$$\ln x = b \quad \text{if and only if} \quad e^b = x.$$

Letting $x = 2$, and $b = 0.6931\ldots$, we have

$$\ln 2 = 0.6931\ldots \text{ if and only if } e^{0.6931\ldots} = 2.$$

5. Since $y = e^x$ is the inverse of $y = \ln x$, we have

$$e^{\ln x} = x.$$

Therefore, $e^{\ln x} = 4$ if $x = 4$.

11. (a) The graph is symmetric with respect to the y-axis since

$$y = e^{-(-x)^2} = e^{-x^2}.$$

(b) The y-intercept is $(0, 1)$.

(c) The x-axis is a horizontal asymptote since

$$\lim_{x \to \infty} e^{-x^2} = \lim_{x \to \infty} \frac{1}{e^{x^2}} = 0.$$

(d) The graph lies entirely above the x-axis, since for all x

$$0 < e^{-x^2}.$$

The following table shows solution points of the equation. By plotting these points and using the information given above we have the graph shown in the figure.

x	0	0.5	1	2
y	1	0.607	0.368	0.135

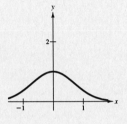

17. The graphs of

$$f(x) = e^x - 1 \quad \text{and} \quad g(x) = \ln(x + 1)$$

are shown in the figure. Since the graph of g is the reflection of the graph of f in the line $y = x$, $g = f^{-1}$.

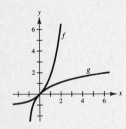

33. $y = e^{\sqrt{x}}$

$$\frac{dy}{dx} = e^{\sqrt{x}} \frac{d}{dx}\left[\sqrt{x}\right]$$

$$= e^{\sqrt{x}}\left(\frac{1}{2\sqrt{x}}\right) = \frac{e^{\sqrt{x}}}{2\sqrt{x}}$$

37. Using the inverse relationship between the exponential and logarithmic functions yields

$$y = \ln(e^{x^2}) = x^2 \ln e = x^2.$$

Therefore, $dy/dx = 2x$.

45. $y = e^{-x} \ln x$

$$y' = \left(e^{-x}\right)\left(\frac{1}{x}\right) + (\ln x)(-e^{-x}) = e^{-x}\left(\frac{1}{x} - \ln x\right)$$

53. $y = e^x\left(\cos \sqrt{2}x + \sin \sqrt{2}x\right)$

$y' = e^x\left(-\sqrt{2}\sin \sqrt{2}x + \sqrt{2}\cos \sqrt{2}x\right) + e^x\left(\cos \sqrt{2}x + \sin \sqrt{2}x\right)$

$y'' = e^x\left(-2\cos \sqrt{2}x - 2\sin \sqrt{2}x\right) + 2e^x\left(-\sqrt{2}\sin \sqrt{2}x + \sqrt{2}\cos \sqrt{2}x\right) + e^x\left(\cos \sqrt{2}x + \sin \sqrt{2}x\right)$

$\quad = -e^x\left(\cos \sqrt{2}x + \sin \sqrt{2}x\right) + 2e^x\left(\sqrt{2}\cos \sqrt{2}x - \sqrt{2}\sin \sqrt{2}x\right)$

$y'' - 2y' + 3y = -e^x\left(\cos \sqrt{2}x + \sin \sqrt{2}x\right) + 2e^x\left(\sqrt{2}\cos \sqrt{2}x - \sqrt{2}\sin \sqrt{2}x\right) - 2e^x\left(\cos \sqrt{2}x + \sin \sqrt{2}x\right)$

$\qquad\qquad\qquad\qquad - 2e^x\left(\sqrt{2}\cos \sqrt{2}x - \sqrt{2}\sin \sqrt{2}x\right) + 3e^x\left(\cos \sqrt{2}x + \sin \sqrt{2}x\right) = 0$

59. $f(x) = x^2 e^{-x}$

$f'(x) = x^2(-e^{-x}) + e^{-x}(2x) = xe^{-x}(-x + 2)$

$f''(x) = (xe^{-x})(-1) + (-x + 2)[x(-e^{-x}) + e^{-x}(1)]$

$\quad = e^{-x}[-x + (-x + 2)(-x + 1)] = e^{-x}(x^2 - 4x + 2)$

Intercept: $(0, 0)$

Solving the equation $f' = 0$ yields the critical numbers $x = 0$, and $x = 2$. Solve the equation $x^2 - 4x + 2 = 0$ to determine where $f''(x) = 0$. The solutions of this equation are $x = 2 \pm \sqrt{2}$.

Interval	$f(x)$	$f'(x)$	$f''(x)$	Shape of graph
$-\infty < x < 0$		$-$	$+$	decreasing, concave up
$x = 0$	0	0	$+$	relative minimum
$0 < x < 2 - \sqrt{2}$		$+$	$+$	increasing, concave up
$x = 2 - \sqrt{2}$	0.191	$+$	0	point of inflection
$2 - \sqrt{2} < x < 2$		$+$	$-$	increasing, concave down
$x = 2$	0.541	0	$-$	relative maximum
$2 < x < 2 + \sqrt{2}$		$-$	$-$	decreasing, concave down
$x = 2 + \sqrt{2}$	0.384	$-$	0	point of inflection
$2 + \sqrt{2} < x < \infty$		$-$	$+$	decreasing, concave up

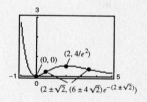

63. $y = \dfrac{L}{1 + ae^{-x/b}} = L(1 + ae^{-x/b})^{-1}, \quad a > 0, \ b > 0, L > 0$

$y' = -L(1 + ae^{-x/b})^{-2}(ae^{-x/b})\left(-\dfrac{1}{b}\right)$

$\quad = \dfrac{aLe^{-x/b}}{b(1 + ae^{-x/b})^2}$

Since a, b, and L are positive, $y' > 0$ and y is an increasing function. The point on the graph where the function is increasing at a maximum rate is the point where the first derivative is maximum. The maximum of the first derivative will occur where its derivative, (the second derivative) is zero.

$y' = \dfrac{aLe^{-x/b}}{b(1 + ae^{-x/b})^2}$

$y'' = \dfrac{aL}{b}\left[\dfrac{(1 + ae^{-x/b})^2(e^{-x/b})(-1/b) - e^{-x/b}(2)(1 + ae^{-x/b})(ae^{-x/b})(-1/b)}{(1 + ae^{-x/b})^4}\right]$

$\quad = \dfrac{aLe^{-x/b}(ae^{-x/b} - 1)}{(1 + ae^{-x/b})^3}$

—CONTINUED—

63. —CONTINUED—

$y'' = 0$ when $ae^{-x/b} - 1 = 0$ or $x = b \ln a$. Note that the sign of y'' is determined by the sign of $ae^{-x/b} - 1$, and that $y'' > 0$ if $x < b \ln a$ and $y'' < 0$ if $x > b \ln a$. Therefore, y' is maximum when $x = b \ln a$ and the corresponding value of y is

$$y = \frac{L}{1 + ae^{-b \ln a/b}} = \frac{L}{1 + ae^{-\ln a}} = \frac{L}{1 + a/a} = \frac{L}{2}.$$

The limits at infinity show that the graph of the function has horizontal asymptotes at $y = 0$ and $y = L$. The shape of the graph is shown in the figure.

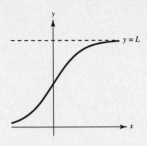

69. (a) The plot of the points $(h, \ln P)$ is shown in the figure. Note that the points appear approximately linear. Use the regression capabilities of a graphing utility to fit a linear model to the points $(h, \ln P)$ yields

$$\ln P = -0.1499h + 9.3018.$$

The graph of this line is also shown in the figure.

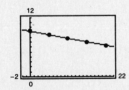

(b) By definition,

$\ln x = b$ if and only if $e^b = x$.

Letting $x = P$, and $b = -0.1499h + 9.3018$, we have

$\ln P = -0.1499h + 9.3018$

$P = e^{-0.1499h + 9.3018} = e^{-0.1499h} \cdot e^{9.3018}$

$\quad = 10,957.7e^{-0.1499h}.$

(c) The plot of the original data and the graph of the model of part (b) are shown in the figure.

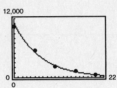

(d) $P = 10,957.7e^{-0.1499h}$

$P'(h) = -1642.56e^{-0.1499h}$

$P'(5) \approx -776$ kilograms per square meter per kilometer

$P'(18) \approx -111$ kilograms per square meter per kilometer

81. If $u = 3/x$, then $du = (-3/x^2) \, dx$. When $x = 1$, $u = 3$ and when $x = 3$, $u = 1$. Therefore,

$$\int_1^3 \frac{e^{3/x}}{x^2} \, dx = -\frac{1}{3} \int_1^3 e^{3/x} \left(-\frac{3}{x^2}\right) dx$$

$$= -\frac{1}{3} \int_3^1 e^u \, du$$

$$= -\frac{1}{3} \left[e^u\right]_3^1 = -\frac{1}{3}(e - e^3) = \frac{e}{3}(e^2 - 1).$$

85. If $u = e^x - e^{-x}$, then $du = (e^x + e^{-x}) \, dx$.

$$\int \frac{e^x + e^{-x}}{e^x - e^{-x}} \, dx = \int \frac{1}{u} \, du$$

$$= \ln|u| + C = \ln|e^x - e^{-x}| + C$$

91. If $u = e^{-x}$, then $du = -e^{-x} \, dx$.

$$\int e^{-x} \tan(e^{-x}) \, dx = -\int \tan u \, du$$

$$= \ln|\cos u| + C$$

$$= \ln|\cos(e^{-x})| + C$$

99. The graph of the region is shown in the figure.

$$A = \int_0^5 e^x \, dx = \left[e^x\right]_0^5 = e^5 - e^0 = e^5 - 1 \approx 147.413$$

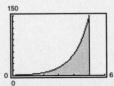

Section 5.5   Bases Other than *e* and Applications

7. From the definition of the exponential and logarithmic functions to the base *a*, it follows that $f(x) = a^x$ and $g(x) = \log_a x$ are inverse functions of each other. Therefore,

$$y = \log_a x \text{ if and only if } a^y = x.$$

(a) $\log_{10} 0.01 = -2$ if and only if $10^{-2} = 0.01$.

(b) $\log_{0.5} 8 = -3$ if and only if $0.5^{-3} = 8$.

13. (a)
$$x^2 - x = \log_5 25$$
$$x^2 - x = \log_5(5^2)$$
$$x^2 - x = 2$$
$$x^2 - x - 2 = 0$$
$$(x - 2)(x + 1) = 0 \Rightarrow x = -1, 2$$

(b) $3x + 5 = \log_2 64$
$$3x + 5 = \log_2(2^6)$$
$$3x + 5 = 6$$
$$3x = 1 \Rightarrow x = \tfrac{1}{3}$$

19. Using the point plotting method of graphing a function yields the following table and graph.

x	-1	0	1	2	3
$h(x) = 5^{x-2}$	$\frac{1}{125}$	$\frac{1}{25}$	$\frac{1}{5}$	1	5

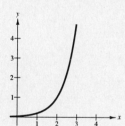

25. The graphs of

$$f(x) = 4x \quad \text{and} \quad g(x) = \log_4 x$$

are shown in the figure. Observe that *g* is a reflection of *f* in the line $y = x$ which implies that $g = f^{-1}$.

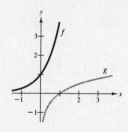

33. Using the Product Rule and Theorem 5.13 the derivative of the function $g(t) = t^2 2^t$ is

$$g'(t) = t^2(\ln 2)2^t + 2t(2^t) = t(2^t)(t \ln 2 + 2).$$

39. Begin by rewriting the logarithmic function.

$$f(x) = \log_2 \frac{x^2}{x - 1}$$
$$= \log_2 x^2 - \log_2(x - 1) = 2 \log_2 x - \log_2(x - 1)$$

Using Theorem 5.13, the derivative is

$$f'(x) = \frac{2}{x \ln 2} - \frac{1}{(x - 1)\ln 2} = \frac{x - 2}{(\ln 2)x(x - 1)}.$$

41. Begin by rewriting the logarithmic function.

$$y = \log_5 \sqrt{x^2 - 1}$$
$$= \log_5(x^2 - 1)^{1/2} = \frac{1}{2} \log_5(x^2 - 1)$$

Using Theorem 5.13 the derivative of the function is

$$\frac{dy}{dx} = \left(\frac{1}{2}\right)\frac{1}{(\ln 5)(x^2 - 1)}(2x) = \frac{x}{(x^2 - 1)\ln 5}.$$

51. (a) $C(t) = 24.95(1.05)^t$

$$C(10) = 24.95(1.05)^{10} \approx \$40.64$$

(c) $\dfrac{dC}{dt} = (\ln 1.05)[P(1.05)^t] = (\ln 1.05)C(t)$

Therefore, dC/dt is proportional to $C(t)$ and the constant of proportionality is $\ln 1.05$.

(b) $C(t) = P(1.05)^t$

$$C'(t) = P \ln(1.05)(1.05)^t$$
$$C'(1) = P \ln(1.05)(1.05)^1 \approx 0.051P$$
$$C'(8) = P \ln(1.05)(1.05)^8 \approx 0.072P$$

61. The balance in the account after 8 years is computed for each option.

(a) A deposit of $20,000 is made now and earns interest 8 years.

$$A = 20,000\left(1 + \frac{0.06}{365}\right)^{(365 \cdot 8)} = \$32,320.21$$

(b) In 8 years the amount is $A = \$30,000$.

(c) The first deposit of $8000 earns interest 8 years. The second deposit of $20,000 in 4 years earns interest 4 years.

$$A = 8000\left(1 + \frac{0.06}{365}\right)^{(365 \cdot 8)} + 20,000\left(1 + \frac{0.06}{365}\right)^{(365 \cdot 4)} = \$38,352.57$$

(d) Three deposits of $9000 each are made. The first earns interest 8 years, the second 4 years, and the final deposit earns no interest.

$$A = 9000\left(1 + \frac{0.06}{365}\right)^{(365 \cdot 8)} + 9000\left(1 + \frac{0.06}{365}\right)^{(365 \cdot 4)} + 9000 = \$34,985.11$$

Therefore, option (c) yields the greatest balance in 8 years.

73. $\displaystyle\int x(5^{x^2})\,dx = -\frac{1}{2}\int 5^{x^2}(-2x)\,dx$

$\displaystyle = -\left(\frac{1}{2}\right)\frac{5^{-x^2}}{\ln 5} + C = \frac{-1}{2 \ln 5}(5^{-x^2}) + C$

81. $\displaystyle P = \int_0^{10} 2000e^{-0.06t}\,dt$

$\displaystyle = \left[\frac{2000}{-0.06}e^{-0.06t}\right]_0^{10}$

$\displaystyle = -33,333.33(e^{-0.6} - 1) \approx \$15,039.61$

Section 5.6   Differential Equations: Growth and Decay

3. $y' = \sqrt{xy}$

$\dfrac{1}{y}y'\,dx = \sqrt{x}\,dx$

$\displaystyle\int \frac{1}{y}y'\,dx = \int \sqrt{x}\,dx$

$\displaystyle\int \frac{1}{y}\,dy = \int x^{1/2}\,dx$

$\ln y = \dfrac{2}{3}x^{3/2} + C_1$

$y = e^{2x^{3/2}/3 + C_1}$

$y = e^{C_1}e^{2x^{3/2}/3} = Ce^{2x^{3/2}/3}$

13. $\dfrac{dy}{dt} = \dfrac{1}{2}t$

$y = \displaystyle\int \frac{1}{2}t\,dt = \frac{1}{4}t^2 + C$

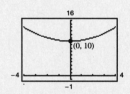

Since the solution passes through the point $(0, 10)$, we have

$$10 = \frac{1}{4}(0^2) + C \implies C = 10.$$

Therefore, $f(t) = (1/4)t^2 + 10$ and its graph is shown in the figure.

25. Let y represent the mass (in grams) of the isotope after t years. Since the rate of decay is proportional to y, apply the Law of Exponential Decay to conclude that $y = y_0e^{kt}$ where y_0 is the initial amount. Since the half life of Pu^{239} is 24,360 years, we have

$$y_0e^{24,360k} = \frac{1}{2}y_0$$

$$e^{24,360k} = \frac{1}{2}$$

$$24,360k = \ln\frac{1}{2}$$

$$k = \frac{-\ln 2}{24,360}.$$

—CONTINUED—

25. —CONTINUED—

Therefore, $y = y_0 e^{(-\ln 2/24,360)t}$. Since $y = 2.1$ when $t = 1000$, we have

$$y_0 e^{(-\ln 2/24,360)1000} = 2.1$$

$$y_0 = 2.1 e^{1000 \ln 2/24,360} \approx 2.16 \text{ grams.}$$

After 10,000 years the amount of the isotope remaining will be

$$y = 2.16 e^{(-\ln 2/24,360)10,000} \approx 1.6 \text{ grams.}$$

31. Since the interest is compounded continuously, use the formula

$$A = Pe^{rt}.$$

To determine the rate, use the fact that the investment doubles in $7\frac{3}{4}$ years.

$$2(750) = 750 e^{r(7.75)}$$

$$2 = e^{7.75r}$$

$$\ln 2 = \ln e^{7.75r} = 7.75r$$

$$r = \frac{\ln 2}{7.75} \approx 0.0894 = 8.94\%$$

The amount after 10 years is given by

$$A = 750 e^{(0.0894)10} \approx \$1833.67.$$

39. Since the population of the city is 4.22 million and 6.49 million in the years 1990 and 2000, respectively, and $t = 0$ corresponds to the year 1990, the model $y = Ce^{kt}$ has solution points $(0, 4.22)$ and $(10, 6.49)$.

$$y = Ce^{kt}$$

$$4.22 = Ce^{k(0)} = C$$

$$y = 4.22 e^{kt}$$

$$6.49 = 4.22 e^{k(10)}$$

$$e^{10k} = \frac{6.49}{4.22}$$

$$\ln e^{10k} = \ln\left(\frac{6.49}{4.22}\right)$$

$$10k = \ln\left(\frac{6.49}{4.22}\right) \implies k = \frac{1}{10}\ln\left(\frac{6.49}{4.22}\right) \approx 0.0430$$

Therefore, the model is $y \approx 4.22 e^{0.0430t}$. When $t = 10$,

$$y \approx 4.22 e^{0.430} \approx 9.97 \text{ million.}$$

47. (a) Since $N(20) = 19$, it follows that

$$19 = 30(1 - e^{20k})$$

$$30 e^{20k} = 11$$

$$e^{20k} = \frac{11}{30}$$

$$k = \frac{\ln(11/30)}{20} \approx -0.0502.$$

Therefore,

$$N(t) = 30(1 - e^{-0.0502t}).$$

(b) To determine the time when the worker will be producing 25 units per day solve the equation

$$25 = 30(1 - e^{-0.0502t})$$

$$e^{-0.0502t} = \frac{1}{6}$$

$$t = \frac{-\ln 6}{-0.0502} \approx 36 \text{ days.}$$

57. Let y be the temperature reading of the thermometer at time t. From Newton's Law of Cooling, we know that the rate of change of y is proportional to the difference between y and $20°$. This can be written as

$$\frac{dy}{dt} = k(y - 20)$$

$$\left(\frac{1}{y - 20}\right)\frac{dy}{dt} = k$$

$$\int \frac{1}{y - 20} \, dy = \int k \, dt$$

$$\ln|y - 20| = kt + C$$

—CONTINUED—

57. —CONTINUED—

Using $y = 72$ when $t = 0$ yields $C = \ln 52$, which implies that

$$kt = \ln|y - 20| - \ln 52.$$

Since $y = 48$ when $t = 1$, we know that

$$k = \ln 28 - \ln 52 = \ln \frac{28}{52} = \ln \frac{7}{13}.$$

Therefore,

$$\ln|y - 20| = \ln \frac{7}{13}t + \ln 52$$

$$y - 20 = e^{[\ln(7/13)]t + \ln 52}$$

$$y = e^{\ln 52} e^{[\ln(7/13)]t} + 20 = 52e^{[\ln(7/13)]t} + 20.$$

After 5 minutes the reading on the thermometer is

$$y = 52e^{5 \ln(7/13)} + 20 \approx 22.35°.$$

Section 5.7   Differential Equations: Separation of Variables

9. Since $y = e^{-2x}$, we have

$$y' = -2e^{-2x}$$

$$y'' = 4e^{-2x}$$

$$y''' = -8e^{-2x}$$

$$y^{(4)} = 16e^{-2x}.$$

Therefore,

$$y^{(4)} - 16y = 16e^{-2x} - 16(e^{-2x}) = 0$$

and the function is a solution to the differential equation.

19. Since $y = Ce^{kx}$ is a solution to the differential equation, begin by substituting the derivative of the solution into the differential equation.

$$y = Ce^{kx}$$

$$\frac{dy}{dx} = Ce^{kx}(k) = k(Ce^{ky}) = ky = 0.07y$$

Therefore, $k = 0.07$.

27. Since $y = C_1 \sin 3x + C_2 \cos 3x$, we have

$$y' = 3C_1 \cos 3x - 3C_2 \sin 3x$$

$$y'' = -9C_1 \sin 3x - 9C_2 \cos 3x.$$

Therefore,

$$y'' + 9y = -9C_1 \sin 3x - 9C_2 \cos 3x + 9(C_1 \sin 3x + C_2 \cos 3x) = 0$$

and the function is a solution to the differential equation. Furthermore, since $y = 2$ and $y' = 1$ when $x = \pi/6$, we have

$$2 = C_1 \sin 3\left(\frac{\pi}{6}\right) + C_2 \cos 3\left(\frac{\pi}{6}\right) \quad \Rightarrow 2 = C_1(1) + C_2(0)$$

$$1 = 3C_1 \cos 3\left(\frac{\pi}{6}\right) - 3C_2 \sin 3\left(\frac{\pi}{6}\right) \quad \Rightarrow 1 = 3C_1(0) - 3C_2(1).$$

Therefore, $C_1 = 2$, $C_2 = -1/3$, and the particular solution is

$$y = 2 \sin 3x - \frac{1}{3} \cos 3x.$$

33. $\dfrac{dy}{dx} = \dfrac{x-2}{x}$

$$y = \int \dfrac{x-2}{x}\, dx$$

$$= \int \left(1 - \dfrac{2}{x}\right) dx$$

$$= x - 2\ln|x| + C = x - \ln x^2 + C$$

43. Begin by separating the variables as follows.

$$(2+x)y' = 3y$$

$$(2+x)\dfrac{dy}{dx} = 3y$$

$$\dfrac{1}{y}\, dy = \dfrac{3}{2+x}\, dx$$

By integration we obtain

$$\int \dfrac{1}{y}\, dy = 3 \int \dfrac{1}{2+x}\, dx$$

$$\ln|y| = 3\ln|2+x| + \ln C$$

$$\ln y = \ln |C(2+x)^3|$$

$$y = C(2+x)^3.$$

51. Begin by separating the variables to obtain

$$y(x+1) + y' = 0$$

$$\dfrac{dy}{dx} = -y(x+1)$$

$$\dfrac{dy}{y} = -(x+1)\, dx.$$

Integration yields

$$\int \dfrac{dy}{y} = -\int (x+1)\, dx$$

$$\ln|y| = -\dfrac{(x+1)^2}{2} + C_1$$

$$y = e^{C_1 - (x+1)^2/2} = Ce^{-(x+1)^2/2}.$$

Since $y = 1$ when $x = -2$, it follows that

$$1 = Ce^{-(-2+1)^2/2} = Ce^{-1/2} \quad \text{or} \quad C = e^{1/2}.$$

Therefore, the particular solution is

$$y = e^{1/2}e^{-(x+1)^2/2} = e^{[-(x+1)^2/2 + 1/2]} = e^{-(x^2+2x)/2}.$$

59. $\dfrac{dy}{dx} = -\dfrac{9x}{16y}$

$$16\, y\, dy = -9\, x\, dx$$

$$\int 16\, y\, dy = \int (-9x)\, dx$$

$$8y^2 = -\dfrac{9}{2}x^2 + C$$

$$\dfrac{9}{2}x^2 + 8y^2 = C$$

Since $(1, 1)$ is a solution point, we have

$$\dfrac{9}{2}(1^2) + 8(1)^2 = \dfrac{25}{2} = C.$$

The required equation is

$$\dfrac{9}{2}x^2 + 8y^2 = \dfrac{25}{2} \quad \text{or} \quad 9x^2 + 16y^2 = 25.$$

67. $f(x, y) = 2\ln\dfrac{x}{y}$

$$f(tx, ty) = 2\ln\dfrac{tx}{ty} = 2\ln\dfrac{x}{y} = t^0 f(x, y)$$

Therefore, f is homogeneous of degree 0.

73. Letting $y = vx$, yields

$$y' = \frac{xy}{x^2 - y^2}$$

$$v + x\frac{dv}{dx} = \frac{x(vx)}{x^2 - v^2x^2} = \frac{v}{1 - v^2}$$

$$x\frac{dv}{dx} = \frac{v}{1 - v^2} - v = \frac{v^3}{1 - v^2}$$

$$x\,dv = \frac{v^3}{1 - v^2}\,dx.$$

Separating variables, we obtain

$$\frac{1 - v^2}{v^3}\,dv = \frac{dx}{x}$$

$$\int\left(v^{-3} - \frac{1}{v}\right)dv = \int\frac{dx}{x}$$

$$\frac{v^{-2}}{-2} - \ln|v| = \ln|x| + \ln|C_1|$$

$$\frac{1}{-2x^2} = \ln|v| + \ln|x| + \ln|C_1| = \ln|C_1\,vx|$$

$$\frac{x^2}{-2y^2} = \ln|C_1 y|$$

$$e^{-x^2/2y^2} = C_1 y \implies y = Ce^{-x^2/2y^2}.$$

77. Letting $y = vx$, yields

$$\left[x\sec\left(\frac{vx}{x}\right) + vx\right]dx - x(v\,dx + x\,dv) = 0$$

$$x\sec v\,dx + vx\,dx - vx\,dx - x^2\,dx = 0$$

$$x\sec v\,dx = x^2\,dv$$

$$\frac{x}{x^2}\,dx = \frac{dv}{\sec v}.$$

Integration yields

$$\int\frac{dx}{x} = \int\cos v\,dv$$

$$\ln|x| = \sin v + C_1 = \sin\frac{y}{x} + C_1.$$

Since $y = 0$ when $x = 1$, we have

$$0 = \sin(0) + C_1 = C_1$$

and it follows that

$$\ln|x| = \sin\frac{y}{x} \implies x = e^{\sin(y/x)}.$$

87. (a) The differential equation is

$$\frac{dS}{dt} = kS(L - S).$$

To verify that

$$S = \frac{L}{1 + Ce^{-kt}}$$

is a solution, find and simplify both sides of the differential equation to show they are equal.

$$S = \frac{L}{1 + Ce^{-kt}} = L(1 + Ce^{-kt})^{-1}$$

$$\frac{dS}{dt} = -L(1 + Ce^{-kt})^{-2}(Ce^{-kt})(-k) = \frac{kCLe^{-kt}}{(1 + Ce^{-kt})^2}$$

$$kS(L - S) = k\left(\frac{L}{1 + Ce^{-kt}}\right)\left(L - \frac{L}{1 + Ce^{-kt}}\right)$$

$$= kL\left[\frac{1}{1 + Ce^{-kt}} - \frac{1}{(1 + Ce^{-kt})^2}\right]$$

$$= \frac{kCLe^{-kt}}{(1 + Ce^{-kt})^2}$$

Therefore, $dS/dt = kS(L - S)$. Since $L = 100$, it follows that

$$S = \frac{100}{1 + Ce^{-kt}}.$$

Since $S = 10$ when $t = 0$, we have

$$10 = \frac{100}{1 + Ce^{-k(0)}} = \frac{100}{1 + C}.$$

—CONTINUED—

87 —CONTINUED—

Hence, $C = 9$. Substituting $C = 9$ and $S = 20$ when $t = 1$ yields

$$20 = \frac{100}{1 + 9e^{-k}}$$

$$20 + 180e^{-k} = 100$$

$$180e^{-k} = 80$$

$$e^{-k} = \frac{80}{180} \implies k = -\ln\frac{4}{9} \approx 0.8109.$$

Therefore, the model is

$$S = \frac{100}{1 + 9e^{-0.8109t}}.$$

(b) $\dfrac{dS}{dt} = kS(L - S)\left(\ln\dfrac{4}{9}\right)S(100 - S)$

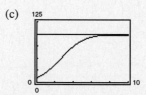

$$\frac{d^2S}{dt^2} = \ln\left(\frac{4}{9}\right)\left[S\left(\frac{dS}{dt}\right) + (100 - S)\frac{dS}{dt}\right]$$

$$= \ln\left(\frac{4}{9}\right)(100 - 2S)\frac{dS}{dt}$$

The second derivative is zero when $S = 50$ or $dS/dt = 0$. Choosing $S = 50$, we have

$$50 = \frac{100}{1 + 9e^{-0.8109t}}$$

$$1 + 9e^{-0.8109t} = 2$$

$$e^{-0.8109t} = \frac{1}{9}$$

$$t = \frac{-\ln 9}{-0.8109} \approx 2.7 \text{ months.}$$

(c)

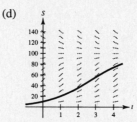

(d)

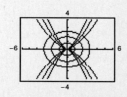

(e) Sales will decrease toward the line $S = L$.

97. First, solve for C in the given equation and obtain

$$C = \frac{y^2}{x^3}.$$

Then, differentiating $y^2 = Cx^3$ implicitly with respect to x and substituting the expression for C given above, we have

$$2yy' = 3Cx^2 = 3\left(\frac{y^2}{x^3}\right)x^2 = \frac{3y^2}{x}.$$

Therefore,

$$\frac{dy}{dx} = \frac{3y}{2x}. \quad \text{Slope of given family}$$

Since dy/dx represents the slope of the given family of curves at (x, y), it follows that the orthogonal family has the negative reciprocal slope, and we write

$$\frac{dy}{dx} = -\frac{2x}{3y}. \quad \text{Slope of orthogonal family}$$

Now, find the orthogonal family by separating variables and integrating to obtain

$$3\int y\, dy = -2\int x\, dx$$

$$\frac{3}{2}y^2 = -x^2 + K$$

$$2x^2 + 3y^2 = K.$$

Section 5.8   Inverse Trigonometric Functions and Differentiation

7. Since $y = \arccos\left(\frac{1}{2}\right)$ if and only if $\cos y = \frac{1}{2}$, then $y = \pi/3$ in the interval $[0, \pi]$. Thus

$$\arccos \frac{1}{2} = \frac{\pi}{3}.$$

19. (a) Begin by sketching a triangle to represent θ, "the angle whose tangent is $\frac{3}{4}$." Then

$$\theta = \arctan \frac{3}{4}$$

and

$$\sin\left(\arctan \frac{3}{4}\right) = \sin \theta = \frac{3}{5}.$$

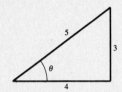

(b) Begin by sketching a triangle to represent θ, "the angle whose sine is $\frac{4}{5}$." Then

$$\theta = \arcsin \frac{4}{5}$$

and

$$\sec\left(\arcsin \frac{4}{5}\right) = \sec \theta = \frac{5}{3}.$$

25. Begin by sketching a triangle to represent θ, "the angle whose secant is x." Then

$$\theta = \text{arcsec } x$$

and

$$\sin(\text{arcsec } x) = \sin \theta = \frac{\sqrt{x^2 - 1}}{x}.$$

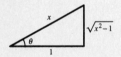

29. Begin by sketching a triangle to represent θ, "the angle whose tangent is $x/\sqrt{2}$." Then

$$\theta = \arctan \frac{x}{\sqrt{2}}$$

and

$$\csc\left(\arctan \frac{x}{\sqrt{2}}\right) = \csc \theta = \frac{\sqrt{x^2 + 2}}{x}.$$

41. Taking the sine of each member of the equation, yields

$$\sin\left(\arcsin \sqrt{2x}\right) = \sin\left(\arccos \sqrt{x}\right)$$

$$\sqrt{2x} = \sqrt{1 - x} \quad \text{(see figure)}$$

$$2x = 1 - x$$

$$3x = 1 \implies x = \frac{1}{3}.$$

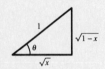

43. $f(x) = 2 \arcsin(x - 1)$

$$f'(x) = 2 \frac{1}{\sqrt{1 - (x - 1)^2}}(1) = \frac{2}{\sqrt{2x - x^2}}$$

51. From the figure we know that

$$h(t) = \sin(\arccos t) = \sqrt{1 - t^2}.$$

Therefore,

$$h'(t) = \frac{1}{2}(1 - t^2)^{-1/2}(-2t) = \frac{-t}{\sqrt{1 - t^2}}.$$

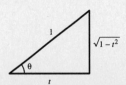

53. $f(x) = \frac{1}{2}\left[\frac{1}{2}\ln\left(\frac{x+1}{x-1}\right) + \arctan x\right]$

$\qquad = \frac{1}{2}\left[\frac{1}{2}\ln(x+1) - \frac{1}{2}\ln(x-1) + \arctan x\right]$

$f'(x) = \frac{1}{2}\left[\frac{1}{2(x+1)} - \frac{1}{2(x-1)} + \frac{1}{1+x^2}\right]$

$\qquad = \frac{1}{2}\left[\frac{(x-1)-(x+1)}{2(x^2-1)} + \frac{1}{1+x^2}\right]$

$\qquad = \frac{1}{2}\left[\frac{-1}{x^2-1} + \frac{1}{1+x^2}\right]$

$\qquad = \frac{1}{2}\left[\frac{1}{1-x^2} + \frac{1}{1+x^2}\right]$

$\qquad = \frac{1}{2}\left[\frac{1+x^2+1-x^2}{1-x^4}\right] = \frac{1}{1-x^4}$

55. $f(x) = x\arcsin x + \sqrt{1-x^2}$

$f'(x) = x\left(\frac{1}{\sqrt{1-x^2}}\right) + \arcsin x + \frac{1}{2}(1-x^2)^{-1/2}(-2x)$

$\qquad = \frac{x}{\sqrt{1-x^2}} + \arcsin x + \frac{-x}{\sqrt{1-x^2}} = \arcsin x$

57. Begin by evaluating the function f and its first and second derivatives at $x = 1/2$.

$$f(x) = \arcsin x \qquad\qquad f\left(\frac{1}{2}\right) = \frac{\pi}{6}$$

$$f'(x) = \frac{1}{\sqrt{1-x^2}} \qquad\qquad f'\left(\frac{1}{2}\right) = \frac{2\sqrt{3}}{3}$$

$$f''(x) = \frac{x}{(1-x)^{3/2}} \qquad\qquad f''\left(\frac{1}{2}\right) = \frac{4\sqrt{3}}{9}$$

Therefore,

$$P_1(x) = \frac{\pi}{6} + \frac{2\sqrt{3}}{3}\left(x - \frac{1}{2}\right)$$

$$P_2(x) = \frac{\pi}{6} + \frac{2\sqrt{3}}{3}\left(x - \frac{1}{2}\right) + \frac{2\sqrt{3}}{9}\left(x - \frac{1}{2}\right)^2.$$

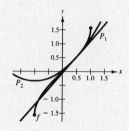

The graphs of f, P_1, and P_2, produced by a graphing utility, are shown in the figure.

Section 5.9   Inverse Trigonometric Functions and Integration

1. If $a = 1$ and $u = 3x$, then $du = 3\,dx$. Thus,

$$\int_0^{1/6} \frac{1}{\sqrt{1-9x^2}}\,dx = \frac{1}{3}\int_0^{1/6}\left(\frac{1}{\sqrt{1-(3x)^2}}\right)(3)\,dx$$

$$= \frac{1}{3}\left[\arcsin(3x)\right]_0^{1/6}$$

$$= \frac{1}{3}\left[\arcsin\left(\frac{1}{2}\right) - \arcsin 0\right]$$

$$= \frac{1}{3}\left(\frac{\pi}{6}\right) = \frac{\pi}{18}.$$

7. Since the degree of the numerator is greater than the degree of the denominator, we divide to obtain

$$\frac{x^3}{x^2+1} = x - \frac{x}{x^2+1}.$$

Therefore,

$$\int \frac{x^3}{x^2+1}\,dx = \int\left(x - \frac{x}{x^2+1}\right)dx$$

$$= \int x\,dx - \frac{1}{2}\int\frac{2x}{x^2+1}\,dx$$

$$= \frac{x^2}{2} - \frac{1}{2}\ln(x^2+1) + C$$

$$= \frac{1}{2}x^2 - \frac{1}{2}\ln(x^2+1) + C.$$

13. If $u = \arcsin x$, then $du = \dfrac{1}{\sqrt{1 - x^2}}\, dx$. Thus,

$$\int_0^{1/\sqrt{2}} \frac{\arcsin x}{\sqrt{1 - x^2}}\, dx = \int_0^{1/\sqrt{2}} (\arcsin x)^1 \frac{1}{\sqrt{1 - x^2}}\, dx$$

$$= \left[\frac{(\arcsin x)^2}{2} \right]_0^{1/\sqrt{2}}$$

$$= \frac{1}{2} \left\{ \left[\arcsin\left(\frac{1}{\sqrt{2}}\right) \right]^2 - [\arcsin(0)]^2 \right\}$$

$$= \frac{1}{2}\left[\left(\frac{\pi}{4}\right)^2 - (0)^2 \right] = \frac{\pi^2}{32} \approx 0.308.$$

17. If $a = 2$ and $u = e^{2x}$, then $du = 2e^{2x}\, dx$. Thus

$$\int \frac{e^{2x}}{4 + e^{4x}}\, dx = \frac{1}{2} \int \frac{1}{2^2 + (e^{2x})^2}(2e^{2x})\, dx$$

$$= \frac{1}{2} \int \frac{du}{a^2 + u^2}$$

$$= \frac{1}{2a} \arctan\frac{u}{a} + C = \frac{1}{4} \arctan\frac{e^{2x}}{2} + C.$$

23. $\displaystyle\int \frac{2x}{x^2 + 6x + 13}\, dx = \int \frac{(2x + 6) - 6}{x^2 + 6x + 13}\, dx$

$$= \int \frac{2x + 6}{x^2 + 6x + 13}\, dx - \int \frac{6}{x^2 + 6x + 13}\, dx$$

$$= \int \frac{2x + 6}{x^2 + 6x + 13}\, dx - \int \frac{6}{(x^2 + 6x + 9) + 4}\, dx$$

$$= \int \frac{2x + 6}{x^2 + 6x + 13}\, dx - 6\int \frac{1}{(x + 3)^2 + 2^2}\, dx$$

$$= \ln(x^2 + 6x + 13) - 3\arctan\left(\frac{x + 3}{2}\right) + C$$

25. $\displaystyle\int \frac{1}{\sqrt{-x^2 - 4x}}\, dx = \int \frac{1}{\sqrt{-(x^2 + 4x)}}\, dx$

$$= \int \frac{1}{\sqrt{4 - (x^2 + 4x + 4)}}\, dx$$

$$= \int \frac{1}{\sqrt{2^2 - (x + 2)^2}}\, dx$$

$$= \arcsin\left(\frac{x + 2}{2}\right) + C$$

29. $\displaystyle\int_2^3 \frac{2x - 3}{\sqrt{4x - x^2}}\, dx = \int_2^3 \frac{(2x - 4) + 1}{\sqrt{4x - x^2}}\, dx$

$$= \int_2^3 \frac{2x - 4}{\sqrt{4x - x^2}}\, dx + \int_2^3 \frac{1}{\sqrt{4x - x^2}}\, dx$$

$$= -\int_2^3 (4x - x^2)^{-1/2}(4 - 2x)\, dx + \int_2^3 \frac{1}{\sqrt{4 - (x^2 - 4x + 4)}}\, dx$$

$$= -\int_2^3 (4x - x^2)^{-1/2}(4 - 2x)\, dx + \int_2^3 \frac{1}{\sqrt{2^2 - (x - 2)^2}}\, dx$$

$$= \left[-\frac{(4x - x^2)^{1/2}}{1/2} + \arcsin\left(\frac{x - 2}{2}\right) \right]_2^3$$

$$= -2\sqrt{3} + \frac{\pi}{6} - (-4 + 0) = 4 - 2\sqrt{3} + \frac{\pi}{6} \approx 1.059$$

37. If $u = \sqrt{e^t - 3}$, then $e^t = u^2 + 3$, $t = \ln(u^2 + 3)$, and $dt = 2u/(u^2 + 3)\, du$. Therefore,

$$\int \sqrt{e^t - 3}\, dt = \int \frac{2u^2}{u^2 + 3}\, du.$$

Since the numerator and denominator are of equal degree, divide to obtain

$$2\int \frac{u^2}{u^2 + 3}\, du = 2\int \left[1 - \frac{3}{u^2 + 3}\right] du$$

$$= 2\left[\int du - 3\int \frac{1}{(\sqrt{3})^2 + u^2}\, du\right]$$

$$= 2\left[u - 3\left(\frac{1}{\sqrt{3}}\right) \arctan\left(\frac{u}{\sqrt{3}}\right) + C\right]$$

$$= 2\sqrt{e^t - 3} - 2\sqrt{3} \arctan\left(\frac{\sqrt{e^t - 3}}{\sqrt{3}}\right) + C.$$

41. From the figure, we can see that the area is

$$A = \int_1^3 \frac{1}{x^2 - 2x + 5}\, dx$$

$$= \int_1^3 \frac{1}{(x^2 - 2x + 1) + 5 - 1}\, dx$$

$$= \int_1^3 \frac{1}{(x - 1)^2 + 2^2}\, dx$$

$$= \frac{1}{2}\left[\arctan \frac{x - 1}{2}\right]_1^3$$

$$= \frac{1}{2}[\arctan 1 - \arctan 0] = \frac{\pi}{8}.$$

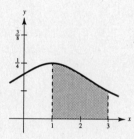

Section 5.10   Hyperbolic Functions

3. (a) $\operatorname{csch}(\ln 2) = \dfrac{1}{\sinh(\ln 2)} = \dfrac{2}{e^{\ln 2} - e^{-\ln 2}} = \dfrac{2}{2 - (1/2)} = \dfrac{4}{3}$

(b) $\coth(\ln 5) = \dfrac{\cosh(\ln 5)}{\sinh(\ln 5)} = \dfrac{e^{\ln 5} + e^{-\ln 5}}{e^{\ln 5} - e^{-\ln 5}} = \dfrac{5 + (1/5)}{5 - (1/5)} = \dfrac{13}{12}$

19. $y = \ln\left(\tanh \dfrac{x}{2}\right)$

$$y' = \frac{1}{\tanh(x/2)}\left[\frac{1}{2} \operatorname{sech}^2\left(\frac{x}{2}\right)\right]$$

$$= \frac{1}{2}\left[\frac{\cosh(x/2)}{\sinh(x/2)}\right]\left[\frac{1}{\cosh^2(x/2)}\right]$$

$$= \frac{1}{2\sinh(x/2)\cosh(x/2)} = \frac{1}{\sinh x} = \operatorname{csch} x$$

25. Using logarithmic differentiation yields

$$y = x^{\cosh x}$$

$$\ln y = \ln(x^{\cosh x}) = (\cosh x)(\ln x)$$

$$\frac{y'}{y} = (\cosh x)\left(\frac{1}{x}\right) + (\ln x)(\sinh x)$$

$$y' = y\left[\frac{\cosh x}{x} + (\ln x)(\sinh x)\right]$$

$$= \frac{y}{x}[\cosh x + x(\sinh x)\ln x].$$

35. $f(x) = \tanh x$ $\qquad\qquad f(1) \approx 0.76$

$f'(x) = \text{sech}^2 x$ $\qquad\qquad f'(1) \approx 0.42$

$f''(x) = -2\,\text{sech}^2 x \tanh x \quad f''(1) \approx 0.64$

$\qquad P_1(x) = f(a) + f'(a)(x - a)$

$\qquad\qquad = f(1) + f'(1)(x - 1)$

$\qquad\qquad = 0.76 + 0.42(x - 1)$

$\qquad P_2(x) = f(a) + f'(a)(x - a) + \frac{1}{2}f''(a)(x - a)^2$

$\qquad\qquad = f(1) + f'(1)(x - 1) + \frac{1}{2}f''(1)(x - 1)^2$

$\qquad\qquad = 0.76 + 0.42(x - 1) - 0.32(x - 1)^2$

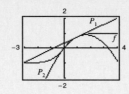

The graphs of f, P_1, and P_2 are shown in the figure.

39. If $u = \cosh(x - 1)$, then $du = \sinh(x - 1)\,dx$.

$$\int \cosh^2(x - 1) \sinh(x - 1)\,dx = \int u^2\,du$$

$$= \frac{1}{3}u^3 + C = \frac{1}{3}\cosh^3(x - 1) + C.$$

45. If $u = \dfrac{1}{x}$, then $du = -\dfrac{1}{x^2}\,dx$ and

$$\int \frac{\text{csch}(1/x)\,\coth(1/x)}{x^2}\,dx = -\int \text{csch}\,\frac{1}{x}\,\coth\,\frac{1}{x}\left(-\frac{1}{x^2}\right)dx$$

$$= -\int \text{csch}\,u\,\coth u\,du$$

$$= \text{csch}\,u + C = \text{csch}\,\frac{1}{x} + C.$$

59. $y = 2x\sinh^{-1}(2x) - \sqrt{1 + 4x^2}$

$$y' = 2x\left[\frac{1}{\sqrt{1 + (2x)^2}}\right](2) + 2\sinh^{-1}(2x) - \frac{1}{2}(1 + 4x^2)^{-1/2}(8x)$$

$$= \frac{4x}{\sqrt{1 + 4x}} + 2\sinh^{-1}(2x) - \frac{4}{\sqrt{1 + 4x^2}} = 2\sinh^{-1}(2x)$$

61. Note that the domain of this function restricts x to the interval $(0, a)$ where $0 < a$.

$$y = a\,\text{sech}^{-1}\left(\frac{x}{a}\right) - \sqrt{a^2 - x^2}$$

$$\frac{dy}{dx} = a\left[\frac{-1}{|x/a|\sqrt{1 - (x/a)^2}}\right]\left(\frac{1}{a}\right) - \frac{1}{2}(a^2 - x^2)^{-1/2}(-2x)$$

$$= \frac{-1}{|x/a|\sqrt{(a^2 - x^2)/a^2}} + \frac{x}{\sqrt{a^2 - x^2}}$$

$$= \frac{-a^2}{x\sqrt{a^2 - x^2}} + \frac{x}{\sqrt{a^2 - x^2}}$$

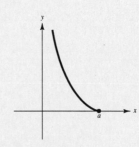

(Note that the absolute value signs can be deleted since $0 < x$.)

$$\frac{dy}{dx} = \frac{-a^2 + x^2}{x\sqrt{a^2 - x^2}} = \frac{-(a^2 - x^2)}{x\sqrt{a^2 - x^2}} = \frac{-\sqrt{a^2 - x^2}}{x}$$

63. If $u = e^x$, then $du = e^x \, dx$ and we have

$$\int \frac{1}{\sqrt{1 + e^{2x}}} \, dx = \int \frac{e^x}{e^x \sqrt{1 + (e^x)^2}} \, dx$$

$$= \int \frac{du}{u \sqrt{1 + u^2}}$$

$$= -\ln\left(\frac{1 + \sqrt{1 + e^{2x}}}{e^x}\right) + C$$

$$= -\operatorname{csch}^{-1}(e^x) + C.$$

65. If $u = \sqrt{x}$, then $du = 1/(2\sqrt{x}) \, dx$ and we have

$$\int \frac{1}{\sqrt{x}\sqrt{1 + x}} \, dx = 2\int \frac{1}{\sqrt{1 + (\sqrt{x})^2}}\left(\frac{1}{2\sqrt{x}}\right) dx$$

$$= 2\int \frac{du}{\sqrt{1 + u^2}}$$

$$= 2 \sinh^{-1}\sqrt{x} + C$$

$$= 2 \ln\left(\sqrt{x} + \sqrt{1 + x}\right) + C.$$

77. The graph of the region is shown in the figure.

$$A = \int_0^2 \frac{5x}{\sqrt{x^4 + 1}} \, dx = \frac{5}{2}\int_0^2 \frac{1}{\sqrt{(x^2)^2 + 1}}(2x) \, dx$$

$$= \frac{5}{2}\left[\ln\left(x^2 + \sqrt{x^4 + 1}\right)\right]_0^2$$

$$= \frac{5}{2}\ln\left(4 + \sqrt{17}\right) \approx 5.237$$

85. If $y = \cosh^{-1}x$, then $\cosh y = x$. Differentiating implicitly with respect to x yields the following.

$$\cosh y = x$$

$$(\sinh y)\frac{dy}{dx} = 1$$

$$\frac{dy}{dx} = \frac{1}{\sinh y} = \frac{1}{\sqrt{\cosh^2 y - 1}} = \frac{1}{\sqrt{x^2 - 1}}$$

Review Exercises for Chapter 5

5. $\ln 3 + \dfrac{1}{3}\ln(4 - x^2) - \ln x = \ln 3 + \ln(4 - x^2)^{1/3} - \ln x$

$$= \ln\left(3\sqrt[3]{4 - x^2}\right) - \ln x = \ln\left(\frac{3\sqrt[3]{4 - x^2}}{x}\right)$$

15. $y = \dfrac{1}{b^2}[\ln(a + bx) + a(a + bx)^{-1}]$

$$\frac{dy}{dx} = \frac{1}{b^2}\left[\left(\frac{1}{a + bx}\right)(b) + a(-1)(a + bx)^{-2}(b)\right]$$

$$= \frac{1}{b^2}\left[\frac{b}{a + bx} - \frac{ab}{(a + bx)^2}\right]$$

$$= \frac{1}{b^2}\left[\frac{b(a + bx) - ab}{(a + bx)^2}\right] = \frac{x}{(a + bx)^2}$$

25. $\displaystyle\int_0^{\pi/3} \sec\theta \, d\theta = \Big[\ln|\sec\theta + \tan\theta|\Big]_0^{\pi/3}$

$$= \ln\left(2 + \sqrt{3}\right) - \ln(1 + 0)$$

$$= \ln\left(2 + \sqrt{3}\right)$$

33. (a) $f(x) = \ln\sqrt{x}$

$$y = \tfrac{1}{2}\ln x$$

$$2y = \ln x$$

$$x = e^{2y}$$

$$y = e^{2x} \implies f^{-1}(x) = e^{2x}$$

(c) $f^{-1}(f(x)) = f^{-1}\left(\ln\sqrt{x}\right) = e^{2\ln\sqrt{x}} = e^{\ln x} = x$

$\qquad f(f^{-1}(x)) = f(e^{2x}) = \ln\sqrt{e^{2x}} = \ln\sqrt{(e^x)^2} = \ln e^x = x$

(b) The graph of f and f^{-1} are shown in the figure. Note that the graph of f^{-1} is a reflection of the graph of f in the line $y = x$.

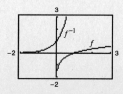

43. Since $g(t) = t^2 e^t$, begin by using the Product Rule.

$$g'(t) = t^2 \frac{d}{dt}[e^t] + e^t \frac{d}{dt}[t^2] = t^2 e^t + e^t(2t) = te^t(t + 2)$$

51. Begin by sketching a triangle to represent θ, "the angle whose sine is x." This yields

$$\theta = \arcsin x$$

$$y = \tan(\arcsin x) = \tan \theta = \frac{x}{\sqrt{1 - x^2}}.$$

Therefore,

$$\frac{dy}{dx} = \frac{\sqrt{1 - x^2}(1) - x(1/2)\left(1/\sqrt{1 - x^2}\right)(-2x)}{1 - x^2}$$

$$= \frac{(1 - x^2) + x^2}{(1 - x^2)\sqrt{1 - x^2}} = (1 - x^2)^{-3/2}.$$

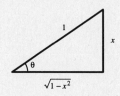

57. Let $u = \sqrt{x}$ in the second term of $y = 2x - \cosh\sqrt{x}$.

$$\frac{dy}{dx} = \frac{d}{dx}[2x] - \frac{d}{du}[\cosh u]\frac{du}{dx}$$

$$= 2 - \sinh\sqrt{x}\left(\frac{1}{2\sqrt{x}}\right) = 2 - \frac{\sinh\sqrt{x}}{2\sqrt{x}}$$

59. Using the Product Rule on the first term and differentiating implicitly yields the following.

$$y \ln x + y^2 = 0$$

$$y\left(\frac{1}{x}\right) + (\ln x)y' + 2yy' = 0$$

$$y'(2y + \ln x) = -\frac{y}{x}$$

$$y' = \frac{-y}{x(2y + \ln x)}$$

67. $\displaystyle\int \frac{e^{4x} - e^{2x} + 1}{e^x} \, dx = \int (e^{3x} - e^x + e^{-x}) \, dx$

$$= \frac{e^{3x}}{3} - e^x - e^{-x} + C$$

$$= \frac{e^{4x} - 3e^{2x} - 3}{3e^x} + C$$

71. If we let $u = e^{2x}$, then $du = 2e^{2x} \, dx$. Thus,

$$\int \frac{1}{e^{2x} + e^{-2x}} \, dx = \int \left(\frac{1}{e^{2x} + e^{-2x}}\right)\left(\frac{e^{2x}}{e^{2x}}\right) dx$$

$$= \int \frac{e^{2x}}{e^{4x} + 1} \, dx$$

$$= \frac{1}{2}\int \left[\frac{1}{1 + (e^{2x})^2}\right](2e^{2x}) \, dx$$

$$= \frac{1}{2}\int \frac{du}{1 + u^2} = \frac{1}{2}\arctan(e^{2x}) + C.$$

79. If $u = x^2$, then $du = 2x \, dx$.

$$\int \frac{x}{\sqrt{x^4 + 1}} \, dx = \frac{1}{2}\int \frac{1}{\sqrt{(x^2)^2 - 1}}(2x) \, dx$$

$$= \frac{1}{2}\ln\left(x^2 + \sqrt{x^4 + 1}\right) + C$$

83. $\dfrac{dy}{dx} = \dfrac{x^2 + 3}{x}$

$$y = \int \frac{x^2 + 3}{x} \, dx$$

$$= \int \left(x + \frac{3}{x}\right) dx = \frac{1}{2}x^2 + 3\ln|x| + C$$

93. (a) The probability that the product is less than 25 is

$$P = \frac{1}{100}\left(25 + \int_{2.5}^{10} \frac{25}{x} \, dx\right) = \frac{1}{100}\left(25 + \Big[25 \ln x\Big]_{2.5}^{10}\right)$$

$$= \frac{1}{4}(1 + \ln 4) \approx 0.60.$$

(b) The probability that the product is less than 50 is

$$P = \frac{1}{100}\left(50 + \int_{5}^{10} \frac{50}{x} \, dx\right) = \frac{1}{100}\left(50 + \Big[50 \ln x\Big]_{5}^{10}\right)$$

$$= \frac{1}{2}(1 + \ln 2) \approx 0.85.$$

CHAPTER 6
Applications of Integration

CHAPTER 6
Applications of Integration

Section 6.1   Area of a Region Between Two Curves

Solutions to Selected Odd-Numbered Exercises

15. The points of intersection f and g are found by solving

$$f(x) = g(x)$$

$$x^2 + 2x + 1 = 3x + 3$$

$$x^2 - x - 2 = 0$$

$$(x - 2)(x + 1) = 0 \implies x = -1, 2.$$

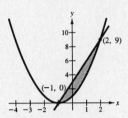

Since $x^2 + 2x + 1 \le 3x + 3$ for $-1 \le x \le 2$, we have

$$\text{Area} = \int_{-1}^{2} \left[(3x + 3) - (x^2 + 2x + 1) \right] dx$$

$$= \int_{-1}^{2} (-x^2 + x + 2)\, dx$$

$$= \left[\frac{-x^3}{3} + \frac{x^2}{2} + 2x \right]_{-1}^{2}$$

$$= \left(\frac{-8}{3} + 2 + 4 \right) - \left(\frac{1}{3} + \frac{1}{2} - 2 \right)$$

$$= \frac{-16 + 12 + 24 - 2 - 3 + 12}{6} = \frac{27}{6} = \frac{9}{2}.$$

23. Using horizontal representative rectangles, we have

$$\text{Area} = \int_{-1}^{2} (y^2 + 1)\, dy = \left[\frac{y^3}{3} + y \right]_{-1}^{2}$$

$$= \left(\frac{8}{3} + 2 \right) - \left(\frac{-1}{3} - 1 \right)$$

$$= \frac{8 + 6 + 1 + 3}{3} = 6.$$

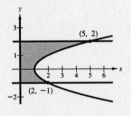

33. The graph of the region is shown in the figure. To find the points of intersection, solve the equation $f(x) = g(x)$.

$$\frac{1}{1 + x^2} = \frac{x^2}{2}$$

$$x^4 + x^2 - 2 = 0$$

$$(x^2 + 2)(x^2 - 1) = 0$$

$$(x^2 + 2)(x + 1)(x - 1) = 0 \implies x = \pm 1$$

Therefore, the points of intersection are $\left(-1, \frac{1}{2} \right)$ and $\left(1, \frac{1}{2} \right)$. Using the integration capabilities of a graphing utility and symmetry, we have

$$\text{Area} = \int_{-1}^{1} [f(x) - g(x)]\, dx$$

$$= 2 \int_{0}^{1} \left(\frac{1}{1 + x^2} - \frac{x^2}{2} \right) dx = \frac{\pi}{2} - \frac{1}{3} \approx 1.237.$$

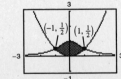

37. Since *f* and *g* are symmetric to the origin, the area of the region bounded by their graphs for
$-\pi/3 \le x \le \pi/3$ is twice the area of the region bounded by their graphs for $0 \le x \le \pi/3$. Since
$\tan x \le 2 \sin x$ for $0 \le x \le \pi/3$, we have

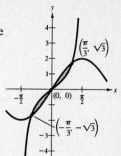

$$\text{Area} = 2\int_0^{\pi/3} (2 \sin x - \tan x)\, dx$$

$$= 2\left[-2 \cos x + \ln|\cos x|\right]_0^{\pi/3}$$

$$= 2\left[-2\left(\frac{1}{2}\right) + \ln\left(\frac{1}{2}\right) - (-2)\right]$$

$$= 2(1 - \ln 2) \approx 0.614.$$

43. The graph of the region is shown in the figure. Using the integration capabilities of a
graphing utility, we have

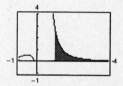

$$\text{Area} = \int_1^3 \left(\frac{1}{x^2}e^{1/x} - 0\right) dx = \int_1^3 \frac{1}{x^2}e^{1/x}\, dx \approx 1.323.$$

47. From the figure, observe that the triangular region is bounded by $y = (c/b)x$, $y = [c/(b-a)](x-a)$, and $y = 0$. Therefore,
the area is given by

$$\text{Area} = \int_0^b \frac{c}{b}x\, dx + \int_b^a \left(\frac{c}{b-a}\right)(x-a)\, dx$$

$$= \frac{c}{b}\left[\frac{x^2}{2}\right]_0^b + \left(\frac{c}{b-a}\right)\left[\frac{x^2}{2} - ax\right]_b^a$$

$$= \frac{1}{2}bc + \left(\frac{c}{b-a}\right)\left[\left(\frac{a^2}{2} - a^2\right) - \left(\frac{b^2}{2} - ab\right)\right]$$

$$= \frac{1}{2}bc + \frac{1}{2}(a-b)c = \frac{1}{2}ac.$$

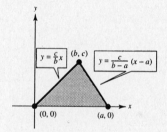

49. Since $f(x) = x^3$, $f'(x) = 3x^2$, and $f'(1) = 3$ (slope of the tangent line), the equation of the tangent line to the graph of *f* is
given by

$$y - 1 = 3(x - 1)$$

$$y = 3x - 2.$$

The *x*-coordinates of the points of intersection of the tangent line and the function are the solutions to the equation

$$x^3 = 3x - 2$$

$$x^3 - 3x + 2 = 0$$

$$(x - 1)(x^2 + x - 2) = 0$$

$$(x - 1)^2(x + 2) = 0 \implies x = -2, 1.$$

Therefore, the points of intersection are given by $(1, 1)$ and $(-2, -8)$. (See the figure.)

$$\text{Area} = \int_{-2}^1 [x^3 - (3x - 2)]\, dx$$

$$= \int_{-2}^1 (x^3 - 3x + 2)\, dx$$

$$= \left[\frac{1}{4}x^4 - \frac{3}{2}x^2 + 2x\right]_{-2}^1$$

$$= \left(\frac{1}{4} - \frac{3}{2} + 2\right) - (4 - 6 - 4) = \frac{27}{4}$$

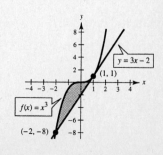

57. The reduction in revenue is approximated by the area between the two models over the interval $0 \leq t \leq 5$.

$$\int_0^5 [(7.21 + 0.58t) - (7.21 + 0.45t)] \, dt = \int_0^5 0.13t \, dt = \left[\frac{0.13t^2}{2} \right]_0^5 = \$1.625 \text{ billion}$$

67. Solving the equations simultaneously yields the point of equilibrium to be (80, 10). Therefore,

$$\text{Consumer Surplus} = \int_0^{80} [(50 - 0.5x) - 10] \, dx$$

$$= \left[40x - 0.25x^2 \right]_0^{80} = 1600$$

$$\text{Producer Surplus} = \int_0^{80} (10 - 0.125x) \, dx$$

$$= \left[10x - 0.0625x^2 \right]_0^{80} = 400.$$

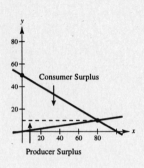

Section 6.2   Volume: The Disc Method

5. From the figure we have

$$R(x) = x^2 \qquad \text{Outer radius}$$

$$r(x) = x^3. \qquad \text{Inner radius}$$

Now, integrating between 0 and 1, we have

$$V = \pi \int_a^b ([R(x)]^2 - [r(x)]^2) \, dx$$

$$= \pi \int_0^1 (x^4 - x^6) \, dx$$

$$= \pi \left[\frac{x^5}{5} - \frac{x^7}{7} \right]_0^1 = \pi \left(\frac{1}{5} - \frac{1}{7} \right) = \frac{2\pi}{35}.$$

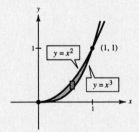

11. (a) From the figure, we have

$$R(x) = y = \sqrt{x}.$$

$$V = \pi \int_a^b ([R(x)]^2 - [r(x)]^2) \, dx$$

$$= \pi \int_0^4 (\sqrt{x})^2 \, dx$$

$$= \pi \int_0^4 x \, dx = \pi \left[\frac{x^2}{2} \right]_0^4 = \frac{16\pi}{2} = 8\pi$$

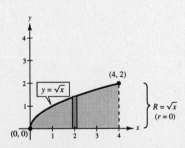

(b) From the figure we have

$$R(y) = 4 \qquad \text{Outer radius}$$

$$r(y) = y^2. \qquad \text{Inner radius}$$

$$V = \pi \int_a^b ([R(y)]^2 - [r(y)]^2) \, dy$$

$$= \pi \int_0^2 [4^2 - (y^2)^2] \, dy$$

$$= \pi \int_0^2 (16 - y^4) \, dy$$

$$= \left[16y - \frac{y^5}{5} \right]_0^2$$

$$= \pi \left[32 - \frac{32}{5} \right] = \frac{128\pi}{5}$$

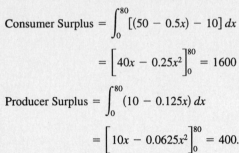

—CONTINUED—

11. —CONTINUED—

(c) From the figure, we have

$$R(y) = 4 - y^2.$$

$$V = \pi \int_a^b ([R(y)]^2 - [r(y)]^2) \, dy$$

$$= \pi \int_0^2 (4 - y^2)^2 \, dy$$

$$= \pi \int_0^2 (16 - 8y^2 + y^4) \, dy$$

$$= \pi \left[16y - \frac{8y^3}{3} + \frac{y^5}{5} \right]_0^2$$

$$= \pi \left[32 - \frac{64}{3} + \frac{32}{5} \right] = \frac{256\pi}{15}$$

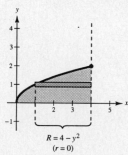

$R = 4 - y^2$
$(r = 0)$

(d) From the figure, we have

$$R(y) = 6 - y^2 \qquad \text{Outer radius}$$

$$r(y) = 2. \qquad \text{Inner radius}$$

$$V = \pi \int_a^b ([R(y)]^2 - [r(y)]^2) \, dy$$

$$= \pi \int_0^2 [(6 - y^2)^2 - 2^2] \, dy$$

$$= \pi \int_0^2 (32 - 12y^2 + y^4) \, dy$$

$$= \pi \left[32y - 4y^3 + \frac{y^5}{5} \right]_0^2$$

$$= \pi \left[64 - 32 + \frac{32}{5} \right] = \frac{192\pi}{5}$$

$R = 6 - y^2$

$r = 2$

13. The points of intersection of the graphs of the two functions are $(0, 0)$ and $(2, 4)$.

(a) From the figure, we have

$$R(x) = 4x - x^2 \qquad \text{Outer radius}$$

$$r(x) = x^2. \qquad \text{Inner radius}$$

$$V = \pi \int_a^b ([R(x)]^2 - [r(x)]^2) \, dx$$

$$= \pi \int_0^2 [(4x - x^2)^2 - (x^2)^2] \, dx$$

$$= \pi \int_0^2 (16x^2 - 8x^3) \, dx$$

$$= \pi \left[\frac{16}{3} x^3 - 2x^4 \right]_0^2$$

$$= 8\pi \left[\frac{16}{3} - 4 \right]$$

$$= \frac{32\pi}{3}$$

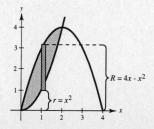

$R = 4x - x^2$

$r = x^2$

(b) From the figure, we have

$$R(y) = 6 - (4x - x^2) \qquad \text{Outer radius}$$

$$r(y) = 6 - x^2. \qquad \text{Inner radius}$$

$$V = \pi \int_a^b ([R(x)]^2 - [r(x)]^2) \, dx$$

$$= \pi \int_0^2 [(6 - x^2)^2 - (6 - 4x + x^2)^2] \, dx$$

$$= 8\pi \int_0^2 (x^3 - 5x^2 + 6x) \, dx$$

$$= 8\pi \left[\frac{x^4}{4} - \frac{5}{3} x^3 + 3x^2 \right]_0^2$$

$$= 32\pi \left(1 - \frac{10}{3} + 3 \right)$$

$$= \frac{64\pi}{3}$$

$r = 6 - (4x - x^2)$

$R = 6 - x^2$

21. From the figure, we have

$R(y) = 6 - y^2$ Outer radius

$r(y) = 2.$ Inner radius

$V = \pi \int_{-2}^{2} [(6 - y^2)^2 - (2)^2] \, dy$

$= 2\pi \int_{0}^{2} (y^4 - 12y^2 + 32) \, dy = 2\pi \left[\dfrac{y^5}{5} - 4y^3 + 32y \right]_{0}^{2} = \dfrac{384\pi}{5}$

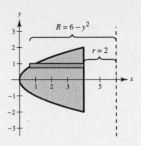

27. $V = \pi \int_{0}^{1} (e^{-x})^2 \, dx$

$= \pi \int_{0}^{1} e^{-2x} \, dx = \left[-\dfrac{\pi}{2} e^{-2x} \right]_{0}^{1} = \dfrac{\pi}{2} \left(1 - \dfrac{1}{e^2} \right) \approx 1.358$

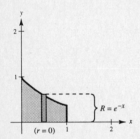

35. From the figure we have $R(x) = e^{x/2} + e^{-x/2}$ and $r(x) = 0$. Using a graphing utility to approximate the definite integral yields

$V = \pi \int_{-1}^{2} (e^{x/2} + e^{-x/2}) \, dx \approx 49.02.$

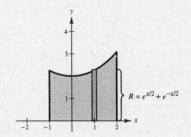

43. Let $y = \sqrt{r^2 - x^2}$ and let the region bounded by $y = \sqrt{r^2 - x^2}$ and $y = 0$ be revolved about the x-axis. The resulting solid of revolution is a sphere with volume

$V = \pi \int_{-r}^{r} \left(\sqrt{r^2 - x^2} \right)^2 \, dx$

$= \pi \int_{-r}^{r} (r^2 - x^2) \, dx$

$= \pi \left[r^2 x - \dfrac{x^3}{3} \right]_{-r}^{r}$

$= \pi \left[\left(r^3 - \dfrac{r^3}{3} \right) - \left(-r^3 + \dfrac{r^3}{3} \right) \right] = \dfrac{4\pi r^3}{3}.$

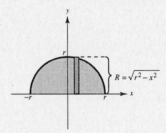

51. The total volume of the sphere is

$V = \dfrac{4}{3} \pi r^3 = \dfrac{4\pi (50)^3}{3} = \dfrac{500,000\pi}{3} \text{ft}^3.$

The volume of the portion filled with water (see figure) is

$\dfrac{1}{4} V = \dfrac{125,000\pi}{3} = \pi \int_{-50}^{y_0} \left(\sqrt{2500 - y^2} \right)^2 \, dy$

$= \pi \int_{-50}^{y_0} (2500 - y^2) \, dy = \pi \left[2500y - \dfrac{y^3}{3} \right]_{-50}^{y_0}$

$= \pi \left[\left(2500y_0 - \dfrac{y_0^3}{3} \right) - \left(-125,000 + \dfrac{125,000}{3} \right) \right] = \pi \left[2500y_0 - \dfrac{y_0^3}{3} + \dfrac{250,000}{3} \right].$

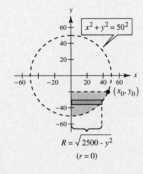

Simplifying yields the equation $y_0^3 - 7500y_0 - 125,000 = 0.$

Using the root-finding capabilities of a graphing utility yields the approximate root $y_0 = -17.36$. When the tank is one-fourth full the approximate depth of the water is $[-17.36 - (-50)] = 32.64$ feet. By symmetry, the tank is three-fourth full when the depth of the water is 67.36 feet.

57. The base of the solid is shown in the figure. Since the cross sections are taken perpendicular to the y-axis, the base of each cross section is given by $(1 - x) = (1 - \sqrt[3]{y})$.

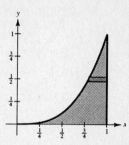

(a) The cross sections are squares whose sides are given by

$$s = \left(1 - \sqrt[3]{y}\right).$$

Thus,

$$A(y) = s^2 = \left(1 - \sqrt[3]{y}\right)^2$$

and

$$V = \int_0^1 \left(1 - \sqrt[3]{y}\right)^2 dy$$

$$= \int_0^1 \left(1 - 2y^{1/3} + y^{2/3}\right) dy$$

$$= \left[y - \frac{2y^{4/3}}{4/3} + \frac{y^{5/3}}{5/3}\right]_0^1 = 1 - \frac{3}{2} + \frac{3}{5} = \frac{1}{10}.$$

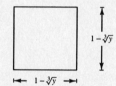

(b) The cross sections are semicircles whose radii are given by

$$r = \left(\frac{1}{2}\right)\left(1 - \sqrt[3]{y}\right).$$

Thus,

$$A(y) = \left(\frac{1}{2}\right)\pi\left[\left(\frac{1}{2}\right)\left(1 - \sqrt[3]{y}\right)\right]^2 = \frac{\pi}{8}\left(1 - \sqrt[3]{y}\right)^2$$

and

$$V = \frac{\pi}{8}\int_0^1 \left(1 - \sqrt[3]{y}\right)^2 dy$$

$$= \frac{\pi}{8}\left(\frac{1}{10}\right) = \frac{\pi}{80}. \qquad \text{From part (a)}$$

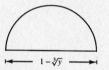

(c) The cross sections are equilateral triangles whose sides are given by

$$1 - \sqrt[3]{y}.$$

Thus,

$$A(y) = \frac{1}{2}bh = \frac{1}{2}\left(1 - \sqrt[3]{y}\right)\left[\frac{\sqrt{3}}{2}\left(1 - \sqrt[3]{y}\right)\right]$$

$$= \frac{\sqrt{3}}{4}\left(1 - \sqrt[3]{y}\right)^2$$

and

$$V = \frac{\sqrt{3}}{4}\int_0^1 \left(1 - \sqrt[3]{y}\right)^2 dy$$

$$= \frac{\sqrt{3}}{4}\left(\frac{1}{10}\right) = \frac{\sqrt{3}}{40}. \qquad \text{From part (a)}$$

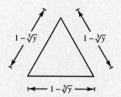

(d) The cross sections are semiellipses whose heights are twice the lengths of their bases. Thus $a = 2b$, where

$$b = 1 - \sqrt[3]{y}.$$

Thus,

$$A(y) = \left(\frac{1}{2}\right)\pi(a)\left(\frac{b}{2}\right) = \frac{\pi}{2}b^2 = \frac{\pi}{2}\left(1 - \sqrt[3]{y}\right)^2$$

and

$$V = \frac{\pi}{2}\int_0^1 \left(1 - \sqrt[3]{y}\right)^2 dy$$

$$= \frac{\pi}{2}\left(\frac{1}{10}\right) = \frac{\pi}{20}. \qquad \text{From part (a)}$$

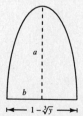

61. (a) Since $\theta = 45°$, the cross sections are isosceles right triangles for which the base and height are equal. Thus,

$$A(x) = \frac{1}{2}bh$$

$$= \frac{1}{2}\left(\sqrt{r^2 - x^2}\right)\left(\sqrt{r^2 - x^2}\right) = \frac{1}{2}(r^2 - x^2)$$

and

$$V = \frac{1}{2}\int_{-r}^{r} (r^2 - x^2)\, dx$$

$$= \frac{1}{2}\left[r^2x - \frac{x^3}{3}\right]_{-r}^{r}$$

$$= \frac{1}{2}\left[\left(r^3 - \frac{r^3}{3}\right) - \left(-r^3 + \frac{r^3}{3}\right)\right] = \frac{2r^3}{3} \text{ in}^3.$$

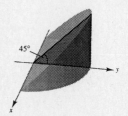

(b) For the arbitrary angle θ, $(0 < \theta < 90°)$, the cross sections are right triangles with base $\sqrt{r^2 - x^2}$ and height $\sqrt{r^2 - x^2} \tan \theta$. Thus,

$$A(x) = \frac{1}{2}bh$$

$$= \frac{1}{2}\left(\sqrt{r^2 - x^2}\right)\left(\sqrt{r^2 - x^2}\right)\tan \theta$$

$$= \frac{1}{2}(r^2 - x^2)\tan \theta.$$

Since the integration is with respect to x, we can use the result of part (a) and determine that the volume is

$$V = \frac{2r^3}{3}(\tan \theta).$$

As θ increases, the volume of the wedge increases.

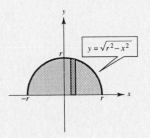

Section 6.3   Volume: The Shell Method

7. The distance from the center of the rectangle to the axis of revolution is $p(x) = x$, and the height of the rectangle is $h(x) = (4x - x^2) - (x^2)$.

$$V = 2\pi \int_a^b p(x)h(x)\, dx$$

$$= 2\pi \int_0^2 x[(4x - x^2) - (x^2)]\, dx$$

$$= 2\pi \int_0^2 x(4x - 2x^2)\, dx$$

$$= 4\pi \int_0^2 (2x^2 - x^3)\, dx$$

$$= 4\pi\left[\frac{2x^3}{3} - \frac{x^4}{4}\right]_0^2 = 4\pi\left[\frac{16}{3} - 4\right] = \frac{16\pi}{3}$$

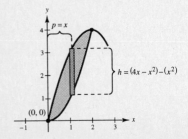

13. The distance from the center of the rectangle to the axis of revolution is $p(y) = y$, and the height of the rectangle is $h(y) = 2 - y$.

$$V = 2\pi \int_a^b p(y)h(y)\, dy$$

$$= 2\pi \int_0^2 y(2 - y)\, dy$$

$$= 2\pi \int_0^2 (2y - y)^2\, dy$$

$$= 2\pi\left[y^2 - \frac{y^3}{3}\right]_0^2 = 2\pi\left[4 - \frac{8}{3}\right] = \frac{8\pi}{3}$$

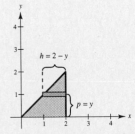

17. The distance from the center of the rectangle to the axis of revolution is $p(x) = 4 - x$, and the height of the rectangle is $h(x) = (4x - x^2) - (x^2)$.

$$V = 2\pi \int_a^b p(x)h(x)\, dx$$

$$= 2\pi \int_0^2 (4 - x)[(4x - x^2) - (x^2)]\, dx$$

$$= 2\pi \int_0^2 (4 - x)(4x - 2x^2)\, dx$$

$$= 4\pi \int_0^2 (8x - 6x^2 + x^3)\, dx$$

$$= 4\pi \left[4x^2 - 2x^3 + \frac{x^4}{4} \right]_0^2 = 4\pi[16 - 16 + 4] = 16\pi$$

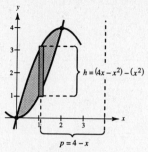

21. (a) Disc Method

$$V = \pi \int_a^b ([R(x)]^2 - [r(x)]^2)\, dx$$

$$= \pi \int_0^2 [(x^3)^2 - (0)^2]\, dx$$

$$= \pi \int_0^2 x^6\, dx$$

$$= \pi \left[\frac{x^7}{7} \right]_0^2 = \frac{128\pi}{7}$$

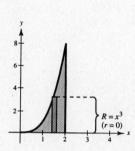

(b) Shell Method

$$V = 2\pi \int_a^b p(x)h(x)\, dx$$

$$= 2\pi \int_0^2 x(x^3)\, dx$$

$$= 2\pi \int_0^2 x^4\, dx$$

$$= 2\pi \left[\frac{x^5}{5} \right]_0^2 = \frac{64\pi}{5}$$

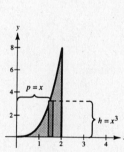

(c) Shell Method

$$V = 2\pi \int_a^b p(x)h(x)\, dx$$

$$= 2\pi \int_0^2 (4 - x)x^3\, dx$$

$$= 2\pi \int_0^2 (4x^3 - x^4)\, dx$$

$$= 2\pi \left[x^4 - \frac{x^5}{5} \right]_0^2 = \frac{96\pi}{5}$$

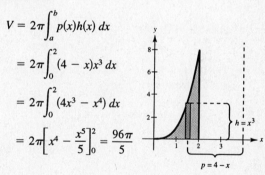

25. (a) The graph of the region is shown in the figure.

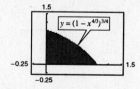

(b) Solving for y in the equation $x^{4/3} + y^{4/3} = 1$ yields $y = \pm(1 - x^{4/3})^{3/4}$. Since the required region is in the first quadrant, $h(x) = (1 - x^{4/3})^{3/4}$. Also, $p(x) = x$. Using a graphing utility to approximate the definite integral yields

$$V = 2\pi \int_a^b p(x)h(x)\, dx$$

$$= 2\pi \int_0^1 x(1 - x^{4/3})^{3/4}\, dx \approx 1.506.$$

33. The total volume of the solid is given by

$$V = 2\pi \int_0^2 x\left(2 - \frac{x^2}{2}\right) dx$$

$$= 2\pi \int_0^2 \left(2x - \frac{x^3}{2}\right) dx$$

$$= 2\pi \left[x^2 - \frac{x^4}{8}\right]_0^2 = 2\pi(4 - 2) = 4\pi.$$

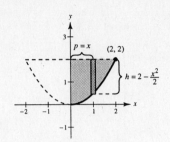

If a hole drilled in the center with a radius of x_0 removes $\frac{1}{4}$ of this volume, we have

$$\left(\frac{3}{4}\right)V = 2\pi \int_{x_0}^2 x\left(2 - \frac{x^2}{2}\right) dx$$

$$3\pi = 2\pi \left[x^2 - \frac{x^4}{8}\right]_{x_0}^2$$

$$3\pi = 2\pi \left[(4 - 2) - \left(x_0^2 - \frac{x_0^4}{8}\right)\right]$$

$$3 = 4 - 2x_0^2 + \left(\frac{x_0^4}{4}\right)$$

$$x_0^4 - 8x_0^2 + 4 = 0$$

$$x_0^2 = \frac{8 \pm \sqrt{64 - 16}}{2}$$

$$x_0 = \sqrt{4 - 2\sqrt{3}}.$$

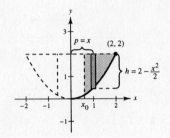

(Since $0 < x_0 < 2$, you are not interested in $x_0 = \sqrt{4 + 2\sqrt{3}} \approx 2.7$.) Finally, since

radius $= x_0 = \sqrt{4 - 2\sqrt{3}} = 0.732$

we have

diameter $= 2x_0 = 2\sqrt{4 - 2\sqrt{3}} \approx 1.464.$

43. (a) The distance from the center of the rectangle to the axis of revolution is $p(x) = x$ and the height $h(x)$ is the depth of the water. Therefore,

$$V = 2\pi \int_0^{200} xh(x)\, dx$$

$$\approx \frac{2\pi(200)}{3(8)}[0 + 4(25)(19) + 2(50)(19) + 4(75)(17) + 2(100)(15) + 4(125)(14)$$

$$+ 2(150)(10) + 4(175)(6) + 0]$$

$$\approx 1,366,593 \text{ cubic feet.}$$

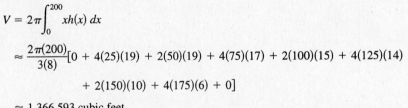

(b) Using the regression capabilities of a graphing utility yields the following quadratic model for the data.

$$h(x) = -0.000561x^2 + 0.0189x + 19.39$$

A plot of the data and a graph of the model are given in the figure.

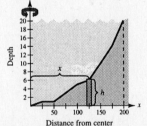

(c) Using the model of part (b) and the integration capabilities of a graphing utility, we obtain the following approximation of the volume of the pond.

$$V \approx 2\pi \int_0^{200} x(-0.000561x^2 + 0.0189x + 19.39)\, dx$$

$$\approx 1,343,345 \text{ cubic feet}$$

(d) Since there are approximately 7.48 gallons per cubic foot of water, the approximate volume of the pond, in gallons, is

$$1,343,345(7.48) \approx 10,048,221 \text{ gallons.}$$

Section 6.4   Arc Length and Surfaces of Revolution

7. $y = \dfrac{x^4}{8} + \dfrac{1}{4x^2}$

$y' = \dfrac{x^3}{2} - \dfrac{1}{2x^3}$

$s = \displaystyle\int_1^2 \sqrt{1 + (y')^2}\, dx = \int_1^2 \sqrt{1 + \left(\dfrac{x^3}{2} - \dfrac{1}{2x^3}\right)^2}\, dx$

$= \displaystyle\int_1^2 \sqrt{1 + \dfrac{x^6}{4} - \dfrac{1}{2} + \dfrac{1}{4x^6}}\, dx$

$= \displaystyle\int_1^2 \sqrt{\dfrac{x^6}{4} + \dfrac{1}{2} + \dfrac{1}{4x^6}}\, dx$

$= \displaystyle\int_1^2 \sqrt{\left(\dfrac{x^3}{2} + \dfrac{1}{2x^3}\right)^2}\, dx$

(Note that $0 < (x^3/2) + 2/(2x^3)$ for $1 \le x \le 2$.)

$s = \displaystyle\int_1^2 \left(\dfrac{x^3}{2} + \dfrac{1}{2x^3}\right) dx = \left[\dfrac{x^4}{8} - \dfrac{1}{4x^2}\right]_1^2$

$= \left(\dfrac{16}{8} - \dfrac{1}{16}\right) - \left(\dfrac{1}{8} - \dfrac{1}{4}\right) = \dfrac{32 - 1 - 2 + 4}{16} = \dfrac{33}{16}$

17. (a) The required graph of $y = 2\arctan x$ is shown in the figure.

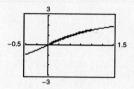

(b) $y = 2\arctan x$

$y' = \dfrac{2}{1 + x^2}$

$s = \displaystyle\int_0^1 \sqrt{1 + (y')^2}\, dx$

$= \displaystyle\int_0^1 \sqrt{1 + \left(\dfrac{2}{1 + x^2}\right)^2}\, dx$

(c) Using the integration capabilities of a graphing utility, $s \approx 1.871$.

25. The y-intercept of $y = \frac{1}{3}(x^{3/2} - 3x^{1/2} + 2)$ is $\left(0, \frac{2}{3}\right)$. Therefore, the fleeing object traveled from $(0, 0)$ to $\left(0, \frac{2}{3}\right)$, a distance of $\frac{2}{3}$.

$y = \dfrac{1}{3}(x^{3/2} - 3x^{1/2} + 2)$

$y' = \dfrac{1}{3}\left[\dfrac{3}{2}x^{1/2} - \dfrac{3}{2}x^{-1/2}\right]$

$= \dfrac{1}{2}\left(\sqrt{x} - \dfrac{1}{\sqrt{x}}\right) = \dfrac{x - 1}{2\sqrt{x}}$

$1 + (y')^2 = 1 + \left(\dfrac{x - 1}{2\sqrt{x}}\right)^2 = \dfrac{4x + (x^2 - 2x + 1)}{4x}$

$= \dfrac{x^2 + 2x + 1}{4x} = \dfrac{(x + 1)^2}{4x}$

Therefore, the distance traveled by the pursuer is given by

$s = \displaystyle\int_0^1 \sqrt{1 + (y')^2}\, dx$

$= \dfrac{1}{2}\displaystyle\int_0^1 \dfrac{x + 1}{\sqrt{x}}\, dx$

$= \dfrac{1}{2}\displaystyle\int_0^1 (x^{1/2} + x^{-1/2})\, dx$

$= \dfrac{1}{2}\left[\dfrac{2}{3}x^{3/2} + 2x^{1/2}\right]_0^1 = \dfrac{4}{3}.$

Thus the pursuer traveled a distance of $\frac{4}{3}$, twice the distance of the fleeing object.

33. $y = \dfrac{x^3}{6} + \dfrac{1}{2x}$

$y' = \dfrac{1}{2}x^2 - \dfrac{1}{2x^2} = \dfrac{x^4 - 1}{2x^2}$

$1 + (y')^2 = 1 + \left(\dfrac{x^4 - 1}{2x^2}\right)^2$

$= \dfrac{4x^4 + (x^8 - 2x^4 + 1)}{4x^2}$

$= \dfrac{x^8 + 2x^4 + 1}{4x^4} = \left(\dfrac{x^4 + 1}{2x^2}\right)^2$

$S = 2\pi\displaystyle\int_1^2 y\sqrt{1 + (y')^2}\, dx$

$= 2\pi\displaystyle\int_1^2 \left(\dfrac{x^3}{6} + \dfrac{1}{2x}\right)\left(\dfrac{x^4 + 1}{2x^2}\right) dx$

$= 2\pi\displaystyle\int_1^2 \left(\dfrac{x^5}{12} + \dfrac{x}{3} + \dfrac{1}{4x^3}\right) dx$

$= 2\pi\left[\dfrac{x^6}{72} + \dfrac{x^2}{6} - \dfrac{1}{8x^2}\right]_1^2 = \dfrac{47\pi}{16}$

35. $y = \sqrt[3]{x} + 2$

$$y' = \frac{1}{3}x^{-2/3}$$

$$S = 2\pi \int_1^8 x\sqrt{1 + (y')^2}\, dx$$

$$= 2\pi \int_1^8 x\sqrt{1 + \left(\frac{1}{3x^{2/3}}\right)^2}\, dx$$

$$= 2\pi \int_1^8 x\sqrt{\frac{9x^{4/3} + 1}{9x^{4/3}}}\, dx$$

$$= 2\pi \int_1^8 \frac{x}{3x^{2/3}}\sqrt{9x^{4/3} + 1}\, dx$$

$$= \frac{2\pi}{3}\left(\frac{1}{12}\right)\int_1^8 (9x^{4/3} + 1)^{1/2}(12x^{1/3})\, dx$$

$$= \frac{\pi}{18}\left[\frac{(9x^{4/3} + 1)^{3/2}}{3/2}\right]_1^8$$

$$= \frac{\pi}{27}\left[145\sqrt{145} - 10\sqrt{10}\right] \approx 199.48$$

39. The distance between the y-axis and the graph of the line $y = hx/r$ is

$$r(y) = g(y) = \frac{ry}{h}$$

and since $g'(y) = r/h$, the surface area is given by

$$S = 2\pi \int_0^h r(y)\sqrt{1 + [g'(y)]^2}\, dy$$

$$= 2\pi \int_0^h \left(\frac{ry}{h}\right)\sqrt{1 + \left(\frac{r}{h}\right)^2}\, dy$$

$$= \frac{2\pi r\sqrt{r^2 + h^2}}{h^2}\int_0^h y\, dy$$

$$= \frac{2\pi r\sqrt{r^2 + h^2}}{h^2}\left[\frac{1}{2}y^2\right]_0^h = \pi r\sqrt{r^2 + h^2}.$$

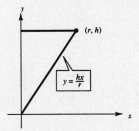

41. $y = \sqrt{9 - x^2}$

$$y' = \frac{-x}{\sqrt{9 - x^2}}$$

$$S = 2\pi \int_0^2 x\sqrt{1 + \left(\frac{-x}{\sqrt{9 - x^2}}\right)^2}\, dx$$

$$= 2\pi \int_0^2 x\sqrt{\frac{9}{9 - x^2}}\, dx$$

$$= 2\pi \int_0^2 3x(9 - x^2)^{-1/2}\, dx$$

$$= -3\pi \int_0^2 (9 - x^2)^{-1/2}(-2x)\, dx$$

$$= -3\pi \left[\frac{(9 - x^2)^{1/2}}{1/2}\right]_0^2$$

$$= -6\pi\left(\sqrt{5} - 3\right) = 6\pi\left(3 - \sqrt{5}\right) \approx 14.40$$

Section 6.5   Work

9. Let $F(x)$ be the force required to stretch a spring x units. By Hooke's Law, $F(x) = kx$. Since a force of 250 newtons stretches the spring 30 centimeters, it follows that

$$250 = k(30) \quad \text{or} \quad k = \frac{25}{3}.$$

Therefore, the work done by stretching the spring from 20 centimeters to 50 centimeters is

$$W = \int_{20}^{50} \underbrace{F(x)}_{\text{(force)}} \underbrace{dx}_{\text{(distance)}}$$

$$= \int_{20}^{50} \frac{25}{3} x \, dx$$

$$= \left[\frac{25}{6} x^2 \right]_{20}^{50}$$

$$= \frac{25}{6}(2500 - 400)$$

$$= 8750 \text{ newton} \cdot \text{centimeters} = 87.5 \text{ newton} \cdot \text{meters}.$$

17. Because the weight of a body varies inversely as the square of its distance from the center of the earth, the force $F(x)$ exerted by gravity is

$$F(x) = \frac{C}{x^2}.$$

Because the satellite weighs 10 tons on the surface of the earth and the radius of the earth is approximately 4000 miles, you have

$$10 = \frac{C}{(4000)^2} \implies C = 160,000,000.$$

(a) $W = \displaystyle\int_{4000}^{15,000} \frac{160,000,000}{x^2} \, dx$

$\quad = \left[-\dfrac{160,000,000}{x} \right]_{4000}^{15,000}$

$\quad = 2.93 \times 10^4 \text{ mi} \cdot \text{tons} \approx 3.10 \times 10^{11} \text{ ft} \cdot \text{lb}$

(b) $W = \displaystyle\int_{4000}^{26,000} \frac{160,000,000}{x^2} \, dx$

$\quad = \left[-\dfrac{160,000,000}{x} \right]_{4000}^{26,000}$

$\quad = 3.38 \times 10^4 \text{ mi} \cdot \text{tons} \approx 3.57 \times 10^{11} \text{ ft} \cdot \text{lb}$

23. A disc of water at height y must be lifted $(6 - y)$ feet to the top of the tank and has volume of $\pi x^2 \Delta y$. To find x in terms of y, solve the equation

$$\frac{6 - 0}{4 - 0} = \frac{y - 0}{x - 0}$$

$$x = \frac{2y}{3}.$$

Thus, the work done in moving the water over the top of the tank is

$$W = \int_0^6 \underbrace{(6 - y)}_{\text{(distance)}} \underbrace{\left[62.4\pi \left(\frac{2y}{3} \right)^2 dy \right]}_{\text{(force: weight of water)}}$$

$$= \frac{4}{9}(62.4)\pi \int_0^6 (6y^2 - y^3) \, dy$$

$$= \frac{83.2}{3}\pi \left[2y^3 - \frac{y^4}{4} \right]_0^6$$

$$= \frac{83.2}{3}\pi(432 - 324) = 2995.2\pi \text{ ft} \cdot \text{lb}.$$

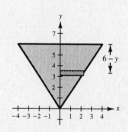

25. To fill the tank with water through a hole in the bottom, all the water is not moved to a height of 6 ft. Some of the water must be moved 6 ft, some 5 ft, some 4 ft, and so on. In general, the "disc" of water that must be moved y feet has a volume of

$$\pi x^2 \Delta y = \pi \left(\sqrt{36 - y^2}\right)^2 \Delta y = \pi(36 - y^2)\Delta y \text{ cubic feet.}$$

The weight of the disc of water is

$$62.4(\pi)(36 - y^2)\Delta y \text{ pounds.}$$

Thus, the work done in filling the tank from the bottom is

$$W = \int_0^6 \underbrace{(y)}_{\text{(distance)}} \underbrace{[62.4\pi(36 - y^2)\, dy]}_{\text{(force: weight of water)}}$$

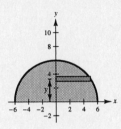

$$= 62.4\pi \int_0^6 (36y - y^3)\, dy$$

$$= 62.4\pi \left[18y^2 - \frac{y^4}{4} \right]_0^6$$

$$= 62.4\pi(648 - 324) = 20{,}217.6\pi \text{ ft} \cdot \text{lb.}$$

27. A layer of gasoline at height y (see figure) is lifted $\left(\frac{13}{2} - y\right)$ feet and has a volume $V = lwh = 4(2x)\Delta y$. To find x in terms of y, solve for x in the equation of the circle representing a cross-section of the tank and obtain

$$x^2 + y^2 = \frac{9}{4}$$

$$x^2 = \frac{9}{4} - y^2$$

$$x = \sqrt{\frac{9}{4} - y^2}.$$

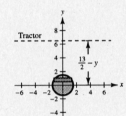

Thus the work done is

$$W = \int_{-1.5}^{1.5} \underbrace{\left(\frac{13}{2} - y\right)}_{\text{(distance)}} \underbrace{\left[42(4)\left(2\sqrt{\frac{9}{4} - y^2}\right) dy\right]}_{\text{(force: weight of water)}}$$

$$= 336\left[\frac{13}{2}\int_{-1.5}^{1.5} \sqrt{\frac{9}{4} - y^2}\, dy - \int_{-1.5}^{1.5} y\sqrt{\frac{9}{4} - y^2}\, dy \right].$$

The first integral represents the area of a semicircle of radius $\frac{3}{2}$ and the second integral is zero since the integrand is odd and the limits of integration are symmetric to the origin. Therefore,

$$W = 336\left(\frac{13}{2}\right)\left(\frac{1}{2}\right)(\pi)\left(\frac{3}{2}\right)^2 = 2457\pi \text{ ft} \cdot \text{lb.}$$

33. A small piece of chain of length Δy at height y must be moved so that it is y feet from the top. Therefore, the distance moved (as seen in the figure) is $15 - 2y$. (For example, the chain at an initial height of 7.5 is moved 0 ft.) The weight of a piece of chain of length Δy is

$$\frac{3 \text{ pounds}}{\text{foot}}\Delta y \text{ feet} = 3\Delta y \text{ pounds.}$$

Finally, since you are only moving a chain that has an initial height between 0 and 7.5 ft, we have

$$W = \int_0^{7.5} \underbrace{(15 - 2y)}_{\text{(distance)}} \underbrace{(3\, dy)}_{\text{(force)}}$$

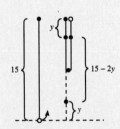

$$= 3\left[15y - y^2 \right]_0^{7.5} = 3(112.5 - 56.25) = 168.75 \text{ ft} \cdot \text{lb.}$$

39. Let $(x, 4)$ be a point on the line segment from $(-2, 4)$ to $(1, 4)$. The distance between the point $(x, 4)$ and $(2, 4)$ is $2 - x$. Since the two electrons repel each other with a force that varies inversely as the square of the distance between them, it follows that

$$F(x) = \frac{k}{(2 - x)^2}.$$

$$W = \int_{-2}^{1} \frac{k}{(2 - x)^2} \, dx$$

$$= \left[\frac{k}{2 - x} \right]_{-2}^{1} = k\left(1 - \frac{1}{4} \right) = \frac{3k}{4} \text{ units of work}$$

Section 6.6   Moments, Centers of Mass, and Centroids

7. The moment produced by each child is the mass of the child times the child's distance from the fulcrum. From the figure we have

left moment $= \frac{50}{32}x$ and right moment $= \frac{75}{32}(10 - x)$.

To balance the seesaw the moments must be equal.

$$\frac{50}{72}x = \frac{75}{32}(10 - x)$$

$$2x = 3(10 - x)$$

$$5x = 30 \implies x = 6 \text{ feet}$$

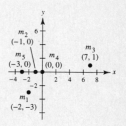

11. $\bar{x} = \dfrac{m_1 x_1 + m_2 x_2 + m_3 x_3 + m_4 x_4 + m_5 x_5}{m_1 + m_2 + m_3 + m_4 + m_5} = \dfrac{3(-2) + 4(-1) + 2(7) + 1(0) + 6(-3)}{3 + 4 + 2 + 1 + 6} = -\dfrac{7}{8}$

$\bar{y} = \dfrac{m_1 y_1 + m_2 y_2 + m_3 y_3 + m_4 y_4 + m_5 y_5}{m_1 + m_2 + m_3 + m_4 + m_5} = \dfrac{3(-3) + 4(0) + 2(1) + 1(0) + 6(0)}{3 + 4 + 2 + 1 + 6} = -\dfrac{7}{16}$

Therefore, $(\bar{x}, \bar{y}) = \left(-\dfrac{7}{8}, -\dfrac{7}{16} \right)$.

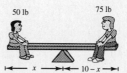

21. Since the region is symmetric with respect to the x-axis, we know that

$$M_x = 0 \quad \text{and} \quad \bar{y} = \frac{M_x}{m} = 0.$$

To find $\bar{x}$, observe that x is a function of y, and use the formula

$$\bar{x} = \frac{\displaystyle\int_a^b \left[\frac{f(y) + g(y)}{2} \right] [f(y) - g(y)] \, dy}{m} = \frac{M_y}{m}$$

where $f(y) = 4 - y^2$, $g(y) = 0$, $a = -2$, and $b = 2$.

$$m = \rho \int_{-2}^{2} (4 - y^2) \, dy = \rho \left[4y - \frac{y^3}{3} \right]_{-2}^{2} = \frac{32\rho}{3}$$

$$M_y = \frac{\rho}{2} \int_{-2}^{2} (4 - y^2)^2 \, dy$$

$$= \frac{\rho}{2} \int_{-2}^{2} (16 - 8y^2 + y^4) \, dy = \frac{\rho}{2} \left[16y - \frac{8y^3}{3} + \frac{y^5}{5} \right]_{-2}^{2}$$

$$= \frac{\rho}{2} \left[\left(32 - \frac{64}{3} + \frac{32}{5} \right) - \left(-32 + \frac{64}{3} - \frac{32}{5} \right) \right]$$

$$= \frac{\rho}{2} \left(\frac{512}{15} \right) = \frac{256\rho}{15}$$

$$\bar{x} = \frac{M_y}{m} = \frac{256\rho/15}{32\rho/3} = \frac{8}{5}$$

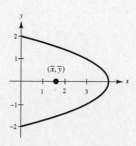

Therefore, $(\bar{x}, \bar{y}) = (8/5, 0)$.

25. The two graphs intersect at the points $(0, 0)$ and $(1, 1)$. The area is given by

$$A = \int_0^1 [f(x) - g(x)]\, dx$$

$$= \int_0^1 (x - x^2)\, dx = \left[\frac{1}{2}x^2 - \frac{1}{3}x^3\right]_0^1 = \frac{1}{6}.$$

$$M_x = \int_0^1 \left[\frac{f(x) + g(x)}{2}\right][f(x) - g(x)]\, dx$$

$$= \int_0^1 \left(\frac{x + x^2}{2}\right)(x - x^2)\, dx = \frac{1}{15}$$

$$M_y = \int_0^1 x[f(x) - g(x)]\, dx$$

$$= \int_0^1 x(x - x^2)\, dx = \frac{1}{12}$$

33. The equation of the line containing $(-a, 0)$ and (b, c) is $y = \left(\dfrac{c}{b + a}\right)(x + a)$.

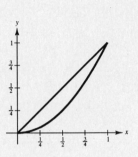

The equation of the line containing $(a, 0)$ and (b, c) is $y = \left(\dfrac{c}{b - a}\right)(x - a)$.

Since the area of the triangle is $A = \left(\dfrac{1}{2}\right)(2a)(c) = ac$,

we have

$$\bar{x} = \frac{\displaystyle\int_{-a}^{b} x\left(\frac{c}{b + a}\right)(x + a)\, dx + \int_{b}^{a} x\left(\frac{c}{b - a}\right)(x - a)\, dx}{ac}$$

$$= \frac{1}{ac}\left[\frac{c}{b + a}\int_{-a}^{b}(x^2 + ax)\, dx + \frac{c}{b - a}\int_{a}^{b}(x^2 - ax)\, dx\right]$$

$$= \frac{1}{ac}\left(\frac{c}{b + a}\left[\frac{x^3}{3} + \frac{ax^2}{2}\right]_{-a}^{b} + \frac{c}{b - a}\left[\frac{x^3}{3} - \frac{ax^2}{2}\right]_{b}^{a}\right)$$

$$= \frac{1}{ac}\left[\frac{c}{b + a}\left(\frac{b^3}{3} + \frac{ab^2}{2} + \frac{a^3}{3} - \frac{a^3}{2}\right) + \frac{c}{b - a}\left(\frac{a^3}{3} - \frac{a^3}{2} - \frac{b^3}{3} + \frac{ab^2}{2}\right)\right]$$

$$= \frac{2b^3 + 3ab^2 - a^3}{6a(b + a)} + \frac{-2b^3 + 3ab^2 - a^3}{6a(b - a)}$$

$$= \frac{(2b^2 + ab - a^2)(a + b)}{6a(b + a)} + \frac{(-2b^2 + ab + a^2)(b - a)}{6a(b - a)}$$

$$= \frac{2ab}{6a} = \frac{b}{3}$$

$$\bar{y} = \frac{\dfrac{1}{2}\displaystyle\int_{-a}^{b}\left[\frac{c}{b + a}(x + a)\right]^2 dx + \frac{1}{2}\int_{b}^{a}\left[\frac{c}{b - a}(x - a)\right]^2 dx}{ac}$$

$$= \frac{1}{2ac}\left\{\frac{c^2}{(b + a)^2}\left[\frac{(x + a)^3}{3}\right]_{-a}^{b} + \frac{c^2}{(b - a)^2}\left[\frac{(x - a)^3}{3}\right]_{b}^{a}\right\}$$

$$= \frac{1}{2ac}\left\{\frac{c^2}{(b + a)^2}\left[\frac{(b + a)^3}{3}\right] - \frac{c^2}{(b - a)^2}\left[\frac{(b - a)^3}{3}\right]\right\}$$

$$= \frac{1}{2ac}\left[\frac{c^2(b + a)}{3} - \frac{c^2(b - a)}{3}\right] = \frac{c}{3}.$$

From Exercise 66, Section P.2, we know that the point $(b/3, c/3)$ is the intersection of the medians of the triangle.

41. (a) From the symmetry of the glass it follows that $\bar{x} = 0$. To approximate the mass of the glass, use its symmetry with respect to the y-axis and Simpson's Rule with $n = 4$ to obtain

$$m = 2\rho \int_0^{40} y\,dx$$

$$\approx 2\rho \left[\frac{40 - 0}{3(4)}\right][30 + 4(29) + 2(26) + 4(20) + 0] \approx 1853.33\rho.$$

To approximate M_x, use the symmetry of the glass and Simpson's Rule with $n = 4$ to obtain

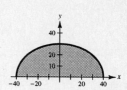

$$M_x = 2\rho \int_0^{40} \left(\frac{y}{2}\right) y\,dx = \rho \int_0^{40} y^2\,dx$$

$$\approx \rho \left[\frac{40 - 0}{3(4)}\right][30^2 + 4(29^2) + 2(26^2) + 4(20^2) + 0^2]$$

$$\approx 24{,}053.33\rho.$$

Therefore, $\bar{y} = \dfrac{M_x}{m} \approx \dfrac{24{,}053.33\rho}{1853.33\rho} \approx 12.98.$

(b) Using the regression capabilities to find a fourth-degree polynomial for the data yields

$$y = (-1.02 \times 10^{-5})x^4 - 0.0019x^2 + 29.28.$$

(c) Use the integration capabilities of a graphing utility to approximate the following integrals where y is the model in part (b).

$$m = 2\rho \int_0^{40} y\,dx \approx 1843.54\rho$$

$$M_x = 2\rho \int_0^{40} \left(\frac{y}{2}\right) y\,dx = \rho \int_0^{40} y^2\,dx \approx 23{,}697.68\rho$$

Therefore. $\bar{y} = \dfrac{M_x}{m} \approx \dfrac{23{,}697.68\rho}{1843.54\rho} \approx 12.85.$

43. Although a coordinate system may be introduced in many different ways, the one shown in the figure is a natural choice. Since both the circle and square have a uniform density, their masses are proportional to their areas, π and 4, respectively. (For simplicity, assume the density to be 1 unit of mass per 1 unit of area.) Again, because of the uniform density, both the circle and the square have their centers of mass at their geometrical centers, $(3, 0)$ and $(1, 0)$, respectively. Therefore, we can find the center of mass of the plate by considering a mass of π centered at $(3, 0)$ and a mass of 4 centered at $(1, 0)$. Thus,

$$\bar{x} = \frac{\pi(3) + 4(1)}{\pi + 4} = \frac{3\pi + 4}{\pi + 4} \quad \text{and} \quad \bar{y} = \frac{\pi(0) + 4(0)}{\pi + 4} = 0.$$

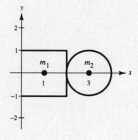

Section 6.7   Fluid Pressure and Fluid Force

7. The force against a representative rectangle of length $2x$ is

$$\Delta F = (\text{density})(\text{depth})(\text{area}) = (62.4)(3 - y)(2x\,\Delta y)$$

$$= (62.4)(3 - y)(2)\left(\frac{1}{3}y + 1\right)\Delta y.$$

Since y ranges from 0 to 3, the total force is

$$F = \int_0^3 (62.4)(3 - y)(2)\left(\frac{1}{3}y + 1\right) dy = 124.8\int_0^3 \left(3 - \frac{1}{3}y^2\right) dy$$

$$= 124.8\left[3y - \frac{1}{9}y^3\right]_0^3 = 748.8 \text{ lb.}$$

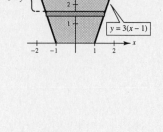

13. The force against a representative rectangle of length x is

$$\Delta F = (\text{density})(\text{depth})(\text{area})$$

$$= (1000)(12 - y)\left(-\frac{2}{3}y + 6\right)\Delta y$$

Since y ranges from 0 to 9, the total force is

$$F = \int_0^9 1000(12 - y)\left(-\frac{2}{3}y + 6\right) dy = 1000\int_0^9 \left(\frac{2}{3}y^2 - 14y + 72\right) dy$$

$$= 1000\left[\frac{2}{9}y^3 - 7y^2 + 72y\right]_0^9 = 243{,}000 \text{ kilograms.}$$

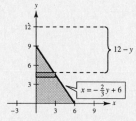

19. The force against a representative rectangle of length $2x$ is

$$\Delta F = (\text{density})(\text{depth})(\text{area}) = (42)(-y)(2x\,\Delta y)$$

$$= (42)(-y)(2)\left(\frac{1}{2}\right)(9 - 4y^2)^{1/2}\,\Delta y = -42y(9 - 4y^2)^{1/2}\,\Delta y.$$

Since y ranges from $-\frac{3}{2}$ to 0, the total force is

$$F = \int_{-3/2}^0 (-42)y(9 - 4y^2)^{1/2}\, dy = \frac{42}{8}\int_{-3/2}^0 (9 - 4y^2)^{1/2}(-8y)\, dy$$

$$= \frac{21}{4}\left(\frac{2}{3}\right)\left[(9 - 4y^2)^{3/2}\right]_{-3/2}^0 = 94.5 \text{ lb.}$$

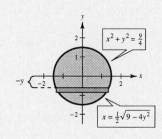

Review Exercises for Chapter 6

3. The graph of the region is shown in the figure. The area is given by

$$\text{Area} = \int_{-1}^1 \frac{1}{x^2 + 1}\, dx$$

$$= 2\int_0^1 \frac{1}{x^2 + 1}\, dx = 2\left[\arctan\right]_0^1$$

$$= 2\arctan 1 = \frac{\pi}{2}.$$

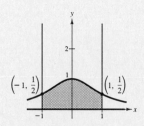

13. Solving the equation

$$\sqrt{x} + \sqrt{y} = 1$$

for y yields

$$y = \left(1 - \sqrt{x}\right)^2 \text{ for } 0 \le x \le 1.$$

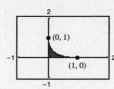

The graph of the region is shown in the figure. Using the integration capabilities of a graphing utility yields the following area.

$$\text{Area} = \int_0^1 \left(1 - \sqrt{x}\right)^2 dx = \frac{1}{6}$$

15. To find the y-intercepts of the graph, solve the equation

$$y^2 - 2y = 0$$

$$y(y - 2) = 0 \implies y = 0, 2.$$

The graph of the equation is a parabola opening to the right as shown in the figure. To find the area of the region by using vertical representative rectangles, begin by solving the equation for y.

$$x = y^2 - 2y \implies 0 = y^2 - 2y - x$$

Using the Quadratic Formula with $a = 1$, $b = -2$, and $c = -x$, we have

$$y = \frac{-(-2) \pm \sqrt{(-2)^2 - 4(1)(-x)}}{2(1)}$$

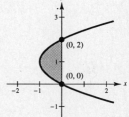

$$= \frac{1}{2}\left[2 \pm \sqrt{4 + 4x}\right] = 1 \pm \sqrt{1 + x}.$$

Therefore, the top half and bottom halves of the parabola are given by $f(x) = 1 + \sqrt{1 + x}$ and $g(x) = 1 - \sqrt{1 + x}$, respectively. Thus, the area of the region is given by

$$A = \int_{-1}^0 \left[\left(1 + \sqrt{1 + x}\right) - \left(1 - \sqrt{1 + x}\right)\right] dx = \int_{-1}^0 \left(2\sqrt{1 + x}\right) dx.$$

Using horizontal representative rectangles and the fact that $y^2 - 2y \le 0$ for $0 \le y \le 2$, we have

$$A = \int_0^2 \left[0 - (y^2 - 2y)\right] dy = \int_0^2 (-y^2 + 2y) \, dy = \left[\frac{-y^3}{3} + y^2\right]_0^2 = -\frac{8}{3} + 4 = \frac{4}{3}.$$

23. (a) Revolving the first quadrant portion of the ellipse about the y-axis will generate one-half the solid. Solving the equation for y yields.

$$y = \frac{3}{4}\sqrt{16 - x^2}.$$

Therefore, using the symmetry and the Shell Method we have

$$V = 2\pi \int_{-4}^4 x f(x) \, dx$$

$$= 4\pi \int_0^4 x\left(\frac{3}{4}\right)\sqrt{16 - x^2} \, dx$$

$$= 3\pi \int_0^4 x\sqrt{16 - x^2} \, dx$$

$$= -\frac{3\pi}{2} \int (16 - x^2)^{1/2}(-2x) \, dx$$

$$= -\frac{3\pi}{2}\left(\frac{2}{3}\right)\left[(16 - x^2)^{3/2}\right]_0^4$$

$$= -\pi[0 - 16^{3/2}] = 64\pi.$$

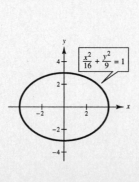

(b) Revolving the first quadrant portion of the ellipse about the x-axis will generate one-half the solid. Solving the equation for y^2 yields.

$$y^2 = \frac{9}{16}(16 - x^2).$$

Therefore, using the symmetry and the Disc Method we have

$$V = \pi \int_{-4}^4 [f(x)]^2 \, dx$$

$$= 2\pi \int_0^4 \frac{9}{16}(16 - x^2) \, dx$$

$$= \frac{9\pi}{8}\left[16x - \frac{1}{3}x^3\right]_0^4$$

$$= \frac{9\pi}{8}\left[64 - \frac{64}{3}\right] = 48\pi.$$

33. Since $f(x) = \frac{4}{5}x^{5/4}$ and $f'(x) = x^{1/4}$, we have

$$s = \int_0^4 \sqrt{1 + [f'(x)]^2}\, dx = \int_0^4 \sqrt{1 + \sqrt{x}}\, dx.$$

Let $u = \sqrt{1 + \sqrt{x}}$. Then $u^2 = 1 + \sqrt{x}, x = (u^2 - 1)^2$, and $dx = 2(u^2 - 1)(2u)\, du$. If $x = 0$, then $u = 1$, and if $x = 4$, then $u = \sqrt{3}$. Thus,

$$s = \int_0^4 \sqrt{1 + \sqrt{x}}\, dx$$

$$= \int_1^{\sqrt{3}} u(2)(u^2 - 1)(2u)\, du$$

$$= 4\int_0^{\sqrt{3}} (u^4 - u^2)\, du$$

$$= 4\left[\frac{u^5}{5} - \frac{u^3}{3}\right]_1^{\sqrt{3}} = \frac{8}{15}(6\sqrt{3} + 1) \approx 6.076.$$

41. A disk of water at height y must be lifted $(175 - y)$ feet and has a volume of $\pi\left(\frac{1}{3}\right)^2 \Delta y$. Thus, the work done in lifting the water to the top of the well is

$$W = \int_0^{150} \underbrace{(175 - y)}_{\text{(distance)}} \underbrace{\left[62.4\pi\left(\frac{1}{3}\right)^2 dy\right]}_{\text{(force: weight of water)}}$$

$$= \frac{62.4\pi}{9}\int_0^{150} (175 - y)\, dy$$

$$= \frac{62.4\pi}{9}\left[175y - \frac{1}{2}y^2\right]_0^{150}$$

$$= 104,000\pi \text{ ft} \cdot \text{lb} \approx 163.4 \text{ ft} \cdot \text{ton}.$$

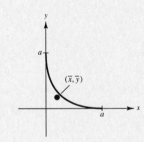

47. Solving the equation $\sqrt{x} + \sqrt{y} = \sqrt{a}$ for y yields $y = \left(\sqrt{a} - \sqrt{x}\right)^2$.

$$A = \int_0^a \left(\sqrt{a} - \sqrt{x}\right)^2 dx$$

$$= \int_0^a \left(a - 2\sqrt{a}x^{1/2} + x\right) dx$$

$$= \left[ax - \frac{4}{3}\sqrt{a}x^{3/2} + \frac{1}{2}x^2\right]_0^a = \frac{a^2}{6}$$

$$\bar{x} = \frac{1}{(a^2/6)}\int_0^a x\left(\sqrt{a} - \sqrt{x}\right)^2 dx$$

$$= \frac{6}{a^2}\int_0^a \left(ax - 2\sqrt{a}x^{3/2} + x^2\right) dx$$

$$= \frac{6}{a^2}\left[\frac{ax^2}{2} - \frac{4}{5}\sqrt{a}x^{5/2} + \frac{1}{3}x^3\right]_0^a = \frac{1}{5}a$$

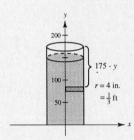

By symmetry, we have $\bar{y} = \frac{a}{5}$. Therefore, $(\bar{x}, \bar{y}) = \left(\frac{a}{5}, \frac{a}{5}\right)$.

51. Since the region is symmetrical to the x-axis, $\bar{y} = 0$. The mass of the region is the sum of the mass of the trapezoid and the mass of the semicircle.

$$m = \left(\frac{2 + 4}{2}\right)6\rho + \left(\frac{1}{2}\right)(\pi)(2)^2\rho = 18\rho + 2\pi\rho = 2\rho(9 + \pi)$$

The moment of the region about the y-axis is the sum of the moment of the trapezoid about the y-axis and the moment of the semicircle about the y-axis.

$$M_y = 2\rho\int_0^6 x\left(\frac{1}{6}x + 1\right) dx + 2\rho\int_6^8 x\sqrt{4 - (x - 6)^2}\, dx$$

—CONTINUED—

51. —CONTINUED—

Evaluating the first integral, we have

$$2\rho \int_0^6 x\left(\frac{1}{6}x + 1\right) dx = 2\rho \int_0^6 \left(\frac{1}{6}x^2 + x\right) dx$$

$$= 2\rho \left[\frac{1}{18}x^3 + \frac{1}{2}x^2\right]_0^6 = 2\rho(12 + 18) = 60\rho.$$

To evaluate the second integral, let $u = x - 6$. Then $x = u + 6$, $dx = du$, $u = 0$ when $x = 6$, and $u = 2$ when $x = 8$.

$$2\rho \int_6^8 x\sqrt{4 - (x - 6)^2}\, dx = 2\rho \int_0^2 (u + 6)\sqrt{4 - u^2}\, du$$

$$= 2\rho \int_0^2 u\sqrt{4 - u^2}\, du + 12\rho \int_0^2 \sqrt{4 - u^2}\, du$$

$$= 2\rho \left[\left(-\frac{1}{2}\right)\left(\frac{2}{3}\right)(4 - u^2)^{3/2}\right]_0^2 + 12\rho\left[\frac{\pi(2)^2}{4}\right]$$

$$= \frac{16\rho}{3} + 12\pi\rho = \frac{4\rho(4 + 9\pi)}{3}$$

Note that the second integral is the area of one-fourth of a circle of radius 2.

Therefore, $M_y = 60\rho + \dfrac{4\rho(4 + 9p)}{3}$, and

$$\bar{x} = \frac{M_y}{m} = \frac{60\rho + \dfrac{4\rho(4 + 9\pi)}{3}}{2\rho(9 + \pi)} = \frac{2(9\pi + 49)}{3(\pi + 9)}.$$

The centroid of the blade is $\left(\dfrac{2(9\pi + 49)}{3(\pi + 9)}, 0\right)$.

C H A P T E R 7
Integration Techniques, L'Hôpital's Rule, and Improper Integrals

CHAPTER 7
Integration Techniques, L'Hôpital's Rule, and Improper Integrals

Section 7.1   Basic Integration Rules
Solutions to Selected Odd-Numbered Exercises

5. If we let $u = 3x - 2$, then $du = 3 \, dx$ and

$$\int (3x - 2)^4 \, dx = \frac{1}{3} \int (3x - 2)^4 (3) \, dx = \frac{1}{3} \int u^4 \, du.$$

11. If we let $u = t^2$, then $du = 2t \, dt$ and

$$\int t \sin t^2 \, dt = \frac{1}{2} \int \sin(t^2)(2t) \, dt = \frac{1}{2} \int \sin u \, du.$$

19. If we let $u = -t^3 + 9t + 1$, then

$$du = (-3t^2 + 9) \, dt = -3(t^2 - 3) \, dt.$$

Thus,

$$\int \frac{t^2 - 3}{-t^3 + 9t + 1} \, dt = -\frac{1}{3} \int \frac{1}{-t^3 + 9t + 1} [-3(t^2 - 3)] \, dt$$

$$= -\frac{1}{3} \int \frac{1}{u} \, du = -\frac{1}{3} \ln|-t^3 + 9t + 1| + C.$$

27. If we let $u = 2\pi x^2$, then $du = 4\pi x \, dx$ and

$$\int x \cos 2\pi x^2 \, dx = \frac{1}{4\pi} \int (\cos 2\pi x^2)(4\pi x) \, dx$$

$$= \frac{1}{4\pi} \int \cos u \, du$$

$$= \frac{1}{4\pi} \sin 2\pi x^2 + C.$$

37. $\displaystyle \int \frac{2t - 1}{t^2 + 4} \, dt = \int \frac{2t}{t^2 + 4} \, dt - \int \frac{1}{t^2 + 2^2} \, dt$

$$= \ln(t^2 + 4) - \frac{1}{2} \arctan \frac{t}{2} + C$$

39. If we let $u = 2t - 1$, then $du = 2 \, dt$. Thus,

$$\int \frac{-1}{\sqrt{1 - (2t - 1)^2}} \, dt = \frac{-1}{2} \int \frac{2}{\sqrt{1 - (2t - 1)^2}} \, dt$$

$$= \frac{-1}{2} \int \frac{1}{\sqrt{a^2 - u^2}} \, du$$

$$= -\frac{1}{2} \arcsin(2t - 1) + C.$$

43. By completing the square we have

$$\int \frac{3}{\sqrt{6x - x^2}} \, dx = 3 \int \frac{1}{\sqrt{9 - (9 - 6x + x^2)}} \, dx$$

$$= 3 \int \frac{1}{\sqrt{9 - (x - 3)^2}} \, dx$$

$$= 3 \arcsin \left(\frac{x - 3}{3} \right) + C.$$

51. $(4 + \tan^2 x)y' = \sec^2 x$

$$y' = \frac{\sec^2 x}{4 + \tan^2 x}$$

$$\int y' \, dx = \int \frac{\sec^2 x}{4 + \tan^2 x} \, dx$$

If we let $u = \tan x$, then $du = \sec^2 x \, dx$ and

$$y = \int \frac{\sec^2 x}{4 + \tan^2 x} \, dx$$

$$= \int \frac{1}{a^2 + u^2} \, du$$

$$= \frac{1}{2} \arctan\left(\frac{\tan x}{2}\right) + C.$$

59. Let $a = 2$ and $u = 3x$. Then $du = 3 \, dx$ and

$$\int_0^{2/\sqrt{3}} \frac{1}{4 + 9x^2} \, dx = \frac{1}{3} \int_0^{2/\sqrt{3}} \frac{1}{2^2 + (3x)^2}(3) \, dx$$

$$= \frac{1}{6}\left[\arctan\left(\frac{3x}{2}\right) \right]_0^{2/\sqrt{3}}$$

$$= \frac{1}{6} \arctan \frac{3}{\sqrt{3}} = \frac{\pi}{18}.$$

69. The graph $y^2 = x^2(1 - x^2)$ is shown in the figure. The first quadrant portion of the curve is given by $y = x\sqrt{1 - x^2}$ where $0 \le x \le 1$. By symmetry we have

$$\text{Area} = 4 \int_0^1 x\sqrt{1 - x^2} \, dx$$

$$= -2 \int_0^1 \underbrace{(1 - x^2)^{1/2}}_{u^{1/2}} \underbrace{(-2x) \, dx}_{du} = -2\left[\frac{2}{3}(1 - x^2)^{3/2} \right]_0^1 = \frac{4}{3}.$$

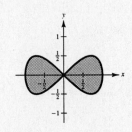

73. (a) By using the Shell Method we have

$$V = 2\pi \int_a^b p(x)h(x) \, dx$$

$$= 2\pi \int_0^b xe^{-x^2} \, dx$$

$$= -\pi \int_0^b e^{-x^2}(-2x \, dx) \qquad (\text{Let } u = -x^2, \, du = -2x \, dx)$$

$$= \left[-\pi e^{-x^2} \right]_0^b = \pi(1 - e^{-b^2}).$$

When $b = 1$, $V = \pi(1 - e^{-1}) \approx 1.986$.

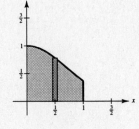

(b) Setting the expression for the volume equal to $4/3$ and solving for b yields

$$\pi(1 - e^{-b^2}) = \frac{4}{3}$$

$$3\pi - 3\pi e^{-b^2} = 4$$

$$e^{-b^2} = \frac{3\pi - 4}{3\pi}$$

$$-b^2 = \ln\left(\frac{3\pi - 4}{3\pi}\right),$$

$$b = \sqrt{\ln\left(\frac{3\pi - 4}{3\pi}\right)^{-1}} = \sqrt{\ln\left(\frac{3\pi}{3\pi - 4}\right)} \approx 0.743.$$

Section 7.2   Integration by Parts

9. Letting $u = x$ and $dv = e^{-2x} \, dx$, we have

$$dv = e^{-2x} \, dx \implies v = \int e^{-2x} \, dx = -\frac{1}{2} e^{-2x}$$

$$u = x \qquad \implies du = dx.$$

Therefore, we have

$$\int u \, dv = uv - \int v \, du$$

$$\int x e^{-2x} \, dx = x\left(-\frac{1}{2} e^{-2x}\right) - \int -\frac{1}{2} e^{-2x} \, dx$$

$$= -\frac{x}{2} e^{-2x} - \frac{1}{4} e^{-2x} + C$$

$$= -\frac{1}{4 e^{2x}} (2x + 1) + C.$$

15. Letting $u = \ln(t + 1)$ and $dv = t \, dt$, we have

$$dv = t \, dt \qquad \implies v = \int t \, dt = \frac{t^2}{2}$$

$$u = \ln(t + 1) \implies du = \frac{1}{t + 1} \, dt.$$

Therefore, we have

$$\int u \, dv = uv - \int v \, du$$

$$\int t \ln(t + 1) \, dt = \frac{t^2}{2} \ln(t + 1) - \frac{1}{2} \int \frac{t^2}{t + 1} \, dt$$

$$= \frac{t^2}{2} \ln(t + 1) - \frac{1}{2} \int \left(t - 1 + \frac{1}{t + 1}\right) dt$$

$$= \frac{t^2}{2} \ln(t + 1) - \frac{1}{2}\left[\frac{t^2}{2} - t + \ln(t + 1)\right] + C$$

$$= \frac{1}{4}[2(t^2 - 1) \ln|t + 1| - t^2 + 2t] + C.$$

23. Letting $u = x$ and $dv = \sqrt{x - 1} \, dx$, we have

$$dv = \sqrt{x - 1} \, dx \implies v = \int \sqrt{x - 1} \, dx = \frac{2}{3}(x - 1)^{3/2}$$

$$u = x \qquad \implies du = dx.$$

Therefore,

$$\int u \, dv = uv - \int v \, du$$

$$\int x \sqrt{x - 1} \, dx = \frac{2}{3} x(x - 1)^{3/2} - \int \frac{2}{3}(x - 1)^{3/2} \, dx$$

$$= \frac{2}{3} x(x - 1)^{3/2} - \frac{4}{15}(x - 1)^{5/2} + C$$

$$= \frac{2}{15}(x - 1)^{3/2}[5x - 2(x - 1)] + C$$

$$= \frac{2}{15}(x - 1)^{3/2}(3x + 2) + C.$$

27. Letting $u = \arctan x$ and $dv = dx$, we have

$$dv = dx \qquad \implies v = \int dx = x$$

$$u = \arctan x \implies du = \frac{1}{1 + x^2} \, dx.$$

Therefore, we have

$$\int \arctan x \, dx = x \arctan x - \int \frac{x}{1 + x^2} \, dx$$

$$= x \arctan x - \frac{1}{2} \ln(1 + x^2) + C.$$

29. Letting $u = e^{2x}$ and $dv = \sin\,dx$, we have

$$dv = \sin x\,dx \implies v = \int \sin x\,dx = -\cos x$$

$$u = e^{2x} \implies du = 2e^{2x}\,dx.$$

$$\int e^{2x} \sin x\,dx = -e^{2x} \cos x + \int \cos x(2e^{2x})\,dx.$$

Using integration by parts again, we have

$$dv = \cos x\,dx \implies v = \int \cos x\,dx = \sin x$$

$$u = 2e^{2x} \implies du = 4e^{2x}\,dx.$$

$$\int e^{2x} \sin x\,dx = -e^{2x} \cos x + 2e^{2x} \sin x - \int 4e^{2x} \sin x\,dx$$

Adding the integral in the right hand member of the equation to both members of the equation yields

$$5\int e^{2x} \sin x\,dx = -e^{2x} \cos x + 2e^{2x} \sin x + C_1.$$

Finally, dividing both members of the equation by 5 yields

$$\int e^{2x} \sin x\,dx = \frac{e^{2x}}{5}(2 \sin x - \cos x) + C.$$

33. Use integration by parts twice.

(1) $dv = \dfrac{1}{\sqrt{2 + 3t}}\,dt \implies v = \displaystyle\int (2 + 3t)^{-1/2}\,dt = \dfrac{2}{3}\sqrt{2 + 3t}$

$\quad u = t^2 \implies du = 2t\,dt$

(2) $dv = \sqrt{2 + 3t}\,dt \implies v = \displaystyle\int (2 + 3t)^{1/2}\,dt = \dfrac{2}{9}(2 + 3t)^{3/2}$

$\quad u = t \implies du = dt$

$$y = \int \frac{t^2}{\sqrt{2 + 3t}}\,dt = \frac{2t^2\sqrt{2 + 3t}}{3} - \frac{4}{3}\int t\sqrt{2 + 3t}\,dt$$

$$= \frac{2t^2\sqrt{2 + 3t}}{3} - \frac{4}{3}\left[\frac{2t}{9}(2 + 3t)^{3/2} - \frac{2}{9}\int (2 + 3t)^{3/2}\,dt\right]$$

$$= \frac{2t^2\sqrt{2 + 3t}}{3} - \frac{8t}{27}(2 + 3t)^{3/2} + \frac{16}{405}(2 + 3t)^{5/2} + C$$

$$= \frac{2\sqrt{2 + 3t}}{405}(27t^2 - 24t + 32) + C$$

43. $dv = \cos\,dx \implies v = \displaystyle\int \cos\,dx = \sin x$

$\quad u = x \implies du = dx$

$$\int x \cos x\,dx = x \sin x - \int \sin x\,dx$$

$$= x \sin x + \cos x + C$$

Finally,

$$\int_0^{\pi/2} x \cos x\,dx = \left[x \sin x + \cos x\right]_0^{\pi/2} = \frac{\pi}{2} - 1.$$

45. Begin by letting $u = x^2$ and $dv = v' \, dx = e^{2x} \, dx$. We next create a table consisting of three columns as follows.

Alternate Signs	u and its Derivatives	v' and its Antiderivatives
$+$	x^2	e^{2x}
$-$	$2x$	$\frac{1}{2}e^{2x}$
$+$	2	$\frac{1}{4}e^{2x}$
$-$	0	$\frac{1}{8}e^{2x}$

Finally, the solution is given by multiplying the signed products of the diagonal entries of the table to obtain

$$\int x^2 e^{2x} \, dx = \frac{1}{2}x^2 e^{2x} - 2x\left(\frac{1}{4}\right)e^{2x} + 2\left(\frac{1}{8}\right)e^{2x} + C$$

$$= \frac{e^{2x}}{4}(2x^2 - 2x + 1) + C.$$

55. (a) $dv = \sqrt{2x - 3} \, dx \implies v = \int (2x - 3)^{1/2} \, dx = \frac{1}{3}(2x - 3)^{3/2}$

$u = 2x \qquad \implies du = 2 \, dx$

$$\int 2x\sqrt{2x - 3} \, dx = \frac{2}{3}x(2x - 3)^{3/2} - \frac{2}{3}\int (2x - 3)^{3/2} \, dx$$

$$= \frac{2}{3}x(2x - 3)^{3/2} - \frac{2}{15}(2x - 3)^{5/2} + C$$

$$= \frac{2}{15}(2x - 3)^{3/2}(3x + 3) + C = \frac{2}{5}(2x - 3)^{3/2}(x + 1) + C$$

(b) $u = \sqrt{2x - 3} \implies x = \frac{1}{2}(u^2 + 3)$ and $dx = u \, du$

$$\int 2x\sqrt{2x - 3} \, dx = \int 2\left(\frac{1}{2}\right)(u^2 + 3)(u)(u) \, du$$

$$= \int (u^4 + 3u^2) \, du$$

$$= \frac{1}{5}u^5 + u^3 + C$$

$$= \frac{1}{5}u^3(u^2 + 5) + C$$

$$= \frac{1}{5}(2x - 3)^{3/2}[(2x - 3) + 5] + C$$

$$= \frac{2}{5}(2x - 3)^{3/2}(x + 1) + C$$

63. $dv = x^n \, dx \implies v = \int x^n \, dx = \frac{x^{n+1}}{n + 1}$

$u = \ln x \implies du = \frac{1}{x} \, dx$

$$\int x^n \ln x \, dx = \frac{x^{n+1}}{n + 1} \ln x - \int \frac{x^{n+1}}{n + 1}\left(\frac{1}{x}\right) dx$$

$$= \frac{x^{n+1}}{n + 1} \ln x - \frac{1}{n + 1}\int x^n \, dx$$

$$= \frac{x^{n+1}}{n + 1} \ln x - \frac{x^{n+1}}{(n + 1)^2} + C$$

$$= \frac{x^{n+1}}{(n + 1)^2}[-1 + (n + 1) \ln x] + C$$

75. (a) From the figure we have

$$A = \int_1^e \ln x \, dx.$$

$$dv = dx \implies v = x$$

$$u = \ln x \implies du = \frac{1}{x} dx.$$

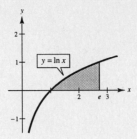

Therefore,

$$A = \int_1^e \ln x \, dx = \Big[x \ln x \Big]_1^e - \int_1^e dx$$

$$= \Big[x \ln x - x \Big]_1^e = 1.$$

(b) Using the Disc Method we have

$$V = \pi \int_1^e (\ln x)^2 \, dx.$$

Let,

$$dv = dx \implies v = x$$

$$u = (\ln x)^2 \implies du = \frac{2 \ln x}{x} dx.$$

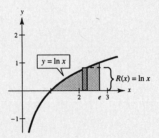

$R(x) = \ln x$

Therefore,

$$\int (\ln x)^2 \, dx = x(\ln x)^2 - 2\int \ln x \, dx$$

$$= x(\ln x)^2 - 2x(\ln x - 1). \quad \text{From part (a)}$$

Finally,

$$V = \pi \int_1^e (\ln x)^2 \, dx$$

$$= \pi \Big[x(\ln x)^2 - 2x(\ln x - 1) \Big]_1^e = \pi(e - 2) \approx 2.257.$$

(c) Using the Shell Method we have

$$V = 2\pi \int_1^e x \ln x \, dx.$$

Let,

$$dv = x \, dx \implies v = \frac{x^2}{2}$$

$$u = \ln x \implies du = \frac{1}{x} dx.$$

$h(x) = \ln x$

$p(x) = x$

Therefore,

$$\int x \ln x \, dx = \frac{x^2}{2} \ln x - \frac{1}{2}\int x \, dx = \frac{x^2}{2} \ln x - \frac{x^2}{4}.$$

Finally,

$$V = 2\pi \int_1^e x \ln x \, dx$$

$$= 2\pi \Big[\frac{x^2}{4}(2 \ln x - 1) \Big]_1^e = \frac{\pi}{2}(e^2 + 1) \approx 13.177.$$

—CONTINUED—

75. —CONTINUED—

(d) $\bar{x} = \dfrac{1}{A}\displaystyle\int_1^e x \ln x \, dx$

$\qquad = \left[\dfrac{x^2}{4}(2 \ln x - 1)\right]_1^e$ From part (c)

$\qquad = \dfrac{1}{4}(e^2 + 1) \approx 2.097$

$\bar{y} = \dfrac{1}{2A}\displaystyle\int_1^e (\ln x)^2 \, dx$

$\qquad = \dfrac{1}{2}\left[x(\ln x)^2 - 2x(\ln x - 1)\right]_1^e$ From part (b)

$\qquad = \dfrac{e - 2}{2} \approx 0.359$

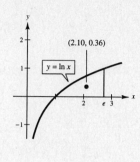

83. Let $u = x, \ dv = \sin\left(\dfrac{n\pi}{2}x\right) dx, \ du = dx, \ v = -\dfrac{2}{n\pi}\cos\left(\dfrac{n\pi}{2}x\right).$

$I_1 = \displaystyle\int_0^1 x \sin\left(\dfrac{n\pi}{2}x\right) dx = \left[\dfrac{-2x}{n\pi}\cos\left(\dfrac{n\pi}{2}x\right)\right]_0^1 + \dfrac{2}{n\pi}\displaystyle\int_0^1 \cos\left(\dfrac{n\pi}{2}x\right) dx$

$\qquad = -\dfrac{2}{n\pi}\cos\left(\dfrac{n\pi}{2}\right) + \left[\left(\dfrac{2}{n\pi}\right)^2 \sin\left(\dfrac{n\pi}{2}x\right)\right]_0^1$

$\qquad = -\dfrac{2}{n\pi}\cos\left(\dfrac{n\pi}{2}\right) + \left(\dfrac{2}{n\pi}\right)^2 \sin\left(\dfrac{n\pi}{2}\right)$

Let $u = (-x + 2), \ dv = \sin\left(\dfrac{n\pi}{2}x\right) dx, \ du = -dx, \ v = -\dfrac{2}{n\pi}\cos\left(\dfrac{n\pi}{2}x\right).$

$I_2 = \displaystyle\int_1^2 (-x + 2) \sin\left(\dfrac{n\pi}{2}x\right) dx = \left[\dfrac{-2(-x + 2)}{n\pi}\cos\left(\dfrac{n\pi}{2}x\right)\right]_1^2 - \dfrac{2}{n\pi}\displaystyle\int_1^2 \cos\left(\dfrac{n\pi}{2}x\right) dx$

$\qquad = \dfrac{2}{n\pi}\cos\left(\dfrac{n\pi}{2}\right) - \left[\left(\dfrac{2}{n\pi}\right)^2 \sin\left(\dfrac{n\pi}{2}x\right)\right]_1^2$

$\qquad = \dfrac{2}{n\pi}\cos\left(\dfrac{n\pi}{2}\right) + \left(\dfrac{2}{n\pi}\right)^2 \sin\left(\dfrac{n\pi}{2}\right)$

$h(I_1 + I_2) = b_n = h\left[\left(\dfrac{2}{n\pi}\right)^2 \sin\left(\dfrac{n\pi}{2}\right) + \left(\dfrac{2}{n\pi}\right)^2 \sin\left(\dfrac{n\pi}{2}\right)\right] = \dfrac{8h}{(n\pi)^2}\sin\left(\dfrac{n\pi}{2}\right)$

(Note that $b_n = 0$ when n is even.)

Section 7.3   Trigonometric Integrals

7. $\displaystyle\int \sin^5 x \cos^2 x\, dx = \int \sin x (\sin^2 x)^2 \cos^2 x\, dx$

$$= \int \sin x (1 - \cos^2 x)^2 \cos^2 x\, dx$$

$$= -\int (\cos^2 x - 2\cos^4 x + \cos^6 x)(-\sin x)\, dx$$

$$= -\frac{1}{3}\cos^3 x + \frac{2}{5}\cos^5 x - \frac{1}{7}\cos^7 x + C$$

11. Use Integration By Part by letting $u = x$ and $dv = \sin^2 x\, dx$. Then $du = dx$ and

$$v = \int \sin^2 x\, dx$$

$$= \int \frac{1 - \cos 2x}{2}\, dx$$

$$= \frac{1}{2}\left(x - \frac{1}{2}\sin 2x\right) = \frac{1}{4}(2x - \sin 2x).$$

Therefore,

$$\int u\, dv = uv - \int v\, du$$

$$\int x \sin 2x\, dx = \frac{1}{4}x(2x - \sin 2x) - \frac{1}{4}\int (2x - \sin 2x)\, dx$$

$$= \frac{1}{4}(2x - \sin 2x) - \frac{1}{4}\left(x^2 + \frac{1}{2}\cos 2x\right) + C$$

$$= \frac{1}{8}(2x^2 - 2x \sin 2x - \cos 2x) + C.$$

23. $\displaystyle\int \tan^5 \frac{x}{4}\, dx = \int \tan^2 \frac{x}{4} \tan^3 \frac{x}{4}\, dx$

$$= \int \left(\sec^2 \frac{x}{4} - 1\right) \tan^3 \frac{x}{4}\, dx$$

$$= \int \tan^3 \frac{x}{4} \sec^2 \frac{x}{4}\, dx - \int \tan^3 \frac{x}{4}\, dx$$

$$= \tan^4 \frac{x}{4} - \int \tan^2 \frac{x}{4} \tan \frac{x}{4}\, dx$$

$$= \tan^4 \frac{x}{4} - \int \left(\sec^2 \frac{x}{4} - 1\right) \tan \frac{x}{4}\, dx$$

$$= \tan^4 \frac{x}{4} - \int \tan \frac{x}{4} \sec^2 \frac{x}{4}\, dx + \int \tan \frac{x}{4}\, dx$$

$$= \tan^4 \frac{x}{4} - 2\tan^2 \frac{x}{4} - 4\ln\left|\cos \frac{x}{4}\right| + C$$

29. $\displaystyle\int \sec^6 4x \tan 4x \, dx = \int (\sec^2 4x)(\sec^2 4x)^2 \tan 4x \, dx$

$$= \frac{1}{4}\int \tan 4x(\tan^2 4x + 1)^2(4 \sec^2 4x) \, dx$$

$$= \frac{1}{4}\int (\tan^5 4x + 2 \tan^3 4x + \tan 4x)(4 \sec^2 4x) \, dx$$

$$= \frac{1}{4}\left[\frac{\tan^6 4x}{6} + \frac{\tan^4 4x}{2} + \frac{\tan^2 4x}{2}\right] + C$$

$$= \frac{1}{24}(\tan^2 4x)(\tan^4 4x + 3 \tan^2 4x + 3) + C$$

or

$$\int \sec^6 4x \tan 4x \, dx = \frac{1}{4}\int \sec^5 4x(4 \sec 4x \tan 4x) \, dx$$

$$= \frac{1}{24} \sec^6 4x + C_1$$

(See Exercise 71 for a comparison of the two methods.)

33. $\displaystyle\frac{dr}{d\theta} = \sin^4 \pi\theta$

$$\int \frac{dr}{d\theta} \, d\theta = \int \sin^4 \pi\theta \, d\theta$$

$$r = \int (\sin^2 \pi\theta)^2 \, d\theta$$

$$= \int \left(\frac{1 - \cos 2\pi\theta}{2}\right)^2 \, d\theta$$

$$= \frac{1}{4}\int (1 - 2\cos 2\pi\theta + \cos^2 2\pi\theta) \, d\theta$$

$$= \frac{1}{4}\int \left(1 - 2\cos 2\pi\theta + \frac{1 + \cos 4\pi\theta}{2}\right) \, d\theta$$

$$= \frac{1}{8}\int (3 - 4\cos 2\pi\theta + \cos 4\pi\theta) \, d\theta$$

$$= \frac{1}{8}\left(3\theta - \frac{2}{\pi}\sin 2\pi\theta + \frac{1}{4\pi}\sin 4\pi\theta\right) + C$$

$$= \frac{1}{32\pi}(12\pi\theta - 8\sin 2\pi\theta + \sin 4\pi\theta) + C$$

39. Using the trigonometric identity

$$\sin u \cos v = \frac{1}{2}[\sin(u + v) + \sin(u - v)],$$

we have

$$\int \sin 3x \cos 2x \, dx = \int \frac{1}{2}(\sin 5x + \sin x) \, dx$$

$$= \frac{1}{2}\left(\frac{1}{5}\right)\int \sin 5x(5) \, dx + \frac{1}{2}\int \sin x \, dx$$

$$= -\frac{1}{10}\cos 5x - \frac{1}{2}\cos x + C$$

$$= -\frac{1}{10}(\cos 5x + 5\cos x) + C.$$

47. $\displaystyle\int \frac{\cot^2 t}{\csc t} \, dt = \int \frac{\csc^2 t - 1}{\csc t} \, dt$

$$= \int (\csc t - \sin t) \, dt$$

$$= \ln|\csc t - \cot t| + \cos t + C$$

55. $\displaystyle\int_0^{\pi/4} \tan^3 x \, dx = \int_0^{\pi/4} (\sec^2 x - 1)(\tan x) \, dx$

$$= \int_0^{\pi/4} \tan x \sec^2 x \, dx - \int_0^{\pi/4} \tan x \, dx$$

$$= \left[\frac{1}{2}\tan^2 x + \ln|\cos x|\right]_0^{\pi/4} = \frac{1}{2}(1 - \ln 2)$$

65. Using a symbolic integration utility to perform the integration yields

$$\int \sec^5 \pi x \tan \pi x \, dx = \frac{1}{5\pi}\sec^5 \pi x + C.$$

The graphs of the antiderivative for $C = 1$ and $C = -2$ are shown in the figure.

71. (a) Method 1: Since there is an odd power of the tangent, write

$$\int \sec^4 3x \tan^3 3x \, dx = \int \sec^3 3x \tan^2 3x (\sec 3x \tan 3x) \, dx$$

$$= \int \sec^3 3x (\sec^2 3x - 1) \sec 3x \tan 3x \, dx$$

$$= \frac{1}{3} \int (\sec^5 3x - \sec^3 3x)(3 \sec 3x \tan 3x \, dx)$$

$$= \frac{1}{3} \left(\frac{1}{6} \sec^6 3x - \frac{1}{4} \sec^4 3x \right) + C.$$

(b) The graphs of Method 1 and Method 2 with $C = 0$ are shown in the figure.

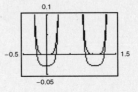

Method 2: Since the power of the secant is even, write

$$\int \sec^4 3x \tan^3 3x \, dx = \int \sec^2 3x \tan^3 3x (\sec^2 3x) \, dx$$

$$= \int (1 + \tan^2 3x) \tan^3 3x (\sec^2 3x) \, dx$$

$$= \frac{1}{3} \int (\tan^3 3x + \tan^5 3x)(3 \sec^2 3x) \, dx$$

$$= \frac{1}{3} \left(\frac{\tan^4 3x}{4} + \frac{\tan^6 3x}{6} \right) + C.$$

(c) Comparing the results of the two methods, we have

$$\frac{1}{3} \left[\frac{1}{4} \tan^4 3x + \frac{1}{6} \tan^6 3x \right] + C = \frac{1}{3} \left[\frac{1}{4} (\sec^2 3x - 1)^2 + \frac{1}{6} (\sec^2 3x - 1)^3 \right] + C$$

$$= \frac{1}{3} \left[\frac{1}{4} (\sec^4 3x - 2 \sec^2 3x + 1) + \frac{1}{6} (\sec^6 3x - 3 \sec^4 3x + 3 \sec^2 3x - 1) \right] + C$$

$$= \frac{1}{3} \left[\frac{1}{6} \sec^6 3x - \frac{1}{4} \sec^4 3x + \frac{1}{4} - \frac{1}{6} \right] + C.$$

Therefore, the results differ only by the constant $\frac{1}{3} \left(\frac{1}{4} - \frac{1}{6} \right)$.

75. (a) $V = \pi \int_0^\pi R^2 \, dx = \pi \int_0^\pi \sin^2 x \, dx$

$$= \pi \int_0^\pi \frac{1 - \cos 2x}{2} \, dx$$

$$= \frac{\pi}{2} \int_0^\pi (1 - \cos 2x) \, dx$$

$$= \frac{\pi}{2} \left[x - \frac{1}{2} \sin 2x \right]_0^\pi = \frac{\pi}{2} \left[\pi - \frac{1}{2}(0) - 0 \right] = \frac{\pi^2}{2}$$

(b) By symmetry $\bar{x} = \frac{\pi}{2}$.

The area of the region is given by

$$A = \int_0^\pi \sin x \, dx = \left[-\cos x \right]_0^\pi = 2.$$

Therefore,

$$\bar{y} = \frac{1}{2A} \int_0^\pi \sin^2 x \, dx$$

$$= \frac{1}{8} \int_0^\pi (1 - \cos 2x) \, dx = \left[\frac{1}{8} \left(x - \frac{1}{2} \sin 2x \right) \right]_0^\pi = \frac{\pi}{8}.$$

Thus,

$$(\bar{x}, \bar{y}) = \left(\frac{\pi}{2}, \frac{\pi}{8} \right).$$

79. Let $dv = \cos^m x \sin x \, dx$ and $u = \sin^{n-1} x$. Then $v = (-\cos^{m+1} x)/(m + 1)$ and $du = (n - 1) \sin^{n-2} x(\cos x) \, dx$. Therefore,

$$\int \cos^m x \sin^n x \, dx = -\frac{\sin^{n-1} x \cos^{m+1} x}{m + 1} + \frac{n - 1}{m + 1} \int \sin^{n-2} x \cos^{m+2} x \, dx$$

$$= -\frac{\sin^{n-1} x \cos^{m+1} x}{m + 1} + \frac{n - 1}{m + 1} \int \sin^{n-2} x \cos^m x(1 - \sin^2 x) \, dx$$

$$= -\frac{\sin^{n-1} x \cos^{m+1} x}{m + 1} + \frac{n - 1}{m + 1} \int \sin^{n-2} x \cos^m x \, dx - \frac{n - 1}{m + 1} \int \sin^n x \cos^m x \, dx.$$

Now observe that the last integral is a multiple of the original. Adding yields

$$\frac{m + n}{m + 1} \int \cos^m x \sin^n x \, dx = -\frac{\sin^{n-1} x \cos^{m+1} x}{m + 1} + \frac{n - 1}{m + 1} \int \cos^m x \sin^{n-2} x \, dx$$

$$\int \cos^m x \sin^n x \, dx = -\frac{\cos^{m+1} x \sin^{n-1} x}{m + n} + \frac{n - 1}{m + n} \int \cos^m x \sin^{n-2} x \, dx.$$

Section 7.4   Trigonometric Substitution

7. Let $x = 5 \sin \theta$. Then $\sqrt{25 - x^2} = 5 \cos \theta$ and $dx = 5 \cos \theta \, d\theta$. Thus,

$$\int \frac{\sqrt{25 - x^2}}{x} \, dx = \int \frac{5 \cos \theta}{5 \sin \theta} 5 \cos \theta \, d\theta$$

$$= 5 \int \frac{\cos^2 \theta}{\sin \theta} \, d\theta$$

$$= 5 \int \frac{1 - \sin^2 \theta}{\sin \theta} \, d\theta$$

$$= 5 \int (\csc \theta - \sin \theta) \, d\theta$$

$$= 5(\ln|\csc \theta - \cot \theta| + \cos \theta) + C$$

$$= 5\left(\ln\left|\frac{5}{x} - \frac{\sqrt{25 - x^2}}{x}\right| + \frac{\sqrt{25 - x^2}}{5}\right) + C$$

$$= 5 \ln\left|\frac{5 - \sqrt{25 - x^2}}{x}\right| + \sqrt{25 - x^2} + C.$$

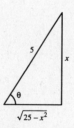

11. Let $x = 2 \sec \theta$. Then $\sqrt{x^2 - 4} = 2 \tan \theta$ and $dx = 2 \sec \theta \tan \theta \, d\theta$. Thus,

$$\int x^3 \sqrt{x^2 - 4} \, dx = \int (2 \sec \theta)^3 (2 \tan \theta)(2 \sec \theta \tan \theta) \, d\theta$$

$$= 32 \int \sec^4 \theta \tan^2 \theta \, d\theta$$

$$= 32 \int \sec^2 \theta \tan^2 \theta \sec^2 \theta \, d\theta$$

$$= 32 \int (\tan^2 \theta + 1) \tan^2 \theta \sec^2 \theta \, d\theta$$

$$= 32 \int (\tan^4 \theta + \tan^2 \theta) \sec^2 \theta \, d\theta$$

$$= 32\left[\frac{1}{5} \tan^5 \theta + \frac{1}{3} \tan^3 \theta\right] + C$$

$$= 32\left[\frac{1}{5}\left(\frac{\sqrt{x^2 - 4}}{2}\right)^5 + \frac{1}{3}\left(\frac{\sqrt{x^2 - 4}}{2}\right)^3\right] + C$$

$$= \frac{1}{15}(x^2 - 4)^{3/2}(3x^2 + 8) + C.$$

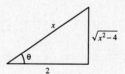

15. Let $x = \tan\theta$. Then $1 + x^2 = \sec^2\theta$ and $dx = \sec^2\theta\, d\theta$. Thus,

$$\int \frac{1}{(1+x^2)^2}\, dx = \int \frac{1}{\sec^4\theta}(\sec^2\theta)\, d\theta = \int \cos^2\theta\, d\theta$$

$$= \frac{1}{2}\int (1 + \cos 2\theta)\, d\theta$$

$$= \frac{1}{2}\left(\theta + \frac{1}{2}\sin 2\theta\right) + C$$

$$= \frac{1}{2}(\theta + \sin\theta\cos\theta) + C$$

$$= \frac{1}{2}\left(\arctan x + \frac{x}{\sqrt{1+x^2}} \cdot \frac{1}{\sqrt{1+x^2}}\right) + C$$

$$= \frac{1}{2}\left(\arctan x + \frac{x}{1+x^2}\right) + C.$$

27. Let $2x = 3\tan\theta$. Then $x = \frac{3}{2}\tan\theta$, $\sqrt{4x^2+9} = 3\sec\theta$, and $dx = \frac{3}{2}\sec^2\theta\, d\theta$.

$$\int \frac{1}{x\sqrt{4x^2+9}}\, dx = \int \frac{(3/2)\sec^2\theta\, d\theta}{(3/2)\tan\theta(3\sec\theta)} = \frac{1}{3}\int \csc\theta\, d\theta$$

$$= \frac{1}{3}\ln|\csc\theta - \cot\theta| + C$$

$$= \frac{1}{3}\ln\left|\frac{\sqrt{4x^2+9}-3}{2x}\right| + C$$

$$= -\frac{1}{3}\ln\left|\frac{2x}{\sqrt{4x^2+9}-3}\right| + C$$

$$= -\frac{1}{3}\left|\ln\left(\frac{2x}{\sqrt{4x^2+9}-3}\right)\left(\frac{\sqrt{4x^2+9}+3}{\sqrt{4x^2+9}+3}\right)\right| + C$$

$$= -\frac{1}{3}\ln\left|\frac{3+\sqrt{4x^2+9}}{2x}\right| + C$$

35. $\displaystyle \int \frac{1}{4+4x^2+x^4}\, dx = \int \frac{1}{(2+x^2)^2}\, dx$

Let $x = \sqrt{2}\tan\theta$. Then $2 + x^2 = 2\sec^2\theta$ and $dx = \sqrt{2}\sec^2\theta\, d\theta$. Thus,

$$\int \frac{1}{(2+x^2)^2}\, dx = \int \frac{\sqrt{2}\sec^2\theta\, d\theta}{(2\sec^2\theta)^2} = \frac{\sqrt{2}}{4}\int \cos^2\theta\, d\theta$$

$$= \frac{\sqrt{2}}{4}\int \frac{1+\cos 2\theta}{2}\, d\theta = \frac{\sqrt{2}}{8}\left(\theta + \frac{1}{2}\sin 2\theta\right) + C$$

$$= \frac{\sqrt{2}}{8}(\theta + \sin\theta\cos\theta) + C$$

$$= \frac{\sqrt{2}}{8}\left[\arctan\frac{x}{\sqrt{2}} + \left(\frac{x}{\sqrt{x^2+2}}\right)\left(\frac{\sqrt{2}}{\sqrt{x^2+2}}\right)\right] + C$$

$$= \frac{1}{4}\left[\frac{x}{x^2+2} + \frac{1}{\sqrt{2}}\arctan\frac{x}{\sqrt{2}}\right] + C.$$

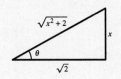

37. Let

$$dv = dx \quad \Rightarrow \quad v = x$$

$$u = \operatorname{arcsec} x \Rightarrow du = \frac{1}{x\sqrt{4x^2-1}}\, dx.$$

—CONTINUED—

37. —CONTINUED—

Therefore,

$$\int \operatorname{arcsec} 2x \, dx = uv - \int v \, du = x \operatorname{arcsec} 2x - \int x\left(\frac{1}{x\sqrt{4x^2-1}}\right) dx$$

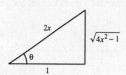

$$= x \operatorname{arcsec} 2x - \frac{1}{2}\int \frac{2}{\sqrt{(2x)^2-1^2}} \, dx \quad (\text{Let } 2x = \sec\theta)$$

$$= x \operatorname{arcsec} 2x - \int \frac{(1/2)\sec\theta \tan\theta \, d\theta}{\tan\theta}$$

$$= x \operatorname{arcsec} 2x - \frac{1}{2}\int \sec\theta \, d\theta$$

$$= x \operatorname{arcsec} 2x - \frac{1}{2}\ln|\sec\theta + \tan\theta| + C$$

$$= x \operatorname{arcsec} 2x - \frac{1}{2}\ln|2x + \sqrt{4x^2-1}| + C.$$

43. If we let $t = \sin\theta$, then $dt = \cos\theta \, d\theta$ and $1 - t^2 = \cos^2\theta$.

$$\int \frac{t^2}{(1-t^2)^{3/2}} \, dt = \int \frac{\sin^2\theta \cos\theta}{\cos^3\theta} \, d\theta$$

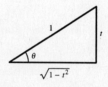

$$= \int \tan^2\theta \, d\theta = \int (\sec^2\theta - 1) \, d\theta$$

$$= \tan\theta - \theta = \frac{t}{\sqrt{1-t^2}} - \arcsin t + C \quad (\text{See figure})$$

(a) Using the antiderivative from above, it follows that

$$\int_0^{\sqrt{3}/2} \frac{t^2}{(1-t^2)^{3/2}} \, dt = \left[\frac{t}{\sqrt{1-t^2}} - \arcsin t\right]_0^{\sqrt{3}/2}$$

$$= \frac{\sqrt{3}/2}{\sqrt{1/4}} - \arcsin \frac{\sqrt{3}}{2}$$

$$= \sqrt{3} - \frac{\pi}{3} \approx 0.685.$$

(b) Using the substitution from above, we have $\theta = 0$ when $t = 0$ and $\theta = \pi/3$ when $t = \sqrt{3}/2$.

$$\int_0^{\sqrt{3}/2} \frac{t^2}{(1-t^2)^{3/2}} \, dt = \int_0^{\pi/3} \frac{\sin^2\theta \cos\theta}{\cos^3\theta} \, d\theta$$

$$= \left[\tan\theta - \theta\right]_0^{\pi/3}$$

$$= \sqrt{3} - \frac{\pi}{3} \approx 0.685$$

53. Shell Method

$$V = 2\pi\int_2^4 x[2\sqrt{1-(x-3)^2}] \, dx = 4\pi\int_2^4 x\sqrt{1-(x-3)^2} \, dx$$

Let $x - 3 = \sin\theta$. Then $\sqrt{1-(x-3)^2} = \cos\theta$ and $dx = \cos\theta \, d\theta$. Also, when $x = 2$, $\sin\theta = -1$ and $\theta = -\pi/2$. When $x = 4$, $\sin\theta = 1$ and $\theta = \pi/2$. Therefore,

$$V = 4\pi\int_{-\pi/2}^{\pi/2} (3 + \sin\theta)(\cos\theta)(\cos\theta) \, d\theta$$

$$= 4\pi\left[\int_{-\pi/2}^{\pi/2} 3\cos^2\theta \, d\theta + \int_{-\pi/2}^{\pi/2} \cos^2\theta \sin\theta \, d\theta\right]$$

$$= 4\pi\left[\int_{-\pi/2}^{\pi/2} \frac{3}{2}(1 + \cos 2\theta) \, d\theta + \int_{-\pi/2}^{\pi/2} \cos^2\theta \sin\theta \, d\theta\right]$$

$$= 4\pi\left[\frac{3}{2}\left(\theta + \frac{1}{2}\sin 2\theta\right) - \frac{1}{3}\cos^3\theta\right]_{-\pi/2}^{\pi/2} = 6\pi^2.$$

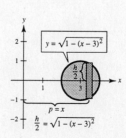

61. $S = 2\pi \int_a^b y\sqrt{1 + (y')^2}\, dx = 2\pi \int_0^{\sqrt{2}} x^2\sqrt{1 + 4x^2}\, dx$

Using trigonometric substitution, let $2x = \tan\theta$. Then $\sqrt{1 + 4x^2} = \sec\theta$, $x^2 = \frac{1}{4}\tan^2\theta$, and $dx = \frac{1}{2}\sec^2\theta\, d\theta$. Therefore,

$$\int x^2\sqrt{1 + 4x^2}\, dx = \int \frac{\tan^2\theta}{4}(\sec\theta)\left(\frac{1}{2}\sec^2\theta\right) d\theta$$

$$= \frac{1}{8}\int \sec^3\theta\,\tan^2\theta\, d\theta$$

$$= \frac{1}{8}\int \sec^3\theta\,(\sec^2\theta - 1)\, d\theta$$

$$= \frac{1}{8}\left[\int \sec^5\theta\, d\theta - \int \sec^3\theta\, d\theta\right].$$

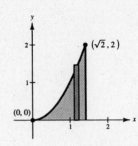

Now using Integration By Parts to evaluate $\int\sec^5\theta\, d\theta$, let

$dv = \sec^2\theta\, d\theta \implies v = \tan\theta$

$u = \sec^3\theta \implies du = 3\sec^3\theta\tan\theta\, d\theta.$

Thus,

$$\int \sec^5\theta\, d\theta = \sec^3\theta\tan\theta - 3\int \sec^3\theta\tan^2\theta\, d\theta$$

$$= \sec^3\theta\tan\theta - 3\int \sec^3\theta(\sec^2\theta - 1)\, d\theta$$

$$= \sec^3\theta\tan\theta - 3\int \sec^5\theta\, d\theta + 3\int \sec^3\theta\, d\theta$$

$$4\int \sec^5\theta\, d\theta = \sec^3\theta\tan\theta + 3\int \sec^3\theta\, d\theta$$

$$\int \sec^5\theta\, d\theta = \frac{1}{4}\left(\sec^3\tan\theta + 3\int \sec^3\theta\, d\theta\right).$$

Hence,

$$\int x^2\sqrt{1 + 4x^2}\, dx = \frac{1}{8}\left[\frac{1}{4}\sec^3\theta\tan\theta + \frac{3}{4}\int \sec^3\theta\, d\theta - \int \sec^3\theta\, d\theta\right]$$

$$= \frac{1}{32}\left(\sec^3\theta\tan\theta - \int \sec^3\theta\, d\theta\right).$$

Using the formula for $\int \sec^3\theta\, d\theta$ from Section 7.2, we have

$$\int x^2\sqrt{1 + 4x^2}\, dx = \frac{1}{32}\left[\sec^2\theta\tan\theta - \frac{1}{2}(\sec\theta\tan\theta + \ln|\sec\theta + \tan\theta|)\right] + C.$$

Finally, when $x = 0$, $\theta = 0$, and when $x = \sqrt{2}$, $\theta = \arctan 2\sqrt{2}$. Therefore,

$$S = 2\pi \int_0^{\sqrt{2}} x^2\sqrt{1 + 4x^2}\, dx$$

$$= \frac{\pi}{16}\left[\sec^3\theta\tan\theta - \frac{1}{2}(\sec\theta\tan\theta + \ln|\sec\theta + \tan\theta|)\right]_0^{\arctan 2\sqrt{2}}$$

$$= \frac{\pi}{16}\left[51\sqrt{2} - \frac{1}{2}\ln\left(2\sqrt{2} + 3\right)\right]$$

$$= \frac{\pi}{32}\left[102\sqrt{2} - \ln\left(2\sqrt{2} + 3\right)\right] \approx 13.989.$$

67. (a) Using the figure and the formula for the slope of a line we have

$$m = \frac{dy}{dx} = \frac{y - \left(y + \sqrt{144 - x^2}\right)}{x - 0} = -\frac{\sqrt{144 - x^2}}{x}.$$

(b) If $x = 12 \sin \theta$ in the following integral, then $dx = 12 \cos \theta \, d\theta$ and $\sqrt{144 - x^2} = 12 \cos \theta$.

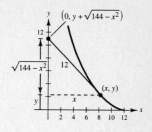

$$y = -\int \frac{\sqrt{144 - x^2}}{x} \, dx$$

$$= -\int \frac{12 \cos \theta}{12 \sin \theta} 12 \cos \theta \, d\theta = -12 \int \frac{1 - \sin^2 \theta}{\sin \theta} \, d\theta$$

$$= -12 \int (\csc \theta - \sin \theta) \, d\theta = -12 \ln|\csc \theta - \cot \theta| - 12 \cos \theta + C$$

$$= -12 \ln\left| \frac{12}{x} - \frac{\sqrt{144 - x^2}}{x} \right| - 12\left(\frac{\sqrt{144 - x^2}}{12} \right) + C$$

$$= -12 \ln\left| \frac{12 - \sqrt{144 - x^2}}{x} \right| - \sqrt{144 - x^2} + C$$

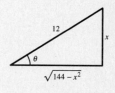

When $x = 12$, $y = 0 \implies C = 0$. Thus,

$$y = -12 \ln\left(\frac{12 - \sqrt{144 - x^2}}{x} \right) - \sqrt{144 - x^2}.$$

Note: $\dfrac{12 - \sqrt{144 - x^2}}{x} > 0$ for $0 < x \le 12$.

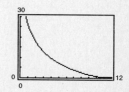

(c) Vertical asymptote: $x = 0$

(d) When the person has reached the point $(0, 12)$, we have

$$y + \sqrt{144 - x^2} = 12 \implies y = 12 - \sqrt{144 - x^2}.$$

Thus,

$$12 - \sqrt{144 - x^2} = -12 \ln\left(\frac{12 - \sqrt{144 - x^2}}{x} \right) = \sqrt{144 - x^2}$$

$$-1 = \ln\left(\frac{12 - \sqrt{144 - x^2}}{x} \right)$$

$$xe^{-1} = 12 - \sqrt{144 - x^2}$$

$$(xe^{-1} - 12)^2 = \left(-\sqrt{144 - x^2} \right)^2$$

$$x^2 e^{-2} - 24xe^{-1} + 144 = 144 - x^2$$

$$x^2(e^{-2} + 1) - 24xe^{-1} = 0$$

$$x[x(e^{-2} + 1) - 24e^{-1}] = 0$$

$$x = 0 \text{ or } x = \frac{24e^{-1}}{e^{-2} + 1} \approx 7.77665.$$

Therefore,

$$s = \int_{7.77665}^{12} \sqrt{1 + \left(-\frac{\sqrt{144 - x^2}}{x} \right)^2} \, dx = \int_{7.77665}^{12} \sqrt{\frac{x^2 + (144 - x^2)}{x^2}} \, dx$$

$$= \int_{7.77665}^{12} \frac{12}{x} \, dx = \left[12 \ln|x| \right]_{7.77665}^{12} = 12(\ln 12 - \ln 7.77665) \approx 5.2 \text{ feet}.$$

Section 7.5   Partial Fractions

13. $\dfrac{x^2 + 12x + 12}{x^3 - 4x} = \dfrac{x^2 + 12x + 12}{x(x - 2)(x + 2)} = \dfrac{A}{x} + \dfrac{B}{x - 2} + \dfrac{C}{x + 2}$

Multiplying by $(x)(x - 2)(x + 2)$, yields

$\quad x^2 + 12x + 12 = A(x - 2)(x + 2) + B(x)(x + 2) + C(x)(x - 2).$

If $x = 0$, then $12 = A(-2)(2)$ or $A = -3$.

If $x = 2$, then $4 + 24 + 12 = B(2)(4)$ or $B = 5$.

If $x = -2$, then $4 - 24 + 12 = C(-2)(-4)$ or $C = -1$.

Thus,

$$\int \frac{x^2 + 12x + 12}{x^3 - 4x}\,dx = \int \left(\frac{-3}{x} + \frac{5}{x - 2} + \frac{-1}{x + 2} \right) dx$$

$$= -3 \ln|x| + 5 \ln|x - 2| - \ln|x + 2| + C$$

$$= \ln \left| \frac{(x - 2)^5}{x^3(x + 2)} \right| + C.$$

25. Since $x^3 - x^2 + x + 3 = (x + 1)(x^2 - 2x + 3)$, we have

$$\frac{x^2 + 5}{x^3 - x^2 + x + 3} = \frac{A}{x + 1} + \frac{Bx + C}{x^2 - 2x + 3}.$$

Multiplying by $(x + 1)(x^2 - 2x + 3)$ yields

$\quad x^2 + 5 = A(x^2 - 2x + 3) + (Bx + C)(x + 1).$

If $x = -1$, then $1 + 5 = A(1 + 2 + 3)$ or $A = 1$.

Therefore,

$\quad x^2 + 5 = x^2 - 2x + 3 + Bx^2 + Bx + Cx + C$

$\quad 2x + 2 = Bx^2 + (B + C)x + C.$

Equating coefficients, yields $B = 0$, $B + C = 2$, and $C = 2$. Finally,

$$\int \frac{x^2}{x^3 - x^2 + x + 3}\,dx = \int \left(\frac{1}{x + 1} + \frac{2}{x^2 - 2x + 3} \right) dx$$

$$= \int \frac{1}{x + 1}\,dx + 2 \int \frac{1}{(x - 1)^2 + 2}\,dx$$

$$= \ln|x + 1| + 2 \left(\frac{1}{\sqrt{2}} \right) \arctan\left(\frac{x - 1}{\sqrt{2}} \right) + C$$

$$= \ln|x + 1| + \sqrt{2} \arctan\left(\frac{x - 1}{\sqrt{2}} \right) + C.$$

29. $\dfrac{x + 1}{x(x^2 + 1)} = \dfrac{A}{x} + \dfrac{Bx + C}{x^2 + 1}$

Multiplying by $(x)(x^2 + 1)$ yields

$\quad x + 1 = A(x^2 + 1) + (Bx + C)x = Ax^2 + A + Bx^2 + Cx$

$\qquad = (A + B)x^2 + Cx + A.$

—CONTINUED—

29. —CONTINUED—

By equating coefficients, you have $A + B = 0$, $C = 1$, and $A = 1$. Thus, $B = -1$, and we obtain

$$\int_1^2 \frac{x+1}{x(x^2+1)} \, dx = \int_1^2 \left(\frac{1}{x} + \frac{-x+1}{x^2+1} \right) dx$$

$$= \int_1^2 \frac{1}{x} \, dx - \frac{1}{2} \int_1^2 \frac{2x}{x^2+1} \, dx + \int_1^2 \frac{1}{x^2+1} \, dx$$

$$= \left[\ln|x| - \frac{1}{2} \ln|x^2+1| + \arctan x \right]_1^2$$

$$= \ln 2 - \frac{1}{2} \ln 5 + \arctan 2 - \ln 1 + \frac{1}{2} \ln 2 - \arctan 1$$

$$= \frac{3}{2} \ln 2 - \frac{1}{2} \ln 5 + \arctan 2 - \frac{\pi}{4}$$

$$= \frac{1}{2} \ln \frac{8}{5} + \arctan 2 - \frac{\pi}{4} \approx 0.557.$$

33. Using a symbolic integration utility to perform the integration yields

$$\int \frac{x^2+x+2}{(x^2+2)^2} \, dx = \frac{\sqrt{2}}{2} \arctan \frac{x}{\sqrt{2}} - \frac{1}{2(x^2+2)} + C.$$

Substitute the solution point $(0, 1)$ into the antiderivative and solve for the constant of integration.

$$y = \frac{\sqrt{2}}{2} \arctan \frac{x}{\sqrt{2}} - \frac{1}{2(x^2+2)} + C$$

$$1 = \frac{\sqrt{2}}{2} \arctan \frac{0}{\sqrt{2}} - \frac{1}{2(0^2+2)} + C$$

$$1 = -\frac{1}{4} + C \implies C = \frac{5}{4}$$

The graph of the function

$$y = \frac{\sqrt{2}}{2} \arctan \frac{x}{\sqrt{2}} - \frac{1}{2(x^2+2)} + \frac{5}{4}$$

is shown in the figure.

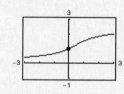

43. If $u = e^x$, then $du = e^x \, dx$, and

$$\int \frac{e^x}{(e^x-1)(e^x+4)} \, dx = \int \frac{du}{(u-1)(u+4)}.$$

Using partial fractions yields

$$\frac{1}{(u-1)(u+4)} = \frac{A}{u-1} + \frac{B}{u+4}.$$

Multiplying by $(u-1)(u+4)$ yields

$$1 = A(u+4) + B(u-1).$$

If $u = 1$, then $1 = A(5)$ or $A = \frac{1}{5}$.

If $u = -4$, then $1 = B(-5)$ or $B = -\frac{1}{5}$.

Therefore,

$$\int \frac{du}{(u-1)(u+4)} = \frac{1}{5} \int \left(\frac{1}{u-1} - \frac{1}{u+4} \right) du$$

$$= \frac{1}{5} (\ln|u-1| - \ln|u+4|) + C$$

$$= \frac{1}{5} \ln \left| \frac{u-1}{u+4} \right| + C$$

$$= \frac{1}{5} \ln \left| \frac{e^x-1}{e^x+4} \right| + C.$$

53. $\dfrac{1}{(x+1)(n-x)} = \dfrac{A}{x+1} + \dfrac{B}{n-x}$

Multiplying by $(x+1)(n-x)$ yields

$$1 = A(n-x) + B(x+1).$$

If $x = -1$, then $1 = A(n+1)$ or $A = \dfrac{1}{n+1}$.

If $x = n$, then $1 = B(n+1)$ or $B = \dfrac{1}{n+1}$.

—CONTINUED—

53. —CONTINUED—

Therefore,

$$\int \frac{1}{(x+1)(n-x)}\,dx = \int k\,dt$$

$$\frac{1}{n+1}\int \left(\frac{1}{x+1} + \frac{1}{n-x}\right)dx = \int k\,dt$$

$$\frac{1}{n+1}[\ln(x+1) - \ln(n-x) = kt + C$$

$$\left(\frac{1}{n+1}\right)\ln\!\left(\frac{x+1}{n-x}\right) = kt + C.$$

When $t = 0$, $x = 0$. Thus,

$$\left(\frac{1}{n+1}\right)\ln\!\left(\frac{1}{n}\right) = C.$$

Therefore,

$$\ln\!\left(\frac{x+1}{n-x}\right) = k(n+1)t + \ln\!\left(\frac{1}{n}\right)$$

$$\frac{x+1}{n-x} = \frac{1}{n}e^{k(n+1)t}$$

$$x = \frac{n[e^{(n+1)kt} - 1]}{e^{(n+1)kt} + n}.$$

Section 7.6   Integration by Tables and Other Integration Techniques

3. Consider the form

$$\int \sqrt{u^2 \pm a^2}\,du = \frac{1}{2}\Big(u\sqrt{u^2 \pm a^2} \pm a^2 \ln\big|u + \sqrt{u^2 \pm a^2}\big|\Big) + C$$

with $a = 1$, $u = e^x$ and $du = e^x\,dx$. Then,

$$\int e^x\sqrt{1 + e^{2x}}\,dx = \int \sqrt{1^2 + (e^x)^2}\,e^x\,dx$$

$$= \frac{1}{2}\Big[e^x\sqrt{e^{2x} + 1} + \ln\big(e^x + \sqrt{e^{2x} + 1}\big)\Big] + C.$$

9. Consider the form

$$\int \frac{1}{1 \pm \cos u}\,du = -\cot u \pm \csc u + C$$

with $u = \sqrt{x}$ and $du = 1/(2\sqrt{x})\,dx$. Then,

$$\int \frac{1}{\sqrt{x}(1 - \cos\sqrt{x})}\,dx = 2\int \frac{1/(2\sqrt{x})}{1 - \cos\sqrt{x}}\,dx$$

$$= 2\big(-\cot\sqrt{x} - \csc\sqrt{x}\big) + C$$

$$= -2\big(\cot\sqrt{x} + \csc\sqrt{x}\big) + C.$$

11. Consider the form

$$\int \frac{1}{1 + e^u}\,du = u - \ln(1 + e^u) + C$$

with $u = 2x$ and $du = 2\,dx$. Then,

$$\int \frac{1}{1 + e^{2x}}\,dx = \frac{1}{2}\int \frac{2}{1 + e^{2x}}\,dx$$

$$= \frac{1}{2}[2x - \ln(1 + e^{2x})] + C$$

$$= x - \frac{1}{2}\ln(1 + e^{2x}) + C.$$

17. (a) Consider the form

$$\int \frac{1}{u^2(a + bu)}\, du = -\frac{1}{a}\left(\frac{1}{u} + \frac{b}{a}\ln\left|\frac{u}{a + bu}\right|\right) + C$$

with $a = 1$, $b = 1$, $u = x$ and $du = dx$. Then,

$$\int \frac{1}{x^2(x + 1)}\, dx = -\left(\frac{1}{x} + \ln\left|\frac{x}{1 + x}\right|\right) + C$$

$$= \ln\left|\frac{x + 1}{x}\right| - \frac{1}{x} + C.$$

(b) $\dfrac{1}{x^2(x + 1)} = \dfrac{A}{x} + \dfrac{B}{x^2} + \dfrac{C}{x + 1}$

Multiplying by $x^2(x + 1)$ yields

$$1 = Ax(x + 1) + B(x + 1) + Cx^2.$$

If $x = 0$, then $1 = B$.

If $x = -1$, then $1 = C$.

If $x = 1$, $B = 1$, and $C = 1$, then $1 = 2A + 2 + 1$ or $A = -1$.

Thus,

$$\int \frac{1}{x^2(x + 1)}\, dx = \int\left(\frac{-1}{x} + \frac{1}{x^2} + \frac{1}{x + 1}\right) dx$$

$$= -\ln|x| - \frac{1}{x} + \ln|x + 1| + C$$

$$= \ln\left|\frac{x + 1}{x}\right| - \frac{1}{x} + C.$$

25. Consider the form

$$\int \frac{1}{u^2\sqrt{u^2 \pm a^2}}\, du = \mp\frac{\sqrt{u^2 \pm a^2}}{a^2 u} + C$$

where $u = x$ and $a = 2$. Then,

$$\int \frac{1}{x^2\sqrt{x^2 - 4}}\, dx = \int \frac{1}{u^2\sqrt{u^2 - a^2}}\, du$$

$$= \frac{\sqrt{u^2 - a^2}}{a^2 u} + C = \frac{\sqrt{x^2 - 4}}{4x} + C.$$

27. Consider the form

$$\int \frac{u}{(a + bu)^2}\, du = \frac{1}{b^2}\left(\frac{a}{a + bu} + \ln|a + bu|\right) + C$$

where $a = 1$, $b = -3$, $u = x$, and $du = dx$. Then,

$$\int \frac{2x}{(1 - 3x)^2}\, dx = 2\int \frac{x}{(1 - 3x)^2}\, dx$$

$$= \frac{2}{9}\left[\frac{1}{1 - 3x} + \ln|1 - 3x|\right] + C.$$

41. Consider the form

$$\int \frac{u}{a + bu}\, du = \frac{1}{b^2}(bu - a\ln|a + bu|) + C$$

where $u = \ln x$, $a = 3$, $b = 2$, and $du = (1/x)\, dx$. Then,

$$\int \frac{\ln x}{x(3 + 2\ln x)}\, dx = \int \frac{\ln x}{3 + 2\ln x}\left(\frac{1}{x}\right) dx$$

$$= \frac{1}{4}[2\ln|x| - 3\ln|3 + 2\ln|x||] + C.$$

47. Consider the form

$$\int \frac{u}{\sqrt{a + bu}}\, du = \frac{-2(2u - bu)}{3b^2}\sqrt{a + bu} + C$$

where $u = x^2$, $a = 4$, $b = -1$, $du = 2x\, dx$, and $\sqrt{a + bu} = \sqrt{4 - x^2}$. Then,

$$\int \frac{x^3}{\sqrt{4 - x^2}}\, dx = \frac{1}{2}\int \frac{x^2}{\sqrt{4 - x^2}}(2x)\, dx$$

$$= \frac{1}{2}\left[\frac{(-2)(8 + x^2)}{3}\right]\sqrt{4 - x^2} + C$$

$$= -\left(\frac{x^2 + 8}{3}\right)\sqrt{4 - x^2} + C.$$

51. Since the numerator and denominator are of the same degree, begin by dividing

$$\frac{u^2}{(a + bu)^2} = \frac{1}{b^2} - \frac{(2a/b)u + (a^2/b^2)}{(a + bu)^2}.$$

Now use partial fractions to obtain

$$\frac{(2a/b)u + (a^2/b^2)}{(a + bu)^2} = \frac{A}{a + bu} + \frac{B}{(a + bu)^2}.$$

Multiplying by $(a + bu)^2$ yields

$$\left(\frac{2a}{b}\right)u + \frac{a^2}{b^2} = A(a + bu) + B = bAu + (aA + B).$$

—CONTINUED—

51. —CONTINUED—

Equating the coefficients of like terms, we have

$$bA = \frac{2a}{b} \quad \text{and} \quad A = \frac{2a}{b^2}$$

$$aA + B = \frac{a^2}{b^2} \quad \text{and} \quad B = \frac{a^2}{b^2} - a\left(\frac{2a}{b^2}\right) = -\frac{a^2}{b^2}.$$

Therefore,

$$\int \frac{u^2}{(a + bu)^2}\, du = \frac{1}{b^2}\int du - \frac{2a}{b^2}\left(\frac{1}{b}\right)\int \frac{b}{a + bu}\, du + \frac{a^2}{b^2}\left(\frac{1}{b}\right)\int \frac{b}{(a + bu)^2}\, du$$

$$= \left(\frac{1}{b^2}\right)u - \frac{2a}{b^3}(\ln|a + bu|) - \frac{a^2}{b^3}\left(\frac{1}{a + bu}\right) + C$$

$$= \frac{1}{b^3}\left[bu - \frac{a^2}{a + bu} - 2a\ln|a + bu|\right] + C.$$

59. Using a symbolic integration utility to perform the integration yields

$$\int \frac{\sqrt{2 - 2x - x^2}}{x + 1}\, dx = \sqrt{2 - 2x - x^2} - \sqrt{3}\ln\left|\frac{\sqrt{3} + \sqrt{2 - 2x - x^2}}{x + 1}\right| + C.$$

Substitute the solution point $\left(0, \sqrt{2}\right)$ into the antiderivative and solve for the constant of integration.

$$y = \sqrt{2 - 2x - x^2} - \sqrt{3}\ln\left|\frac{\sqrt{3} + \sqrt{2 - 2x - x^2}}{x + 1}\right| + C$$

$$\sqrt{2} = \sqrt{2} - \sqrt{3}\ln\left(\sqrt{3} + \sqrt{2}\right) + C \implies C = \sqrt{3}\ln\left(\sqrt{3} + \sqrt{2}\right)$$

The graph of the function

$$y = \sqrt{2 - 2x - x^2} - \sqrt{3}\ln\left|\frac{\sqrt{3} + \sqrt{2 - 2x - x^2}}{x + 1}\right| + \sqrt{3}\ln\left(\sqrt{3} + \sqrt{2}\right)$$

is shown in the figure.

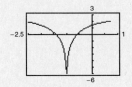

65. Let $u = \dfrac{\sin\theta}{1 + \cos\theta}$. Then,

$$\cos\theta = \frac{1 - u^2}{1 + u^2}, \quad \sin\theta = \frac{2u}{1 + u^2}, \quad \text{and} \quad d\theta = \frac{2\, du}{1 + u^2}.$$

Furthermore, when $\theta = \pi/2$, $u = 1$, and when $\theta = 0$, $u = 0$.

$$\int_0^{\pi/2} \frac{1}{1 + \sin\theta + \cos\theta}\, d\theta = \int_0^1 \frac{[2/(1 + u^2)]\, du}{1 + (2u)/(1 + u^2) + (1 - u^2)/(1 + u^2)}$$

$$= \int_0^1 \frac{1}{u + 1}\, du = \left[\ln|u + 1|\right]_0^1 = \ln 2.$$

Section 7.7   Indeterminate Forms and L'Hôpital's Rule

5. (a) Using the Cancellation Technique from Chapter 1 yields

$$\lim_{x \to 3} \frac{2(x - 3)}{x^2 - 9} = \lim_{x \to 3} \frac{2(x - 3)}{(x + 3)(x - 3)} = \lim_{x \to 3} \frac{2}{x + 3} = \frac{1}{3}.$$

(b) Using L'Hôpital's Rule yields

$$\lim_{x \to 3} \frac{2(x - 3)}{x^2 - 9} = \lim_{x \to 3} \frac{(d/dx)[2(x - 3)]}{(d/dx)[x^2 - 9]} = \lim_{x \to 3} \frac{2}{2x} = \frac{1}{3}.$$

13. Since a direct substitution of $x = 0$ yields the indeterminate form $0/0$, apply L'Hôpital's Rule to obtain

$$\lim_{x \to 0} \frac{\sqrt{4 - x^2} - 2}{x} = \lim_{x \to 0} \frac{(1/2)(4 - x^2)^{-1/2}(-2x)}{1}$$

$$= \lim_{x \to 0} \frac{-x}{\sqrt{4 - x^2}} = \frac{0}{2} = 0.$$

17. Case 1: $n = 1$ (Apply L'Hôpital's Rule once.)

$$\lim_{x \to 0^+} \frac{e^x - (1 + x)}{x} = \lim_{x \to 0^+} \frac{e^x - 1}{1} = 0$$

Case 2: $n = 2$ (Apply L'Hôpital's Rule twice.)

$$\lim_{x \to 0^+} \frac{e^x - (1 + x)}{x^2} = \lim_{x \to 0^+} \frac{e^x - 1}{2x} = \lim_{x \to 0^+} \frac{e^x}{2} = \frac{1}{2}$$

Case 3: $n \geq 3$ (Apply L'Hôpital's Rule twice.)

$$\lim_{x \to 0^+} \frac{e^x - (1 + x)}{x^n} = \lim_{x \to 0^+} \frac{e^x - 1}{nx^{n-1}}$$

$$= \lim_{x \to 0^+} \frac{e^x}{n(n - 1)x^{n-2}} = \infty$$

29. Since direct substitution leads to the indeterminate form ∞/∞, use L'Hôpital's Rule.

$$\lim_{x \to \infty} \frac{\ln x}{x} = \lim_{x \to \infty} \frac{1/x}{1} = \lim_{x \to \infty} \frac{1}{x} = 0$$

33. (a) Since direct substitution yields the indeterminate form $\infty \cdot 0$, use L'Hôpital's Rule.

(b) $\displaystyle \lim_{x \to \infty} x \sin \frac{1}{x} = \lim_{x \to \infty} \frac{\sin(1/x)}{1/x}$

$$= \lim_{x \to \infty} \frac{(-1/x^2) \cos(1/x)}{-1/x^2}$$

$$= \lim_{x \to \infty} \cos \frac{1}{x} = 1$$

(c)

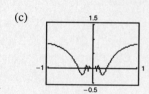

35. (a) 0^∞ is not an indeterminate form.

(b) $\displaystyle \lim_{x \to 0^+} x^{1/x} = 0^\infty = 0$

(c)

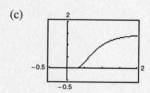

37. (a) Direct substitution yields the indeterminate form ∞^0.

(b) Begin by taking the natural logarithm of both members of the equation

$$y = \lim_{x \to \infty} x^{1/x}.$$

We obtain

$$\ln y = \ln \left[\lim_{x \to \infty} x^{1/x} \right]$$

$$= \lim_{x \to \infty} \left[\frac{1}{x} \ln x \right] = \lim_{x \to \infty} \frac{\ln x}{x} = \lim_{x \to \infty} \frac{1/x}{1} = 0.$$

Finally, as $\ln y \to 0$, we know that $y \to 1$ and conclude that $\displaystyle \lim_{x \to \infty} x^{1/x} = 1$.

(c)

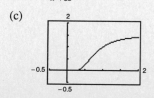

45. (a) The graph of the function is shown in the figure.

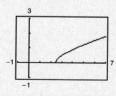

(b) Since direct substitution leads to the indeterminate form $0/0$, use L'Hôpital's Rule.

$$\lim_{x \to 3} \frac{x - 3}{\ln(2x - 5)} = \lim_{x \to 3} \frac{1}{\dfrac{2}{2x - 5}} = \lim_{x \to 3} \frac{2x - 5}{2} = \frac{1}{2}$$

55. $\lim\limits_{x\to\infty} \dfrac{(\ln x)^n}{x^m} = \lim\limits_{x\to\infty}\left[\dfrac{n(\ln x)^{n-1}(1/x)}{mx^{m-1}}\right] = \lim\limits_{x\to\infty}\left[\dfrac{n(\ln x)^{n-1}}{mx^m}\right]$

If $n - 1 \le 0$, this limit is zero. If $n - 1 > 0$, repeat L'Hôpital's Rule to obtain

$\lim\limits_{x\to\infty}\left[\dfrac{n(n-1)(\ln x)^{n-2}}{m^2 x^m}\right].$

Again, if $n - 2 \le 0$, this limit is zero. If $n - 2 > 0$, repeated applications of L'Hôpital's Rule will eventually yield a form where the numerator approaches a finite number and the denominator approaches infinity. Thus, in every case the limit is 0.

69. $\lim\limits_{k\to 0} \dfrac{32}{k}\left(1 - e^{-kt} + \dfrac{v_0 k e^{-kt}}{32}\right) = \lim\limits_{k\to 0}\dfrac{32(1 - e^{-kt})}{k} + \lim\limits_{k\to 0}\left[\dfrac{32}{k}\cdot\dfrac{v_0 k e^{-kt}}{32}\right]$

$= \lim\limits_{k\to 0}\dfrac{32te^{-kt}}{1} + v_0 = 32t + v_0$

(Apply L'Hôpital's Rule to the first limit in the right-hand member of the equation.)

75. $\dfrac{f'(c)}{g'(c)} = \dfrac{f(b) - f(a)}{g(b) - g(a)}$

$\dfrac{\cos c}{-\sin c} = \dfrac{\sin(\pi/2) - \sin 0}{\cos(\pi/2) - \cos 0}$

$-\cot c = -1 \Rightarrow c = \dfrac{\pi}{4}$

Section 7.8   Improper Integrals

3. $\displaystyle\int_0^2 \dfrac{1}{(x-1)^2}\,dx = \lim\limits_{b\to 1^-}\int_0^b (x-1)^{-2}\,dx + \lim\limits_{c\to 1^+}\int_c^2 (x-1)^{-2}\,dx$

$= \lim\limits_{b\to 1^-}\left[\dfrac{-1}{x-1}\right]_0^b + \lim\limits_{c\to 1^+}\left[\dfrac{-1}{x-1}\right]_c^2$

$= \lim\limits_{b\to 1^-}\left[\dfrac{-1}{b-1} - 1\right] + \lim\limits_{c\to 1^+}\left[-1 - \dfrac{-1}{c-1}\right] = \infty$

Therefore, the improper integral diverges.

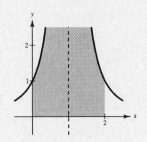

15. Since

$\displaystyle\int e^{-x}\cos x\,dx = \dfrac{1}{2}e^{-x}(-\cos x + \sin x) + C,$

we have

$\displaystyle\int_0^\infty e^{-x}\cos x\,dx = \lim\limits_{b\to\infty}\int_0^b e^{-x}\cos x\,dx$

$= \lim\limits_{b\to\infty}\dfrac{1}{2}\left[\dfrac{(-\cos x + \sin x)}{e^x}\right]_0^b$

$= \lim\limits_{b\to\infty}\dfrac{1}{2}\left[\dfrac{-\cos b + \sin b}{e^b} - (-1)\right]$

$= \dfrac{1}{2}[0 - (-1)] = \dfrac{1}{2}.$

25. $\displaystyle\int_0^8 \dfrac{1}{\sqrt[3]{8-x}}\,dx = \lim\limits_{b\to 8^-}\int_0^b \dfrac{1}{\sqrt[3]{8-x}}\,dx$

$= \lim\limits_{b\to 8^-}\left[-\dfrac{(8-x)^{2/3}}{2/3}\right]_0^b$

$= \lim\limits_{b\to 8^-}\left[\dfrac{3}{2}(8-b)^{2/3} + \dfrac{3}{2}(8-0)^{2/3}\right]$

$= -\dfrac{3}{2}(0) + \dfrac{3}{2}(4) = 6$

31. $\displaystyle\int_2^4 \frac{1}{\sqrt{x^2 - 4}}\, dx = \lim_{a \to 2^+} \int_a^4 \frac{1}{\sqrt{x^2 - 4}}\, dx$

$\displaystyle = \lim_{a \to 2^+} \left[\ln|x + \sqrt{x^2 - 4}| \right]_a^4$

$\displaystyle = \lim_{a \to 2^+} \left[\ln(4 + \sqrt{12}) - \ln|a + \sqrt{a^2 - 4}| \right]$

$\displaystyle = \ln(4 + \sqrt{12}) - \ln(2 + 0)$

$\displaystyle = \ln\left(\frac{4 + 2\sqrt{3}}{2} \right) = \ln(2 + \sqrt{3})$

37. When $n = 1$, the integral converges, since

$\displaystyle\int_0^\infty xe^{-x}\, dx = \lim_{b \to \infty} \int_0^b xe^{-x}\, dx$

$\displaystyle = \lim_{b \to \infty} \left[-e^{-x}(x + 1) \right]_0^b \qquad \text{(Integration by parts)}$

$\displaystyle = \lim_{b \to \infty} \left[-e^{-b}(b + 1) + 1 \right]$

$\displaystyle = 0 + 1 = 1. \qquad \text{(L'Hôpital's Rule)}$

Now assume that the integral converges for $n = k$ and verify that it converges for $n = k + 1$.

$\displaystyle\int_0^\infty x^{k+1}e^{-x}\, dx = \lim_{b \to \infty} \int_0^b x^{k+1}e^{-x}\, dx$

$\displaystyle = \lim_{b \to \infty} \left[-x^{k+1}e^{-x} - \frac{k + 1}{-1} \int_0^b x^k e^{-x}\, dx \right]_0^\infty \qquad \text{(Integration by parts)}$

$\displaystyle = 0 + (k + 1)\int_0^\infty x^k e^{-x}\, dx \qquad \text{(L'Hôpital's Rule)}$

$\displaystyle = (k + 1)\int_0^\infty x^k e^{-x}\, dx$

Therefore, we have shown the integral for $n = k + 1$ converges if the integral for $n = k$ converges. Combining this with the results for $n = 1$, it follows by mathematical induction that the integral converges for any positive integer n.

47. On $[1, \infty)$ we have

$x \le x^2$

$e^x \le e^{x^2}$

$\dfrac{1}{e^x} \ge \dfrac{1}{e^{x^2}}$

$e^{-x} \ge e^{-x^2}.$

Therefore, by Exercise 38 we have

$\displaystyle\int_0^\infty e^{-x^2}\, dx = \int_0^1 e^{-x^2}\, dx + \int_1^\infty e^{-x^2}\, dx$

$\displaystyle \le \int_0^1 e^{-x^2}\, dx + \int_1^\infty e^{-x}\, dx$

$\displaystyle = \int_0^1 e^{-x^2}\, dx + \left[-e^{-x} \right]_1^\infty = \int_0^1 e^{-x^2}\, dx + e^{-1}.$

Therefore, the integral converges.

53. To evaluate $\int e^{-st} \cos at \, dt$ use Integration By Parts.

$$dv = \cos at \, dt \;\Longrightarrow\; v = \int \cos at \, dt = \frac{1}{a}\sin at$$

$$u = e^{-st} \quad\Longrightarrow\; du = -se^{-st}\,dt$$

$$\int e^{-st}\cos at \, dt = \frac{1}{a}e^{-st}\sin at + \frac{s}{a}\int e^{-st}\sin at \, dt$$

Use Integration By Parts again.

$$dv = \sin at \, dt \;\Longrightarrow\; v = \int \sin at \, dt = -\frac{1}{a}\cos at$$

$$u = e^{-st} \quad\Longrightarrow\; du = -se^{-st}\,dt$$

$$\int e^{-st}\cos at \, dt = \frac{1}{a}e^{-st}\sin at + \frac{s}{a}\left(-\frac{1}{a}e^{-st}\cos at - \frac{s}{a}\int e^{-st}\cos at \, dt\right)$$

Solving for the integral in the equation yields

$$\int e^{-st}\cos at \, dt = \frac{e^{-st}}{s^2 + a^2}(a\sin at - s\cos at) + C.$$

Therefore,

$$\int_0^{\infty} e^{-st}\cos at \, dt = \lim_{b\to\infty}\left[\frac{e^{-st}}{s^2 + a^2}(a\sin at - s\cos at)\right]_0^b$$

$$= 0 + \frac{s}{s^2 + a^2} = \frac{s}{s^2 + a^2}, \; s > 0.$$

59. By symmetry the perimeter of this figure is four times the length of the arc in the first quadrant. Thus,

$$s = 4\int_0^1 \sqrt{1 + (y')^2}\, dx.$$

By implicit differentiation we have

$$x^{2/3} + y^{2/3} = 1$$

$$\frac{2}{3}x^{-1/3} + \frac{2}{3}y^{-1/3}y' = 0$$

$$y' = -\frac{y^{1/3}}{x^{1/3}}.$$

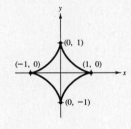

Therefore,

$$s = 4\int_0^1 \sqrt{1 + \left(-\frac{y^{1/3}}{x^{1/3}}\right)^2}\, dx = 4\int_0^1 \sqrt{\frac{x^{2/3} + y^{2/3}}{x^{2/3}}}\, dx$$

$$= 4\lim_{b\to 0^+}\int_b^1 \sqrt{\frac{1}{x^{2/3}}}\, dx = 4\lim_{b\to 0^+}\int_b^1 x^{-1/3}\, dx$$

$$= 4\lim_{b\to 0^+}\left[\frac{x^{2/3}}{2/3}\right]_b^1 = 4\left(\frac{3}{2}\right) = 6.$$

65. (a) $C = 650,000 + \displaystyle\int_0^5 25,000e^{-0.06t}\, dt$

$$= 650,000 - \left[\frac{25,000}{0.06}e^{-0.06t}\right]_0^5 \approx \$757,992.41$$

(b) $C = 650,000 + \displaystyle\int_0^{10} 25,000e^{-0.06t}\, dt$

$$= 650,000 - \left[\frac{25,000}{0.06}e^{-0.06t}\right]_0^{10} \approx \$837,995.15$$

(c) $C = 650,000 + \displaystyle\int_0^{\infty} 25,000e^{-0.06t}\, dt$

$$- 650,000 - \lim_{b\to\infty}\left[\frac{25,000}{0.06}e^{-0.06t}\right]_0^b \approx \$1,066,666.67$$

Review Exercises for Chapter 7

5. $\displaystyle\int \sec^4 \frac{x}{2}\, dx = \int \sec^2 \frac{x}{2} \sec^2 \frac{x}{2}\, dx$

$$= \int \left[\tan^2 \frac{x}{2} + 1 \right] \sec^2 \frac{x}{2}\, dx$$

$$= 2\int \tan^2 \frac{x}{2} \sec^2 \frac{x}{2}\left(\frac{1}{2}\right) dx + 2\int \sec^2 \frac{x}{2}\left(\frac{1}{2}\right) dx$$

$$= \frac{2}{3} \tan^3 \frac{x}{2} + 2 \tan \frac{x}{2} + C$$

$$= \frac{2}{3}\left[\tan^3 \frac{x}{2} + 3 \tan \frac{x}{2} \right] + C$$

9. Since $x^3 - x^2 + x - 1 = (x - 1)(x^2 + 1)$, we have

$$\frac{x^2 + 2x}{(x - 1)(x^2 + 1)} = \frac{A}{x - 1} + \frac{Bx + C}{x^2 + 1}.$$

Multiplying by $(x - 1)(x^2 + 1)$ yields

$$x^2 + 2x = A(x^2 + 1) + (Bx + C)(x - 1).$$

If $x = 1$, then $1 + 2 = A(2) + (B + C)(0)$ or $A = \dfrac{3}{2}$.

Furthermore,

$$x^2 + 2x = Ax^2 + A + Bx^2 - Bx + Cx - C$$

$$= (A + B)x^2 + (C - B)x + (A - C).$$

Now by equating the coefficients, we have

$$1 = A + B = \frac{3}{2} + B \qquad \text{or} \qquad B = -\frac{1}{2}$$

$$2 = C - B = C + \frac{1}{2} \qquad \text{or} \qquad C = \frac{3}{2}.$$

Therefore,

$$\int \frac{x^2 + 2x}{(x - 1)(x^2 + 1)}\, dx = \int \left[\frac{3/2}{x - 1} + \frac{(-x/2) + (3/2)}{x^2 + 1} \right] dx$$

$$= \frac{3}{2}\int \frac{1}{x - 1}\, dx - \frac{1}{2}\int \frac{x}{x^2 + 1}\, dx + \frac{3}{2}\int \frac{1}{x^2 + 1}\, dx$$

$$= \frac{3}{2} \ln|x - 1| - \frac{1}{4} \ln(x^2 + 1) + \frac{3}{2} \arctan x + C$$

$$= \frac{1}{4}[6 \ln|x - 1| - \ln(x^2 + 1) + 6 \arctan x] + C.$$

13. $\displaystyle\int \frac{1}{1 - \sin \theta}\, d\theta = \int \frac{1}{1 - \sin \theta}\left(\frac{1 + \sin \theta}{1 + \sin \theta} \right) d\theta$

$$= \int \frac{1 + \sin \theta}{1 - \sin^2 \theta}\, d\theta = \int \frac{1 + \sin \theta}{\cos^2 \theta}\, d\theta$$

$$= \int \left(\frac{1}{\cos^2 \theta} + \frac{\sin \theta}{\cos \theta \cos \theta} \right) d\theta$$

$$= \int (\sec^2 \theta + \sec \theta \tan \theta)\, d\theta = \tan \theta + \sec \theta + C$$

15. Using Integration By Parts, let

$$dv = \frac{1}{x^2}\,dx \implies v = -\frac{1}{x}$$

$$u = \ln(2x) \implies du = \frac{1}{x}\,dx.$$

$$\int \frac{\ln(2x)}{x^2}\,dx = -\frac{\ln(2x)}{x} - \int \frac{1}{x}\left(-\frac{1}{x}\right)dx$$

$$= -\frac{\ln(2x)}{x} + \int x^{-2}\,dx$$

$$= -\frac{\ln(2x)}{x} - \frac{1}{x} + C$$

$$= -\frac{1}{x}(1 + \ln 2x) + C$$

25. To evaluate $\int \theta \sin\theta \cos\theta\,d\theta$ use Integration By Parts.

$$dv = \sin\theta\cos\theta\,d\theta \implies v = \int \sin\theta\cos\theta\,d\theta = \frac{1}{2}\sin^2\theta$$

$$u = \theta \qquad\qquad \implies du = d\theta$$

$$\int \theta \sin\theta \cos\theta\,du = \frac{1}{2}\theta\sin^2\theta - \frac{1}{2}\int \sin^2\theta\,d\theta$$

$$= \frac{1}{2}\theta\sin^2\theta - \frac{1}{4}\int (1 - \cos 2\theta)\,d\theta$$

$$= \frac{1}{2}\theta\sin^2\theta - \frac{1}{4}\left(\theta - \frac{1}{2}\sin 2\theta\right) + C$$

$$= \frac{1}{2}\theta\sin^2\theta - \frac{1}{4}\theta + \frac{1}{8}\sin 2\theta + C$$

$$= \frac{1}{8}(4\theta\sin^2\theta + \sin 2\theta - 2\theta) + C$$

$$= \frac{1}{8}\left[4\theta\left(\frac{1 - \cos 2\theta}{2}\right) - \sin 2\theta - 2\theta\right] + C$$

$$= \frac{1}{8}(\sin 2\theta - 2\theta\cos 2\theta) + C$$

A second method of evaluating the integral is to use the double angle identity on the integrand and integrate by parts.

$$dv = \sin 2\theta\,d\theta \implies v = \int \sin 2\theta\,d\theta = -\frac{1}{2}\cos\theta$$

$$u = \theta \qquad \implies du = d\theta$$

$$\int \theta\sin\theta\cos\theta\,d\theta = \frac{1}{2}\int \theta\sin 2\theta\,d\theta$$

$$= -\frac{1}{4}\theta\cos 2\theta + \frac{1}{4}\int \cos 2\theta\,d\theta$$

$$= -\frac{1}{4}\theta\cos 2\theta + \frac{1}{8}\sin 2\theta = \frac{1}{8}(\sin 2\theta - 2\theta\cos 2\theta) + C$$

35. Let $u = x^{1/4}$. Then $u^4 = x$, $u^2 = x^{1/2}$, $dx = 4u^3\,du$, and

$$\int \frac{x^{1/4}}{1 + x^{1/2}}\,dx = \int \frac{u}{1 + u^2}(4u^3\,du) = 4\int \frac{u^4}{1 + u^2}\,du$$

$$= 4\int \left(u^2 - 1 + \frac{1}{1 + u^2}\right) du$$

$$= 4\left(\frac{u^3}{3} - u + \arctan u\right) + C$$

$$= \frac{4}{3}(x^{3/4} - 3x^{1/4} + 3\arctan x^{1/4}) + C.$$

43. Using a symbolic integration utility to perform the integration yields

$$\int \frac{1}{1 + \sin \theta + \cos \theta}\,d\theta = \ln\left|\frac{1 + \sin \theta + \cos \theta}{1 + \cos \theta}\right| + C.$$

Substitute the solution point $(0, 0)$ into the antiderivative and solve for the constant of integration.

$$y = \ln\left|\frac{1 + \sin \theta + \cos \theta}{1 + \cos \theta}\right| + C$$

$$0 = \ln\left|\frac{1 + \sin 0 + \cos 0}{1 + \cos 0}\right| + C = \ln 1 + C = C$$

The graph of the function

$$y = \ln\left|\frac{1 + \sin \theta + \cos \theta}{1 + \cos \theta}\right|$$

is shown in the figure.

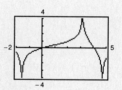

53. (a) Let $x = 2\tan \theta$. Then $\sqrt{4 + x^2} = 2\sec \theta$ and $dx = 2\sec^2 \theta\,d\theta$. Thus,

$$\int \frac{x^3}{\sqrt{4 + x^2}}\,dx = \int \frac{(2\tan \theta)^3}{2\sec \theta}(2\sec^2 \theta\,d\theta)$$

$$= 8\int \tan^3 \theta \sec \theta\,d\theta$$

$$= 8\int (\sec^2 \theta - 1)\tan \theta \sec \theta\,d\theta$$

$$= 8\left(\frac{1}{3}\sec^3 \theta - \sec \theta\right) + C$$

$$= \frac{8}{3}\sec \theta(\sec^2 \theta - 3) + C$$

$$= \frac{\sqrt{4 + x^2}}{3}(x^2 - 8) + C.$$

(b) Let $u^2 = 4 + x^2$. Then, $2u\,du = 2x\,dx$ and we have

$$\int \frac{x^3}{\sqrt{4 + x^3}}\,dx = \int \frac{x^2(x\,dx)}{\sqrt{4 + x^2}}$$

$$= \int \frac{(u^2 - 4)(u\,du)}{u}$$

$$= \int (u^2 - 4)\,du$$

$$= \frac{1}{3}u^3 - 4u + C$$

$$= \frac{u}{3}(u^2 - 12) + C$$

$$= \frac{\sqrt{4 + x^2}}{3}(x^2 - 8) + C.$$

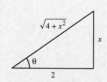

—CONTINUED—

53. **—CONTINUED—**

(c) Let

$$dv = \frac{x}{\sqrt{4 + x^2}} \, dx \implies v = \int \frac{x}{\sqrt{4 + x^2}} \, dx = \sqrt{4 + x^2}$$

$$u = x^2 \qquad \implies du = 2x \, dx.$$

$$\int u \, dv = uv - \int v \, du$$

$$\int \frac{x^3}{\sqrt{4 + x^2}} \, dx = x^2 \sqrt{4 + x^2} - \int 2x \sqrt{4 + x^2} \, dx$$

$$= x^2 \sqrt{4 + x^2} - \frac{2}{3}(4 + x^2)^{3/2} + C$$

$$= \frac{\sqrt{4 + x^2}}{3}(x^2 - 8) + C$$

63. The graph of the region is shown in the figure. Since the region is symmetric to the y-axis, $\bar{x} = 0$. The area of the region is $A = \pi/2$.

$$M_x = \frac{1}{2} \int_{-1}^{1} \left(\sqrt{1 - x^2} \right)^2 dx$$

$$= \int_0^1 (1 - x^2) \, dx = \left[x - \frac{1}{3}x^3 \right]_0^1 = \frac{2}{3}$$

$$\bar{y} = \frac{M_x}{A} = \frac{2/3}{\pi/2} = \frac{4}{3\pi}$$

$$(\bar{x}, \bar{y}) = \left(0, \frac{4}{3\pi} \right)$$

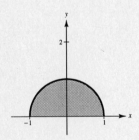

73. Begin by taking the natural logarithm of both members of the equation

$$y = \lim_{x \to \infty} (\ln x)^{2/x}.$$

Then we obtain

$$\ln y = \ln \left[\lim_{x \to \infty} (\ln x)^{2/x} \right]$$

$$= \lim_{x \to \infty} \left[\frac{2}{x} \ln(\ln x) \right]$$

$$= 2 \lim_{x \to \infty} \frac{(1/\ln x)(1/x)}{1} = 0.$$

Finally, as $\ln y \to 0$, we know that $y \to 1$ and conclude that

$$\lim_{x \to \infty} (\ln x)^{2/x} = 1.$$

CHAPTER 8
Infinite Series

CHAPTER 8
Infinite Series

Section 8.1   Sequences

Solutions to Selected Odd-Numbered Exercises

9. Since $a_1 = 3$, and $a_{k+1} = 2(a_k - 1)$, we have

$$a_1 = 3$$
$$a_2 = 2(a_1 - 1) = 2(3 - 1) = 4$$
$$a_3 = 2(a_2 - 1) = 2(4 - 1) = 6$$
$$a_4 = 2(a_3 - 1) = 2(6 - 1) = 10$$
$$a_5 = 2(a_4 - 1) = 2(10 - 1) = 18.$$

23. The difference between two consecutive terms in the sequence 2, 5, 8, 11, ... is 3. Therefore, the sequence can be defined recursively as $a_1 = 2$ and $a_{k+1} = a_k + 3$. The next two terms of the sequence are 14 and 17.

31. $\dfrac{(2n - 1)!}{(2n + 1)!} = \dfrac{(2n - 1)(2n - 2) \cdots 3 \cdot 2 \cdot 1}{(2n + 1)(2n)(2n - 1)(2n - 2) \cdots 3 \cdot 2 \cdot 1} = \dfrac{1}{(2n + 1)(2n)}$

35. $-1, 2, 7, 14, 23, \ldots$

Compare the terms of the sequence $-1, 2, 7, 14, 23, \ldots$ with the sequence of squares

$$1^2, 2^2, 3^2, 4^2, 5^2, \ldots = 1, 4, 9, 16, 25, \ldots.$$

Observe that each term of the given sequence is two less than the sequence of squares. Thus, we write the nth term of the given sequence as $a_n = n^2 - 2$.

39. $2, -1, \dfrac{1}{2}, \dfrac{-1}{4}, \dfrac{1}{8}, \ldots$

First observe that the denominators are powers of 2, where for $n = 4$ the denominator is 2^2. This implies that the nth term has a denominator 2^{n-2}. Note further that the signs alternate starting with a positive sign, and thus the nth term is

$$a_n = (-1)^{n-1}\left(\frac{1}{2^{n-2}}\right) = \frac{(-1)^{n-1}}{2^{n-2}}.$$

45. $1, -\dfrac{1}{1 \cdot 3}, \dfrac{1}{1 \cdot 3 \cdot 5}, -\dfrac{1}{1 \cdot 3 \cdot 5 \cdot 7}, \ldots$

The denominator of the nth term is the product of the first n positive odd integers. Note further that the signs alternate starting with a positive sign, and thus the nth term is

$$a_n = \frac{(-1)^{n-1}}{1 \cdot 3 \cdot 5 \cdots (2n - 1)}.$$

A second form of the nth term is obtained when we multiply the numerator and denominator by the n missing even integers. Then,

$$a_n = \frac{(-1)^{n-1}}{1 \cdot 3 \cdot 5 \cdots (2n - 1)}$$

$$= \frac{(-1)^{n-1} 2 \cdot 4 \cdot 6 \cdot 8 \cdots (2n)}{1 \cdot 2 \cdot 3 \cdot 4 \cdot 5 \cdots (2n - 1)(2n)}$$

$$= \frac{(-1)^{n-1} 2^n (1 \cdot 2 \cdot 3 \cdot 4 \cdots n)}{1 \cdot 2 \cdot 3 \cdot 4 \cdot 5 \cdots (2n - 1)(2n)}$$

$$= \frac{(-1)^{n-1} 2^n n!}{(2n)!}.$$

51. $a_n = (-1)^n\left(\dfrac{n}{n + 1}\right)$

Since

$$\lim_{n \to \infty} a_n = \lim_{n \to \infty} (-1)^n\left(\frac{n}{n + 1}\right) = \pm 1,$$

the limit does not exist and the sequence $\{a_n\}$ diverges.

59. $a_n = \dfrac{(n+1)!}{n!}$

Since

$$\lim_{n\to\infty} a_n = \lim_{n\to\infty} \frac{(n+1)!}{n!}$$

$$= \lim_{n\to\infty} \frac{(n+1)(n)(n-1)\cdots 3\cdot 2\cdot 1}{n(n-1)\cdots 3\cdot 2\cdot 1}$$

$$= \lim_{n\to\infty}(n+1) = \infty,$$

the sequence $\{a_n\}$ diverges.

71. $a_n = (-1)^n\left(\dfrac{1}{n}\right)$

Writing out the first few terms of the sequence we have

$$a_1 = -1, a_2 = \frac{1}{2}, a_3 = -\frac{1}{3}, a_4 = \frac{1}{4}, \cdots.$$

Because of the alternating signs, observe that the terms are neither nondecreasing nor nonincreasing. Therefore, the sequence is *not* monotonic. The sequence is bounded since $-1 \le a_n \le 1$ for all n.

85. (a) If we let A_n be the amount budgeted after n years, we have

$$A_1 = 2.5 - 0.2(2.5) = 0.8(2.5) \text{ billion}$$

$$A_2 = A_1 - 0.2A_1 = 0.8A_1 = 0.8^2(2.5) \text{ billion}$$

$$A_3 = A_2 - 0.2A_2 = 0.8A_2 = 0.8^2(2.5) \text{ billion}$$

$$\vdots$$

$$A_n = 2.5(0.8)^n \text{ billion}.$$

65. $a_n = \left(1 + \dfrac{k}{n}\right)^n$

$$\lim_{n\to\infty} a_n = \lim_{n\to\infty}\left(1 + \frac{k}{n}\right)^n = \lim_{n\to\infty}\left[\left(1 + \frac{k}{n}\right)^{n/k}\right]^k$$

Now let $u = k/n$. Then as n approaches infinity, u approaches zero, and

$$\lim_{n\to\infty} a_n = \lim_{u\to 0}\left[(1+u)^{1/u}\right]^k = e^k.$$

Therefore the sequence converges to e^k.

(b) $A_1 = 2.5(0.8) = \$2$ billion

$A_2 = 2.5(0.8)^2 = \$1.6$ billion

$A_3 = 2.5(0.8)^3 = \$1.28$ billion

$A_4 = 2.5(0.8)^4 = \$1.024$ billion

(c) $\lim\limits_{n\to\infty} 2.5(0.8)^n = 0$ and therefore the sequence converges.

95. $a_{n+2} = a_n + a_{n+1}$

(a)
$a_1 = 1$ $a_7 = 8 + 5 = 13$

$a_2 = 1$ $a_8 = 13 + 8 = 21$

$a_3 = 1 + 1 = 2$ $a_9 = 21 + 13 = 34$

$a_4 = 2 + 1 = 3$ $a_{10} = 34 + 21 = 55$

$a_5 = 3 + 2 = 5$ $a_{11} = 55 + 34 = 89$

$a_6 = 5 + 3 = 8$ $a_{12} = 89 + 55 = 144$

(c) $1 + \dfrac{1}{b_{n-1}} = 1 + \dfrac{1}{\dfrac{a_n}{a_{n-1}}}$

$$= 1 + \frac{a_{n-1}}{a_n}$$

$$= \frac{a_n + a_{n-1}}{a_n} = \frac{a_{n+1}}{a_n} = b_n$$

(b) $b_n = \dfrac{a_{n+1}}{a_n},\ n \ge 1$

$b_1 = \dfrac{1}{1} = 1$ $b_6 = \dfrac{13}{8}$

$b_2 = \dfrac{2}{1} = 2$ $b_7 = \dfrac{21}{13}$

$b_3 = \dfrac{3}{2}$ $b_8 = \dfrac{34}{21}$

$b_4 = \dfrac{5}{3}$ $b_9 = \dfrac{55}{34}$

$b_5 = \dfrac{8}{5}$ $b_{10} = \dfrac{89}{55}$

(d) If $\lim\limits_{n\to\infty} b_n = \rho$, then $\lim\limits_{n\to\infty}\left(1 + \dfrac{1}{b_{n-1}}\right) = \rho.$

Since $\lim\limits_{n\to\infty} b_n = \lim\limits_{n\to\infty} b_{n-1}$ we have $1 + (1/\rho) = \rho.$

$$\rho + 1 = \rho^2$$

$$0 = \rho^2 - \rho - 1$$

$$\rho = \frac{1 \pm \sqrt{1+4}}{2} = \frac{1 \pm \sqrt{5}}{2}$$

Since a_n, and thus b_n, is positive,

$$\rho = \left(1 + \sqrt{5}\right)/2 \approx 1.6180.$$

Section 8.2   Series and Convergence

3. $3 - \dfrac{9}{2} + \dfrac{27}{4} - \dfrac{81}{8} + \dfrac{243}{16} - \cdots$

$S_1 = 3$

$S_2 = 3 - \dfrac{9}{2} = -\dfrac{3}{2} = -1.5$

$S_3 = 3 - \dfrac{9}{2} + \dfrac{27}{4} = \dfrac{21}{4} = 5.25$

$S_4 = 3 - \dfrac{9}{2} + \dfrac{27}{4} - \dfrac{81}{8} = -\dfrac{39}{8} = -4.875$

$S_5 = 3 - \dfrac{9}{2} + \dfrac{27}{4} - \dfrac{81}{8} + \dfrac{243}{16} = \dfrac{165}{16} = 10.3125$

11. $\displaystyle\sum_{n=0}^{\infty} 3\left(\dfrac{3}{2}\right)^n = 3 + \dfrac{9}{2} + \dfrac{27}{4} + \dfrac{81}{8} + \cdots$

The series is geometric with common ratio $r = \dfrac{3}{2}$. Since $|r| \geq 1$, the series diverges.

15. $\displaystyle\sum_{n=1}^{\infty} \dfrac{2^n + 1}{2^{n+1}}$

$\displaystyle\lim_{n\to\infty} a_n = \lim_{n\to\infty} \dfrac{2^n - 1}{2^{n+1}} = \lim_{n\to\infty} \dfrac{1 - (1/2^n)}{2} = \dfrac{1}{2} \neq 0$

Therefore, the series diverges by the nth-Term Test for

17. $\displaystyle\sum_{n=0}^{\infty} 2\left(\dfrac{3}{4}\right)^n = 2 + \dfrac{3}{2} + \dfrac{9}{8} + \dfrac{27}{32} + \dfrac{81}{128} + \cdots$

The series is geometric with $a = 2$ and $r = \dfrac{3}{4}$. Since $|r| < 1$, the series converges.

21. $\displaystyle\sum_{n=1}^{\infty} \dfrac{1}{n(n+1)}$

Using partial fractions, we can write

$a_n = \dfrac{1}{n(n+1)} = \dfrac{1}{n} - \dfrac{1}{n+1}.$

Therefore,

$\displaystyle\sum_{n=1}^{\infty} \dfrac{1}{n(n+1)} = \sum_{n=1}^{\infty} \left(\dfrac{1}{n} - \dfrac{1}{n+1}\right)$

$= \left(1 - \dfrac{1}{2}\right) + \left(\dfrac{1}{2} - \dfrac{1}{3}\right) + \left(\dfrac{1}{3} - \dfrac{1}{4}\right) + \cdots$

where each term after the first term of 1 cancels. Thus, the series converges to 1.

27. (a) $\displaystyle\sum_{n=1}^{\infty} 2(0.9)^{n-1} = \sum_{n=0}^{\infty} 2(0.9)^n = \dfrac{2}{1 - 0.9} = 20$

(b)

n	5	10	20	50	100
S_n	8.1902	13.0264	17.5685	19.8969	19.9995

(c) The graph is shown if the figure.

(d) The terms of the series decrease in magnitude slowly, and the sequence of partial sums approaches the sum of the series relatively slowly.

35. The series

$$1 + 0.1 + 0.01 + 0.001 + \cdots = \sum_{n=0}^{\infty} (0.1)^n$$

is geometric with $a = 1$ and $r = 0.1$. Therefore, the sum is

$$S = \frac{a}{1 - r} = \frac{1}{1 - 0.1} = \frac{10}{9}.$$

41. $\displaystyle\sum_{n=1}^{\infty} \frac{4}{n(n + 2)}$

Using partial fractions, we have

$$\frac{4}{n(n + 2)} = \frac{A}{n} + \frac{B}{n + 2}.$$

Multiplying by $n(n + 2)$, yields

$$4 = A(n + 2) + Bn.$$

If $n = 0$, then $4 = 2A$ or $A = 2$. If $n = -2$, then $4 = -2B$ or $B = -2$. Thus,

$$\sum_{n=1}^{\infty} \frac{4}{n(n + 2)} = \sum_{n=1}^{\infty} \left[\frac{2}{n} - \frac{2}{n + 2} \right]$$

$$= \left[2 - \frac{2}{3} \right] + \left[\frac{2}{2} - \frac{2}{4} \right] + \left[\frac{2}{3} - \frac{2}{5} \right] + \left[\frac{2}{4} - \frac{2}{6} \right] + \left[\frac{2}{5} - \frac{2}{7} \right] + \cdots = 2 + 1 = 3.$$

47. $0.07\overline{75} = 0.075 + 0.00075 + 0.0000075 + \cdots$

$$= 0.075(1 + 0.01 + 0.0001 + \cdots)$$

$$= \sum_{n=0}^{\infty} \frac{75}{1000} \left(\frac{1}{100} \right)^n = \sum_{n=0}^{\infty} \frac{3}{40} \left(\frac{1}{100} \right)^n$$

This is a geometric series with $a = \frac{3}{40}$ and $r = \frac{1}{100}$. Therefore,

$$0.07\overline{75} = \frac{a}{1 - r} = \frac{3/40}{1 - (1/100)} = \frac{5}{66}.$$

53. $\displaystyle\sum_{n=1}^{\infty} \frac{3n - 1}{2n + 1}$

Since

$$\lim_{n\to\infty} a_n = \lim_{n\to\infty} \frac{3n - 1}{2n + 1} = \frac{3}{2} \neq 0,$$

the series diverges by the nth-Term Test for Divergence.

61. (a) The ratios of consecutive terms are x. Therefore, the series is geometric with common ratio x.

(b) Since the first term is $a = 1$ and the common ratio is $r = x$, the series will converge if $|x| < 1$.

$$1 + x + x^2 + x^3 + \cdots = \sum_{n=0}^{\infty} x^n$$

$$= \frac{a}{1 - r} = \frac{1}{1 - x}$$

$$f(x) = \frac{1}{1 - x}, \quad |x| < 1$$

(c) The graph is given in the figure.

71. $P(n) = \frac{1}{2} \left(\frac{1}{2} \right)^n$

$$P(2) = \frac{1}{2} \left(\frac{1}{2} \right)^2 = \frac{1}{8}$$

The series

$$\sum_{n=0}^{\infty} \frac{1}{2} \left(\frac{1}{2} \right)^n$$

is geometric with $a = \frac{1}{2}$ and $r = \frac{1}{2}$. Therefore, it converges and

$$\sum_{n=0}^{\infty} \frac{1}{2} \left(\frac{1}{2} \right)^n = \frac{1/2}{1 - (1/2)} = 1.$$

75. Since the wage keeps doubling every day, the sequence of wages is geometric with $a = 0.01$ and $r = 2$. Therefore, the total wage W after n days is

$$\sum_{n=0}^{n-1} 0.01(2)^n = \frac{0.01(1 - 2^n)}{1 - 2} = 0.01(2^n - 1).$$

(a) When $n = 29$, $W = \$5,368,709.11$.

(b) When $n = 30$, $W = \$10,737,418.23$.

(c) When $n = 31$, $W = \$21,474,836.47$.

77. (a) Use $P = \$50$, $r = 0.03$, and $t = 20$ in the formula

$$A = P\left(\frac{12}{r}\right)\left[\left(1 + \frac{r}{12}\right)^{12t} - 1\right]$$

to obtain

$$A = 50\left(\frac{12}{0.03}\right)\left[\left(1 + \frac{0.03}{12}\right)^{12 \cdot 20} - 1\right]$$

$$= \$16,415.10.$$

(b) Use $P = \$50$, $r = 0.03$, and $t = 20$ in the formula

$$A = \frac{P(e^{rt} - 1)}{e^{r/12} - 1}$$

to obtain

$$A = \frac{50(e^{0.03 \cdot 20} - 1)}{e^{0.03/12} - 1} = \$16,421.83.$$

Section 8.3   The Integral Test and *p*-Series

7. $\dfrac{\ln 2}{2} + \dfrac{\ln 3}{3} + \dfrac{\ln 4}{4} + \dfrac{\ln 5}{5} + \cdots$

First observe that the nth term of the series is $a_n = \dfrac{\ln n}{n}$.

Since $f(x) = \dfrac{\ln x}{x}$ is positive, continuous, and decreasing for $x \geq 2$, we use the Integral Test.

$$\int_2^\infty \frac{\ln x}{x}\, dx = \int_2^\infty (\ln x)\frac{1}{x}\, dx = \left[\frac{(\ln x)^2}{2}\right]_2^\infty = \infty$$

Therefore, the series diverges by the Integral Test.

13. $\displaystyle\sum_{n=2}^{\infty} \frac{1}{n(\ln n)^p}$

$$f(x) = \frac{1}{x(\ln x)^p}$$

The conditions of the Integral Test are satisfied if we let $f(x) = \dfrac{1}{x(\ln x)^p}$.

If $u = \ln x$ then $du = (1/x)\, dx$. Thus,

$$\int_2^\infty \frac{1}{x(\ln x)^p}\, dx = \lim_{b \to \infty} \int_2^b (\ln x)^{-p}\frac{1}{x}\, dx$$

$$= \lim_{b \to \infty} \left[\frac{(\ln x)^{-p+1}}{-p + 1}\right]_2^b$$

$$= \lim_{b \to \infty} \left[\left(\frac{1}{1 - p}\right)\frac{1}{(\ln b)^{p-1}} - \left(\frac{1}{1 - p}\right)\frac{1}{(\ln 2)^{p-1}}\right]$$

$$= 0 + \frac{1}{(p - 1)(\ln 2)^{p-1}} \quad \text{for } p - 1 > 0.$$

Therefore, the given series converges if $p > 1$.

19. Since

$$1 + \frac{1}{2\sqrt{2}} + \frac{1}{3\sqrt{3}} + \frac{1}{4\sqrt{4}} + \frac{1}{5\sqrt{5}} + \cdots = \sum_{n=1}^{\infty} \frac{1}{n\sqrt{n}} = \sum_{n=1}^{\infty} \frac{1}{n^{3/2}},$$

we have a p-series with $p = \frac{3}{2}$. Since $p > 1$, the series converges.

33. $\displaystyle\sum_{n=1}^{\infty} \frac{1}{n^4}$

$$S_6 = 1 + \frac{1}{2^4} + \frac{1}{3^4} + \frac{1}{4^4} + \frac{1}{5^4} + \frac{1}{6^4} \approx 1.0811$$

$$R_6 \le \int_6^{\infty} \frac{1}{x^4}\, dx = \left[-\frac{1}{3x^3} \right]_6^{\infty} \approx 0.0015$$

$$1.0811 \le \sum_{n=1}^{\infty} \frac{1}{n^4} \le 1.0811 + 0.0015 = 1.0826$$

37. $\displaystyle\sum_{n=1}^{\infty} ne^{-n^2}$

$$S_4 = e^{-1} + 2e^{-4} + 3e^{-9} + 4e^{-16} \approx 0.4049$$

$$R_4 \le \int_4^{\infty} xe^{-x^2}\, dx = \left[-\frac{1}{2}e^{-x^2} \right]_4^{\infty} \approx 5.6 \times 10^{-8}$$

$$0.4049 \le \sum_{n=1}^{\infty} ne^{-n^2} \le 0.4049 + (5.6 \times 10^{-8})$$

49. $\displaystyle\sum_{n=1}^{\infty} \frac{1}{2n - 1}$

Let $f(x) = \dfrac{1}{2x - 1}$.

f is positive, continuous, and decreasing for $x \ge 1$. Therefore,

$$\int_1^{\infty} \frac{1}{2x - 1} = \left[\ln\sqrt{2x - 1}\, \right]_1^{\infty} = \infty,$$

and the series diverges by the Integral Test.

55. $\displaystyle\sum_{n=1}^{\infty} \frac{n}{\sqrt{n^2 + 1}}$

$$\lim_{n\to\infty} \frac{n}{\sqrt{n^2 + 1}} = \lim_{n\to\infty} \frac{1}{\sqrt{1 + (1/n^2)}} = 1 \ne 0$$

Therefore, the series diverges by nth-Term Test for Divergence.

Section 8.4   Comparisons of Series

5. $\displaystyle\sum_{n=2}^{\infty} \frac{1}{n - 1}$

This series resembles $\displaystyle\sum_{n=2}^{\infty} \frac{1}{n}$, a divergent p-series.

Term-by-term comparison yields

$$a_n = \frac{1}{n} < \frac{1}{n - 1} = b_n.$$

By the Direct Comparison Test, the series diverges.

11. $\displaystyle\sum_{n=0}^{\infty} \frac{1}{n!}$

Compare the given series to the convergent p-series

$$\sum_{n=1}^{\infty} \frac{1}{n^2}.$$

If $n > 3$, then $n^2 < n!$ and $\dfrac{1}{n!} < \dfrac{1}{n^2}$.

By the Direct Comparison Test, the series converges.

15. $\displaystyle\sum_{n=1}^{\infty} \frac{n}{n^2 + 1}$

The series can be compared with

$$\sum_{n=1}^{\infty} \frac{1}{n}$$

since the degree of the denominator is only one greater than the degree of the numerator. Furthermore, since

$$\lim_{n\to\infty} \frac{n/(n^2 + 1)}{1/n} = \lim_{n\to\infty} \frac{n^2}{n^2 + 1} = 1$$

and since the series

$$\sum_{n=1}^{\infty} \frac{1}{n}$$

diverges, the given series also diverges by the Limit Comparison Test.

23. $\displaystyle\sum_{n=1}^{\infty} \frac{1}{n\sqrt{n^2 + 1}}$

For "large" values of n

$$\frac{1}{n\sqrt{n^2 + 1}} \approx \frac{1}{n\sqrt{n^2}} = \frac{1}{n^2}.$$

By comparing the given series with the convergent p-series

$$\sum_{n=1}^{\infty} \frac{1}{n^2},$$

we have

$$\lim_{n\to\infty} \frac{a_n}{b_n} = \lim_{n\to\infty} \left(\frac{1}{n\sqrt{n^2 + 1}} \right)\left(\frac{n^2}{1} \right)$$

$$= \lim_{n\to\infty} \frac{1}{\sqrt{1 + (1/n)}} = 1.$$

By the Limit Comparison Test, the given series converges.

33. $\displaystyle\sum_{n=1}^{\infty}\frac{n}{2n+3}$

Since

$$\lim_{n\to\infty} a_n = \lim_{n\to\infty}\frac{n}{2n+3} = \frac{1}{2} \neq 0,$$

the series diverges by the *n*th-Term Divergence Test for Divergence.

45. (a) $\displaystyle\sum_{n=1}^{\infty}\frac{1}{(2n-1)^2}$

Since the degree of the numerator is 2 less than the degree of the denominator, the series converges by the Polynomial Test given in Exercise 38.

(c) Since

$$\sum_{n=1}^{2}\frac{1}{(2n-1)^2} + \sum_{n=3}^{\infty}\frac{1}{(2n-1)^2} = \sum_{n=1}^{\infty}\frac{1}{(2n-1)^2}$$

$$= \frac{\pi^2}{8},$$

we have

$$\sum_{n=3}^{\infty}\frac{1}{(2n-1)^2} = \frac{\pi^2}{8} - \sum_{n=1}^{2}\frac{1}{(2n-1)^2}$$

$$= \frac{\pi^2}{8} - \left(1 + \frac{1}{9}\right) = 0.1226.$$

39. $\dfrac{1}{2} + \dfrac{2}{5} + \dfrac{3}{10} + \dfrac{4}{17} + \dfrac{5}{26} + \cdots$

The *n*th term of the series is

$$a_n = \frac{n}{n^2+1}$$

and the series has the form

$$\sum_{n=1}^{\infty}\frac{n}{n^2+1} = \sum_{n=1}^{\infty}\frac{P(n)}{Q(n)}.$$

$P(n)$ has degree $k = 1$ and $Q(n)$ has degree $k = 2$. Hence, it follows from the Polynomial Test (Exercise 38) that the series diverges.

(b)

n	5	10	20	50	100
S_n	1.1839	1.2087	1.2212	1.2287	1.2312

(d) Since

$$\sum_{n=1}^{9}\frac{1}{(2n-1)^2} + \sum_{n=10}^{\infty}\frac{1}{(2n-1)^2} = \sum_{n=1}^{\infty}\frac{1}{(2n-1)^2}$$

$$= \frac{\pi^2}{8},$$

we have

$$\sum_{n=10}^{\infty}\frac{1}{(2n-1)^2} = \frac{\pi^2}{8} - \sum_{n=1}^{9}\frac{1}{(2n-1)^2}$$

$$= \frac{\pi^2}{8} - S_9 = 0.0277.$$

(Note that in parts (c) and (d) we know the series converges because only a finite number of terms have been subtracted from a known convergent series.)

Section 8.5   Alternating Series

11. $\displaystyle\sum_{n=1}^{\infty}\frac{(-1)^{n+1}}{2n-1}$

Observe that

$$\lim_{n\to\infty} a_n = \lim_{n\to\infty}\frac{1}{2n-1} = 0$$

and

$$a_{n+1} = \frac{1}{2(n+1)-1} = \frac{1}{2n+1} < \frac{1}{2n-1} = a_n$$

for all *n*. Therefore, by the Alternating Series Test, the series converges.

17. $\displaystyle\sum_{n=1}^{\infty}\frac{(-1)^{n+1}(n+1)}{\ln(n+1)}$

Using L'Hôpital's Rule we have

$$\lim_{n\to\infty} a_n = \lim_{n\to\infty}\frac{n+1}{\ln(n+1)}$$

$$= \lim_{n\to\infty}\frac{1}{\dfrac{1}{(n+1)}} = \lim_{n\to\infty}(n+1) = \infty \neq 0.$$

Therefore, the series diverges by the *n*th-Term Test for Divergence.

27. $\displaystyle\sum_{n=1}^{\infty} \frac{2(-1)^{n+1}}{e^n + e^{-n}}$

Use differentiation to establish that $a_{n+1} \le a_n$.

$$f(x) = \frac{2}{e^x + e^{-x}}$$

$$f'(x) = \frac{-2(e^x - e^{-x})}{(e^x + e^{-x})^2}$$

Observe that $f'(x)$ is negative for $x > 0$. Hence, f is a decreasing function and it follows that $a_{n+1} \le a_n$ for $n \ge 1$. Since

$$\lim_{n \to \infty} \frac{2}{e^n + e^{-n}} = 0,$$

the series converges by the Alternating Series Test.

29. (a) $\displaystyle\sum_{n=0}^{\infty} \frac{(-1)^n}{n!}$

The series is alternating, and since

$$\lim_{n \to \infty} \frac{1}{n!} = 0 \quad \text{and} \quad \frac{1}{(n+1)!} < \frac{1}{n!}$$

the series converges. By the Alternating Series Remainder Theorem, the error R_N after N terms satisfies $|R_N| \le a_{N+1}$. Therefore, to ensure an error less than 0.001, choose N sufficiently large so that

$$\frac{1}{(N+1)!} \le 0.001.$$

Since $1/6! = 1/720 = 0.0013888$ and $1/7! = 1/5040 < 0.001$, let $N = 6$ and use 7 terms of the series.

(b) Using a graphing utility, we obtain

$$\sum_{n=0}^{6} \frac{(-1)^n}{n!} \approx 0.368.$$

The approximation is within 0.001 of the actual sum $1/e$.

35. $\displaystyle\sum_{n=1}^{\infty} \frac{(-1)^{n+1}}{2n^3 - 1}$

By the Alternating Series Remainder Theorem, the error R_N after N terms satisfies $|R_N| \le a_{N+1}$. Therefore, to insure an error less than 0.001, choose N sufficiently large so that

$$\frac{1}{2(N+1)^3 - 1} \le 0.001$$

$$1000 \le 2(N+1)^3 - 1$$

$$\frac{1001}{2} \le (N+1)^3$$

$$7.9 \approx \sqrt[3]{\frac{1001}{2}} \le N + 1$$

$$6.9 \le N.$$

Therefore, choose $N = 7$.

39. $\displaystyle\sum_{n=1}^{\infty} \frac{(-1)^{n+1}}{\sqrt{n}}$

In this case, the Alternating Series Test verifies that the series converges. However, the series

$$\sum_{n=1}^{\infty} \left| \frac{(-1)^{n+1}}{\sqrt{n}} \right| = \sum_{n=1}^{\infty} \frac{1}{\sqrt{n}}$$

is a divergent p-series with $p = \frac{1}{2}$. Therefore, the given series converges conditionally.

45. $\displaystyle\sum_{n=2}^{\infty} \frac{(-1)^n n}{n^3 - 1}$

The Alternating Series Test verifies that the series converges. Also, the series

$$\sum_{n=2}^{\infty} \left| \frac{(-1)^{n+1} n}{n^3 - 1} \right| = \sum_{n=2}^{\infty} \frac{n}{n^3 - 1}$$

converges by the Limit Comparison Test with the series

$$\sum_{n=2}^{\infty} \frac{1}{n^2}.$$

Therefore, the given series converges absolutely.

Section 8.6 The Ratio and Root Tests

3. Use Mathematical Induction to verify the formula

$$1 \cdot 3 \cdot 5 \cdots (2k - 1) = \frac{(2k)!}{2^k k!}.$$

The formula is valid when $k = 1$ since

$$1 = \frac{(2 \cdot 1)!}{2^1 1!}.$$

Assume that the formula is true when $k = n$ and then show that it is true when $k = n + 1$. Assuming it is true when $k = n$, we have

$$1 \cdot 3 \cdot 5 \cdots (2n - 1) = \frac{(2n)!}{2^n n}.$$

When $k = n + 1$, we have

$$1 \cdot 3 \cdot 5 \cdots (2n - 1)(2n + 1) \doteq [1 \cdot 3 \cdot 5 \cdots (2n - 1)](2n + 1)$$

$$= \frac{(2n)!}{2^n n!} \cdot (2n + 1)$$

$$= \frac{(2n)!(2n + 1)}{2^n n!} \cdot \frac{(2n + 2)}{2(n + 1)}$$

$$= \frac{(2n)!(2n + 1)(2n + 2)}{2^{n+1} n!(n + 1)}$$

$$= \frac{(2n + 2)!}{2^{n+1}(n + 1)!}$$

Therefore, the formula is valid for all $n \geq 1$.

13. $\displaystyle\sum_{n=0}^{\infty} \frac{3^n}{n!}$

$$\lim_{n \to \infty} \left| \frac{a_{n+1}}{a_n} \right| = \lim_{n \to \infty} \left| \frac{3^{n+1}/(n + 1)!}{3^n/n!} \right|$$

$$= \lim_{n \to \infty} \left| \frac{3^{n+1}}{1 \cdot 2 \cdot 3 \cdot 4 \cdots n \cdot (n + 1)} \cdot \frac{1 \cdot 2 \cdot 3 \cdot 4 \cdots n}{3^n} \right|$$

$$= \lim_{n \to \infty} \frac{3}{n + 1} = 0 < 1$$

Therefore, by the Ratio Test, the series converges.

27. $\displaystyle\sum_{n=0}^{\infty} \frac{4n}{3^n + 1}$

$$\lim_{n \to \infty} \left| \frac{a_{n+1}}{a_n} \right| = \lim_{n \to \infty} \left[\frac{4^{n+1}}{3^{n+1} + 1} \cdot \frac{3^n + 1}{4^n} \right]$$

$$= \lim_{n \to \infty} \left[\frac{4(3^n + 1)}{3^{n+1} + 1} \right]$$

$$= 4 \lim_{n \to \infty} \left[\frac{1 + (1/3^n)}{3 + (1/3^n)} \right] = \frac{4}{3} > 1$$

Therefore, by the Ratio Test, the series diverges.

29. $\displaystyle\sum_{n=0}^{\infty} \frac{(-1)^{n+1}n!}{1 \cdot 3 \cdot 5 \cdots (2n+1)}$

$$\lim_{n \to \infty} \left| \frac{a_{n+1}}{a_n} \right| = \lim_{n \to \infty} \left[\frac{(n+1)!}{1 \cdot 3 \cdot 5 \cdots (2n+1)(2n+3)} \cdot \frac{1 \cdot 3 \cdot 5 \cdots (2n+1)}{n!} \right]$$

$$= \lim_{n \to \infty} \frac{(n+1)}{(2n+3)} = \frac{1}{2}$$

Therefore, by the Ratio Test, the series converges.

37. $\displaystyle\sum_{n=1}^{\infty} \left(2\sqrt[n]{n} + 1 \right)^n$

Using the Root Test, we have

$$\lim_{n \to \infty} \sqrt[n]{a_n} = \lim_{n \to \infty} \sqrt[n]{\left(2\sqrt[n]{n} + 1 \right)^n} = \lim_{n \to \infty} \left(2\sqrt[n]{n} + 1 \right).$$

Now let $y = \sqrt[n]{n} = n^{1/n}$. Then $\ln y = (1/n) \ln n$ and by L'Hôpital's Rule, we have

$$\lim_{n \to \infty} \left(\frac{\ln n}{n} \right) = \lim_{n \to \infty} \left(\frac{1/n}{1} \right) = 0.$$

Thus, $\ln y = 0$, $y = e^0 = 1$, and

$$\lim_{n \to \infty} \left(2\sqrt[n]{n} + 1 \right) = 2(1) + 1 = 3.$$

The series diverges by the Root Test.

43. $\displaystyle\sum_{n=1}^{\infty} \frac{3}{n\sqrt{n}} = 3 \sum_{n=1}^{\infty} \frac{1}{n^{3/2}}$

This is a p-series with $p = \frac{3}{2}$. Therefore, the series converges.

49. $\displaystyle\sum_{n=1}^{\infty} \frac{10n+3}{n2^n}$

Compare the given series to the convergent geometric series

$$\sum_{n=0}^{\infty} \frac{1}{2^n}.$$

Since

$$\lim_{n \to \infty} \frac{(10n+3)/(n2^n)}{1/2^n} = \lim_{n \to \infty} \frac{10n+3}{n} = 10,$$

the series converges by the Limit Comparison Test.

51. $\displaystyle\sum_{n=1}^{\infty} \frac{\cos n}{2^n}$

Since $|\cos n| \leq 1$, it follows that

$$\left| \frac{\cos n}{2^n} \right| \leq \frac{1}{2^n}.$$

Since the series

$$\sum_{n=0}^{\infty} \frac{1}{2^n}$$

is a convergent geometric series, the series

$$\sum_{n=1}^{\infty} \left| \frac{\cos(n)}{2^n} \right|$$

converges by the Comparison Test. Finally, by the Absolute Convergence Theorem, the series

$$\sum_{n=1}^{\infty} \frac{\cos n}{2^n}$$

also converges.

57. $\displaystyle\sum_{n=1}^{\infty} \frac{(-3)^n}{3 \cdot 5 \cdot 7 \cdots (2n+1)}$

Use the Ratio Test to test for convergence or divergence and obtain

$$\lim_{n\to\infty}\left|\frac{a_{n+1}}{a_n}\right| = \lim_{n\to\infty}\left|\frac{(-3)^{n+1}}{3 \cdot 5 \cdot 7 \cdots (2n+1)[2(n+1)+1]} \cdot \frac{3 \cdot 5 \cdot 7 \cdots (2n+1)}{(-3)^n}\right|$$

$$= \lim_{n\to\infty}\left|\frac{-3}{2(n+1)+1}\right| = 0 < 1.$$

Therefore, by the Ratio Test the series converges.

Section 8.7   Taylor Polynomials and Approximations

11. $f(x) = \sin x$ \qquad $f(0) = 0$

$f'(x) = \cos x$ \qquad $f'(0) = 1$

$f''(x) = -\sin x$ \qquad $f''(0) = 0$

$f'''(x) = -\cos x$ \qquad $f'''(0) = -1$

$f^{(4)}(x) = \sin x$ \qquad $f^{(4)}(0) = 0$

$f^{(5)}(x) = \cos x$ \qquad $f^{(5)}(0) = 1$

Therefore, the expansion yields

$$P_5(x) = f(0) + f'(0)x + \frac{f''(0)}{2!}x^2 + \frac{f'''(0)}{3!}x^3 + \frac{f^{(4)}(0)}{4!}x^4 + \frac{f^{(5)}(0)}{5!}x^5$$

$$= x - \frac{x^3}{6} + \frac{x^5}{120}.$$

15. $f(x) = \dfrac{1}{x+1}$ \qquad $f(0) = 1$

$f'(x) = \dfrac{-1}{(x+1)^2}$ \qquad $f'(0) = -1$

$f''(x) = \dfrac{2}{(x+1)^3}$ \qquad $f''(0) = 2$

$f'''(x) = \dfrac{-6}{(x+1)^4}$ \qquad $f'''(0) = -6$

$f^{(4)}(x) = \dfrac{24}{(x+1)^5}$ \qquad $f^{(4)}(0) = 24$

Therefore, the expansion yields

$$P_4(x) = f(0) + f'(0)x + \frac{f''(0)}{2!}x^2 + \frac{f'''(0)}{3!}x^3 + \frac{f^{(4)}(0)}{4!}x^4$$

$$= 1 - x + x^2 - x^3 + x^4.$$

19. $f(x) = \ln x$ $f(1) = 0$

$f'(x) = \dfrac{1}{x}$ $f'(1) = 1$

$f''(x) = -\dfrac{1}{x^2}$ $f''(1) = -1$

$f'''(x) = \dfrac{2}{x^3}$ $f'''(1) = 2$

$f^{(4)}(x) = -\dfrac{6}{x^4}$ $f^{(4)}(1) = -6$

Therefore, the expansion yields

$$P_4(x) = f(1) + f'(1)(x-1) + \frac{f''(1)}{2!}(x-1)^2 + \frac{f'''(1)}{3!}(x-1)^3 + \frac{f^{(4)}(1)}{4!}(x-1)^4$$

$$= (x-1) - \frac{1}{2}(x-1)^2 + \frac{1}{3}(x-1)^3 - \frac{1}{4}(x-1)^4.$$

25. (a) $f(x) = \arcsin x$ $f(0) = 0$

$f'(x) = \dfrac{1}{\sqrt{1-x^2}}$ $f'(0) = 1$

$f''(x) = \dfrac{x}{(1-x^2)^{3/2}}$ $f''(0) = 0$

$f'''(x) = \dfrac{2x^2+1}{(1-x^2)^{5/2}}$ $f'''(0) = 1$

$$P_3(x) = f(0) + f'(0)x + \frac{f''(0)}{2!}x^2 + \frac{f'''(0)}{3!}x^3 = x + \frac{x^3}{6}$$

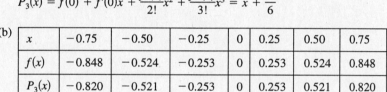

(b)

x	-0.75	-0.50	-0.25	0	0.25	0.50	0.75
$f(x)$	-0.848	-0.524	-0.253	0	0.253	0.524	0.848
$P_3(x)$	-0.820	-0.521	-0.253	0	0.253	0.521	0.820

31. From Exercise 19, we have the following 4th-degree polynomial for $f(x) = \ln x$.

$$f(x) \approx P_4(x) = (x-1) - \frac{1}{2}(x-1)^2 + \frac{1}{3}(x-1)^3 - \frac{1}{4}(x-1)^4$$

$$f(1.2) \approx P_4(1.2) = (1.2-1) - \frac{1}{2}(1.2-1)^2 + \frac{1}{3}(1.2-1)^3 - \frac{1}{4}(1.2-1)^4 \approx 0.1823$$

(The actual functional value accurate to four decimal places is also 0.1823.)

35. Using Taylor's Theorem, we have

$$\arcsin(0.4) \approx 0.4 + \frac{(0.4)^3}{2 \cdot 3} + R_3(x) = 0.4 + \frac{(0.4)^3}{2 \cdot 3} + \frac{f^{(4)}(z)}{4!}x^4$$

where $0 < z < 0.4$. Using a symbolic differentiation utility to find the $f^{(4)}(x)$ and sketch its graph, yields

$$f^{(4)}(x) = \frac{3x(2x^2+3)}{(1-x^2)^{7/2}}.$$

—CONTINUED—

35. **—CONTINUED—**

From the graph of the fourth derivative, it follows that $f^{(4)}(x) < 7.3340$ in the interval $[0, 0.4]$. Therefore,

$$0 < R_3(0.4) = \frac{f^{(4)}(z)}{4!}(0.4)^4 < \frac{7.3340}{4!}(0.4)^4 \approx 7.82 \times 10^{-3}.$$

Since

$$\arcsin(0.4) \approx 0.4 + \frac{(0.4)^3}{2 \cdot 3} = 0.41067,$$

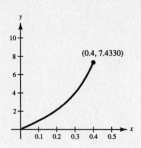

(0.4, 7.4330)

we have

$$0.41067 < \arcsin(0.4) = 0.41067 + R_3(0.4) < 0.41067 + (7.82 \times 10^{-3})$$

$$0.41067 < \arcsin(0.4) < 0.41849.$$

41. From Taylor's Theorem we have

$$e^x = 1 + x + \frac{x^2}{2!} + \frac{x^3}{3!} + R_3$$

where

$$R_3 = \frac{f^{(4)}(z)}{4!}x^4 = \frac{e^z}{4!}x^4.$$

For $z < 0$ we have

$$R_3 = \frac{e^z x^4}{4!} < \frac{x^4}{4!}$$

and we wish to find $x < 0$ such that

$$\frac{x^4}{4!} < 0.001$$

$$x^4 < 24(0.001) = 0.024$$

$$|x| < (0.024)^{1/4} \approx 0.3936.$$

Therefore, for values of x such that $-0.3936 < x\ 0$,

$$e^x \approx 1 + x + \frac{x^2}{2!} + \frac{x^3}{3!}.$$

Section 8.8   Power Series

9. $\displaystyle\sum_{n=1}^{\infty} \frac{(-1)^n x^n}{n}$

Letting $u_n = (-1)^n x^n / n$ produces

$$\lim_{n\to\infty}\left|\frac{u_{n+1}}{u_n}\right| = \lim_{n\to\infty}\left|\frac{(-1)^{n+1}x^{n+1}/(n+1)}{(-1)^2 x^n/n}\right|$$

$$= \lim_{n\to\infty}\left[\frac{x^{n+1}}{n+1}\cdot\frac{n}{x}\right]$$

$$= \lim_{n\to\infty}\left|\frac{nx}{n+1}\right| = |x|.$$

By the Ratio Test, the series converges if $|x| < 1$. Hence, the radius of convergence is 1 and the interval of convergence includes $-1 < x < 1$. When $x = -1$, we have the harmonic series

$$\sum_{n=1}^{\infty}\frac{1}{n}$$

which diverges. When $x = 1$, we have the alternating series

$$\sum_{n=1}^{\infty}\frac{(-1)^n}{n}$$

which converges. Thus, the interval of convergence is $-1 < x \le 1$.

13. $\displaystyle\sum_{n=0}^{\infty} (2n)!\left(\frac{x}{2}\right)^n$

Letting $u_n = (2n)!\left(\dfrac{x}{2}\right)^n$ produces

$$\lim_{n\to\infty}\left|\frac{u_{n+1}}{u_n}\right| = \lim_{n\to\infty}\left|\frac{(2n+2)!/(x/2)^{n+1}}{(2n)!/(x/2)^n}\right|$$

$$= \lim_{n\to\infty}\left|\frac{(2n+2)(2n+1)x}{2}\right| = \infty$$

for any real $x \ne 0$. Therefore, by the Ratio Test, the series converges only when $x = 0$.

17. $\displaystyle\sum_{n=1}^{\infty} \frac{(-1)^{n+1}(x-5)^n}{n5^n}$

Letting $u_n = \dfrac{(-1)^{n+1}(x-5)^n}{n5^n}$ produces

$$\lim_{n\to\infty}\left|\frac{u_{n+1}}{u_n}\right| = \lim_{n\to\infty}\left|\frac{n5^n(x-5)}{(n+1)5^{n+1}}\right|$$

$$= \lim_{n\to\infty}\left|\frac{x-5}{5}\left(\frac{n}{n+1}\right)\right| = \left|\frac{x-5}{5}\right|.$$

By the Ratio Test, the series converges if $|(x-5)/5| < 1$, or $|x-5| < 5$. Hence, the radius of convergence is $R = 5$, and since the series is centered at $x = 5$, the series will converge in the interval $(0, 10)$. Furthermore, when $x = 0$ we have the series

$$\sum_{n=1}^{\infty}\frac{(-1)^{n+1}(-1)^n}{n} = \sum_{n=1}^{\infty}\frac{(-1)^{2n+1}}{n} = -\sum_{n=1}^{\infty}\frac{1}{n}$$

which diverges ($p = 1$). When $x = 10$, we have the series

$$\sum_{n=1}^{\infty}\frac{(-1)^{n+1}}{n}$$

which converges by the Alternating Series Test. Hence, the interval of convergence of the given series is $0 < x \le 10$.

27. $\sum_{n=1}^{\infty} \dfrac{k(k + 1)(k + 2) \cdots (k + n - 1)x^n}{n!}$ $(k \geq 1)$

Since

$$\lim_{n \to \infty} \left| \frac{u_{n+1}}{u_n} \right| = \lim_{n \to \infty} \left| \frac{k(k + 1) \cdots (k + n - 1)(k + n)x^{n+1}}{(n + 1)!} \cdot \frac{n!}{k(k + 1) \cdots (k + n - 1)x^n} \right|$$

$$= \lim_{n \to \infty} \left| \frac{(k + n)x}{n + 1} \right| = |x|,$$

the radius of convergence is $R = 1$. Since the series is centered at $x = 0$, it will converge on the interval $(-1, 1)$. To test for convergence at the endpoints, note that for $k \geq 1$,

$$\lim_{n \to \infty} a_n = \lim_{n \to \infty} \left[\left(\frac{k}{1} \right) \left(\frac{k + 1}{2} \right) \left(\frac{k + 2}{3} \right) \cdots \left(\frac{k + n - 1}{n} \right) \right] \neq 0.$$

Thus, for $x = \pm 1$ the series diverges, and the interval of convergence is $-1 < x < 1$.

31. $f(x) = \sum_{n=0}^{\infty} \left(\dfrac{x}{2} \right)^n$

(a) The given series is geometric with $r = x/2$ and converges if

$$\left| \frac{x}{2} \right| < 1 \quad \text{or} \quad -2 < x < 2.$$

(b) $f'(x) = \sum_{n=1}^{\infty} n \left(\dfrac{x}{2} \right)^{n-1} \left(\dfrac{1}{2} \right) = \sum_{n=1}^{\infty} \left(\dfrac{n}{2} \right) \left(\dfrac{x}{2} \right)^{n-1}$

Therefore the series for $f'(x)$ diverges for $x = \pm 2$, and its interval of convergence is $-2 < x < 2$.

(c) $f''(x) = \sum_{n=2}^{\infty} \left(\dfrac{n}{2} \right)(n - 1) \left(\dfrac{x}{2} \right)^{n-2} \left(\dfrac{1}{2} \right)$

$$= \sum_{n=2}^{\infty} \frac{n(n - 1)}{4} \left(\frac{x}{2} \right)^{n-2}$$

Therefore, the series for $f''(x)$ diverges for $x = \pm 2$, and its interval of convergence is $-2 < x < 2$.

(d) The series for $\int f(x)\, dx$ is

$$\sum_{n=0}^{\infty} \frac{2}{n + 1} \left(\frac{x}{2} \right)^{n+1}$$

and it converges (Alternating Series Test) for $x = -2$ and diverges [Limit Comparison Test with $\sum_{n=1}^{\infty} (1/n)$] for $x = 2$. Therefore, its interval of convergence is

$$-2 \leq x < 2.$$

41.

$$y = \sum_{n=0}^{\infty} \frac{x^{2n}}{2^n n!}$$

$$y' = \sum_{n=1}^{\infty} \frac{2nx^{2n-1}}{2^n n!}$$

$$y'' = \sum_{n=1}^{\infty} \frac{2n(2n - 1)x^{2n-2}}{2^n n!}$$

$$y'' - xy' - y = \sum_{n=1}^{\infty} \frac{2n(2n - 1)x^{2n-2}}{2^n n!} - \sum_{n=1}^{\infty} \frac{2nx^{2n}}{2^n n!} - \sum_{n=0}^{\infty} \frac{x^{2n}}{2^n n!}$$

$$= \sum_{n=1}^{\infty} \frac{2n(2n - 1)x^{2n-2}}{2^n n!} - \sum_{n=0}^{\infty} \frac{(2n + 1)x^{2n}}{2^n n!}$$

$$= \sum_{n=0}^{\infty} \left[\frac{(2n + 2)(2n + 1)x^{2n}}{2^{n+1}(n + 1)!} - \frac{(2n + 1)x^{2n}}{2^n n!} \cdot \frac{2(n + 1)}{2(n + 1)} \right]$$

$$= \sum_{n=0}^{\infty} \frac{2(n + 1)x^{2n}[(2n + 1) - (2n + 1)]}{2^{n+1}(n + 1)!}$$

$$= 0$$

Section 8.9   Representation of Functions by Power Series

5. $f(x) = \dfrac{1}{2 - x}$

Writing $f(x)$ in the form $a/(1 - r)$, we have

$$\frac{1}{2 - x} = \frac{1}{-3 - x + 5} = \frac{-1/3}{1 - [(x - 5)/(-3)]} = \frac{a}{1 - r}$$

which implies that $a = -1/3$ and $r = (x - 5)/(-3)$. Therefore,

$$\frac{1}{2 - x} = \sum_{n=0}^{\infty} ar^n = -\frac{1}{3} \sum_{n=0}^{\infty} \left(\frac{x - 5}{-3}\right)^n = \sum_{n=0}^{\infty} \frac{(x - 5)^n}{(-3)^{n+1}}.$$

Since

$$\lim_{n \to \infty} \left|\frac{u_{n+1}}{u_n}\right| = \lim_{n \to \infty} \left|\frac{(x - 5)^{n+1}}{(-3)^{n+2}} \cdot \frac{(-3)^{n+1}}{(x - 5)^n}\right| = \left|\frac{x - 5}{3}\right|,$$

the radius of convergence is $R = 3$. Since the series is centered at $c = 5$, it converges in the interval $(2, 8)$. Finally, since the series diverges at both endpoints, the interval of convergence is $2 < x < 8$.

13. $f(x) = \dfrac{3x}{x^2 + x - 2}$

Using the method for finding partial fractions (Section 7.5), we have

$$\frac{3x}{x^2 + x - 2} = \frac{2}{x + 2} + \frac{1}{x - 1} = \frac{1}{1 - [-x/2]} - \frac{1}{1 - x}.$$

Therefore, we have the difference of the sum of two geometric series. Using the fact that

$$\sum_{n=0}^{\infty} ar^n = \frac{a}{1 - r},$$

we can write

$$\frac{1}{1 - [-x/2]} - \frac{1}{1 - x} = \sum_{n=0}^{\infty} \left(\frac{x}{-2}\right)^n - \sum_{n=0}^{\infty} x^n = \sum_{n=0}^{\infty} \left[\frac{x^n}{(-2)^n} - x^n\right] = \sum_{n=0}^{\infty} \left[\frac{1}{(-2)^n} - 1\right] x^n.$$

By the Ratio Test

$$\lim_{n \to \infty} \left|\frac{u_{n+1}}{u_n}\right| = \lim_{n \to \infty} \left|\frac{[1/(-2)^{n+1}] - 1}{[1/(-2)^n] - 1} \cdot \frac{x^{n+1}}{x_n}\right| = |x|,$$

and the series converges if $|x| < 1$. Finally, when $x = \pm 1$,

$$\lim_{n \to \infty} a_n = \lim_{n \to \infty} \left[\frac{1}{(-2)^n} - 1\right] \neq 0.$$

Thus, the series diverges when $x = \pm 1$ and the interval of convergence is $-1 < x < 1$.

15. $f(x) = \dfrac{2}{1 - x^2}$

Letting $u = x^2$,

$$\frac{2}{1 - x^2} = \frac{2}{1 - u} = 2 \sum_{n=0}^{\infty} u^n = 2 \sum_{n=0}^{\infty} (x^2)^n = 2 \sum_{n=0}^{\infty} x^{2n}.$$

The series will converge if $x^2 < 1$ or $-1 < x < 1$.

21. $f(x) = \ln(x + 1)$

Since

$$\int \frac{1}{x + 1}\, dx = \ln(x + 1) + C,$$

integrate the power series for $1/(x + 1)$ to obtain the series for $\ln(x + 1)$.

$$\frac{1}{x + 1} = \sum_{n=0}^{\infty} (-1)^n x^n$$

$$\ln(x + 1) = \sum_{n=0}^{\infty} \frac{(-1)^n x^{n+1}}{n + 1} + C$$

Substituting $x = 0$ on both sides of this equation, yields $C = 0$. Furthermore, by the Ratio Test

$$\lim_{n \to \infty} \left| \frac{u_{n+1}}{u_n} \right| = \lim_{n \to \infty} \left| \frac{x^{n+2}}{n + 2} \cdot \frac{n + 1}{x^{n+1}} \right| = |x|,$$

and the series converges if $|x| < 1$. At $x = -1$ we have the divergent series

$$\sum_{n=0}^{\infty} \frac{(-1)^{2n+1}}{n + 1} = -\sum_{n=0}^{\infty} \frac{1}{n + 1}.$$

At $x = 1$ we have the convergent alternating series

$$\sum_{n=0}^{\infty} \frac{(-1)^n}{n + 1}.$$

Therefore,

$$\ln(x + 1) = \sum_{n=0}^{\infty} \frac{(-1)^n x^{n+1}}{n + 1}, \text{ for } -1 < x \le 1.$$

41. The following geometric series converges when $|x| < 1$.

$$\sum_{n=0}^{\infty} x^n = \frac{1}{1 - x}$$

$$\frac{1}{(1 - x)^2} = \frac{d}{dx}\left[\frac{1}{1 - x} \right] = \frac{d}{dx}\left[\sum_{n=0}^{\infty} x^n \right] = \sum_{n=1}^{\infty} n x^{n-1}$$

$$E(n) = \sum_{n=1}^{\infty} n\, P(n) = \sum_{n=1}^{\infty} n \left(\frac{1}{2} \right)^n$$

$$= \frac{1}{2} \sum_{n=1}^{\infty} n \left(\frac{1}{2} \right)^{n-1} = \frac{1}{2} \frac{1}{[1 - (1/2)]^2} = 2$$

The probability of obtaining a head on a single toss is $\frac{1}{2}$, and on average, a head will be obtained in two tosses.

37. From Example 5,

$$\arctan x = \sum_{n=0}^{\infty} \frac{(-1)^n x^{2n+1}}{2n + 1}.$$

Therefore,

$$\arctan x^2 = \sum_{n=0}^{\infty} \frac{(-1)^n (x^2)^{(2n+1)}}{2n + 1} = \sum_{n=0}^{\infty} \frac{(-1)^n x^{(4n+2)}}{2n + 1}$$

and

$$\frac{\arctan x^2}{x} = \sum_{n=0}^{\infty} \frac{(-1)^n x^{4n+1}}{2n + 1}.$$

Thus,

$$\int_0^{1/2} \frac{\arctan x^2}{x}\, dx = \left[\sum_{n=0}^{\infty} \frac{(-1)^n x^{4n+2}}{(2n + 1)(4n + 2)} \right]_0^{1/2}$$

$$= \sum_{n=0}^{\infty} \frac{(-1)^n (1/2)^{4n+2}}{(2n + 1)(4n + 2)}$$

$$\approx \left(\frac{1}{2} \right)\left(\frac{1}{2} \right)^2 = \frac{1}{8}.$$

Note that the second term of the series is

$$-\left(\frac{1}{18} \right)\left(\frac{1}{2} \right)^6 < 0.001.$$

Therefore, by the Alternating Series Remainder Theorem, one term is sufficient to approximate the integral.

49. In Example 4 it was shown that

$$\ln x = \sum_{n=0}^{\infty} (-1)^n \frac{(x - 1)^{n+1}}{n + 1} = \sum_{n=1}^{\infty} (-1)^{n-1} \frac{(x - 1)^n}{n}.$$

Letting $x = \frac{7}{5}$ in the series yields

$$\sum_{n=1}^{\infty} (-1)^{n-1} \frac{([7/5] - 1)^n}{n} = \sum_{n=1}^{\infty} (-1)^{n-1} \frac{(2/5)^n}{n}$$

$$= \sum_{n=1}^{\infty} (-1)^{n-1} \frac{2^n}{5^n n}.$$

Therefore, the sum of the series is

$$\ln \frac{7}{5} \approx 0.3365.$$

Section 8.10   Taylor and Maclaurin Series

7. $f(x) = \sin 2x$

Since

$$f(x) = \sin 2x \qquad\qquad f(0) = 0$$
$$f'(x) = 2\cos 2x \qquad\qquad f'(0) = 2$$
$$f''(x) = -4\sin 2x \qquad\qquad f''(0) = 0$$
$$f'''(x) = -8\cos 2x \qquad\qquad f'''(0) = -8 = -2^3$$
$$f^{(4)}(x) = 16\sin 2x \qquad\qquad f^{(4)}(0) = 0$$
$$f^{(5)}(x) = 32\cos 2x \qquad\qquad f^{(5)}(0) = 32 = 2^5,$$

we can see that the signs alternate and that $|f^{(n)}(0)| = 2^n$ if n is odd. Therefore, the Taylor Series is

$$\sin 2x = f(0) + f'(0)x + \frac{f''(0)x^2}{2!} + \frac{f'''(0)x^3}{3!} + \frac{f^{(4)}(0)x^4}{4!} + \cdots$$

$$= \frac{2x}{1!} - \frac{2^3 x^3}{3!} + \frac{2^5 x^5}{5!} - \cdots + \frac{(-1)^n (2x)^{2n+1}}{(2n+1)!} + \cdots$$

$$= \sum_{n=0}^{\infty} \frac{(-1)^n (2x)^{2n+1}}{(2n+1)!}$$

Note that we could have arrived at the same result by substituting $2x$ into the series for $\sin x$ as follows:

$$\sin x = x - \frac{x^3}{3!} + \frac{x^5}{5!} - \frac{x^7}{7!} + \cdots$$

$$\sin(2x) = (2x) - \frac{(2x)^3}{3!} + \frac{(2x)^5}{5!} - \frac{(2x)^7}{7!} + \cdots$$

$$= \sum_{n=0}^{\infty} \frac{(-1)^n (2x)^{2n+1}}{(2n+1)!}$$

13. $f(x) = \dfrac{1}{\sqrt{4 + x^2}}$

Consider f in the form

$$f(x) = \frac{1}{\sqrt{4 + x^2}} = \frac{1}{2\sqrt{1 + (x/2)^2}} = \frac{1}{2}\left[1 + \left(\frac{x}{2}\right)^2\right]^{-1/2}$$

which is similar to the binomial form $(1 + x)^{-k}$. Since

$$(1 + x)^{-k} = 1 - kx + \frac{k(k+1)x^2}{2!} - \frac{k(k+1)(k+2)x^3}{3!} + \cdots$$

we have, for $k = \dfrac{1}{2}$,

$$(1 + x)^{-1/2} = 1 - \frac{1}{2}x + \frac{(1/2)(3/2)x^2}{2!} - \frac{(1/2)(3/2)(5/2)x^3}{3!} + \cdots$$

$$= 1 - \frac{x}{2} + \frac{1 \cdot 3 x^2}{2^2 2!} - \frac{1 \cdot 3 \cdot 5 x^3}{2^3 3!} + \cdots$$

$$= 1 + \sum_{n=1}^{\infty} \frac{(-1)^n 1 \cdot 3 \cdot 5 \cdots (2n-1) x^n}{2^n n!}.$$

—CONTINUED—

13. —CONTINUED—

Now substituting $(x/2)^2$ for x, yields

$$f(x) = \frac{1}{\sqrt{4 + x^2}} = \frac{1}{2}\left[1 + \left(\frac{x}{2}\right)^2\right]^{-1/2}$$

$$= \frac{1}{2}\left[1 + \sum_{n=1}^{\infty} \frac{(-1)^n 1 \cdot 3 \cdot 5 \cdots (2n-1)(x/2)^{2n}}{2^n n!}\right]$$

$$= \frac{1}{2} + \sum_{n=1}^{\infty} \frac{(-1)^n 1 \cdot 3 \cdot 5 \cdots (2n-1)x^{2n}}{2^{3n+1}n!}$$

$$= \frac{1}{2}\left[1 + \sum_{n=1}^{\infty} \frac{(-1)^n 1 \cdot 3 \cdot 5 \cdots (2n-1)x^{2n}}{2^{3n}n!}\right].$$

17. $f(x) = e^{x^2/2}$

Since

$$e^x = 1 + x + \frac{x^2}{2!} + \frac{x^3}{3!} + \frac{x^4}{4!} + \frac{x^5}{5!} + \cdots,$$

we can substitute $x^2/2$ for x and obtain the series

$$e^{x^2/2} = 1 + \frac{x^2}{2} + \frac{(x^2/2)^2}{2!} + \frac{(x^2/2)^3}{3!} + \frac{(x^2/2)^4}{4!} + \cdots$$

$$= 1 + \frac{x^2}{2} + \frac{x^4}{2^2 2!} + \frac{x^6}{2^3 3!} + \frac{x^8}{2^4 4!} + \cdots$$

$$= \sum_{n=0}^{\infty} \frac{x^{2n}}{2^n n!}$$

23. $f(x) = \frac{\sin x}{x}$

Since

$$\sin x = x - \frac{x^3}{3!} + \frac{x^5}{5!} - \frac{x^7}{7!} + \frac{x^9}{9!} - \cdots,$$

we can divide by x to obtain

$$\frac{\sin x}{x} = 1 - \frac{x^2}{3!} + \frac{x^4}{5!} - \frac{x^6}{7!} + \frac{x^8}{9!} - \cdots$$

$$= \sum_{n=0}^{\infty} \frac{(-1)^n x^{2n}}{(2n+1)!}$$

29. Observe the following powers of i:

$$i = \sqrt{-1} \qquad\qquad i^5 = i^4 \cdot i = 1$$
$$i^2 = -1 \qquad\qquad i^6 = i^4 \cdot i^2 = -1$$
$$i^3 = i^2 \cdot i = -i \qquad i^7 = i^4 \cdot i^3 = -i$$
$$i^4 = i^2 \cdot i^2 = 1 \qquad i^8 = 1^4 \cdot i^4 = 1$$

Since

$$e^x = 1 + x + \frac{x^2}{2!} + \frac{x^3}{3!} + \frac{x^4}{4!} + \frac{x^5}{5!} + \cdots,$$

we can substitute (ix) and $(-ix)$ for x and obtain the series for e^{ix} and e^{-ix}.

$$e^{ix} = 1 + (ix) + \frac{(ix)^2}{2!} + \frac{(ix)^3}{3!} + \frac{(ix)^4}{4!} + \frac{(ix)^5}{5!} + \cdots$$

$$= 1 + ix - \frac{x^2}{2!} - \frac{ix^3}{3!} + \frac{x^4}{4!} + \frac{ix^5}{5!} - \cdots$$

$$e^{-ix} = 1 + (-ix) + \frac{(-ix)^2}{2!} + \frac{(-ix)^3}{3!} + \frac{(-ix)^4}{4!} + \frac{(-ix)^5}{5!} + \cdots$$

$$= 1 - ix - \frac{x^2}{2!} + \frac{ix^3}{3!} + \frac{x^4}{4!} - \frac{ix^5}{5!} - \cdots$$

Therefore, subtracting the series for e^{ix} and e^{-ix} yields

$$e^{ix} - e^{-ix} = 2ix - \frac{2ix^3}{3!} + \frac{2ix^5}{5!} - \cdots \quad \text{and}$$

$$\frac{e^{ix} - e^{-ix}}{2i} = x - \frac{x^3}{3!} + \frac{x^5}{5!} - \cdots$$

$$= \sum_{n=0}^{\infty} \frac{(-1)^n x^{2n+1}}{(2n+1)} = \sin x.$$

35. $g(x) = \dfrac{\sin x}{1 + x} = x - x^2 + \dfrac{5x^3}{6} - \dfrac{5x^4}{6} + \cdots$

$$
\begin{array}{r}
x - x^2 + \dfrac{5x^3}{6} - \dfrac{5x^4}{6} + \cdots \\[2mm]
1 + x \overline{\smash{\big)}\ x + 0x^2 - \dfrac{x^3}{6} + 0x^4 + \dfrac{x^5}{120} + \cdots} \\[2mm]
\underline{x + x^2} \\[2mm]
-x^2 - \dfrac{x^3}{6} \\[2mm]
\underline{-x^2 - x^3} \\[2mm]
\dfrac{5x^3}{6} + 0x^4 \\[2mm]
\underline{\dfrac{5x^3}{6} + \dfrac{5x^4}{6}} \\[2mm]
-\dfrac{5x^4}{6} + \dfrac{x^5}{120} \\[2mm]
\underline{-\dfrac{5x^4}{6} - \dfrac{5x^5}{6}} \\[2mm]
\vdots
\end{array}
$$

The graph is shown in the figure.

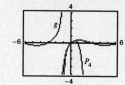

41. $f(x) = \displaystyle\int_0^x (e^{-t^2} - 1)\,dt = \int_0^x \left[\left(\sum_{n=0}^\infty \frac{(-1)^n t^{2n}}{n!} \right) - 1 \right] dt$

$\qquad\qquad = \displaystyle\int_0^x \left[\sum_{n=0}^\infty \frac{(-1)^{n+1} t^{2n+2}}{(n+1)!} \right] dt$

$\qquad\qquad = \left[\displaystyle\sum_{n=0}^\infty \frac{(-1)^{n+1} t^{2n+3}}{(2n+3)(n+1)!} \right]_0^x$

$\qquad\qquad = \displaystyle\sum_{n=0}^\infty \frac{(-1)^{n+1} x^{2n+3}}{(2n+3)(n+1)!}$

49. From Exercise 23 we have

$$\frac{\sin x}{x} = \sum_{n=0}^\infty \frac{(-1)^n x^{2n}}{(2n+1)!}.$$

Therefore,

$$\int_0^1 \frac{\sin x}{x}\,dx = \left[\sum_{n=0}^\infty \frac{(-1)^n x^{2n+1}}{(2n+1)(2n+1)!} \right]_0^1$$

$$= \sum_{n=0}^\infty \frac{(-1)^n}{(2n+1)(2n+1)!}$$

$$= 1 - \frac{1}{3 \cdot 3!} + \frac{1}{5 \cdot 5!} \approx 0.9461.$$

By the Alternating Series Remainder Theorem, we have

$$|R_3| \le a_4 = \frac{1}{7 \cdot 7!} \approx 0.00003 < 0.0001.$$

55. From Exercise 17, we have

$$\frac{1}{\sqrt{2\pi}} \int_0^1 e^{-x^2/2}\,dx = \frac{1}{\sqrt{2\pi}} \int_0^1 \sum_{n=0}^\infty \frac{(-1)^n x^{2n}}{2^n n!}\,dx$$

$$= \frac{1}{\sqrt{2\pi}} \left[\sum_{n=0}^\infty \frac{(-1)^n x^{2n+1}}{2^n n!(2n+1)} \right]_0^1$$

$$= \frac{1}{\sqrt{2\pi}} \sum_{n=0}^\infty \frac{(-1)^n}{2^n n!(2n+1)}$$

$$\approx \frac{1}{\sqrt{2\pi}} \left[1 - \frac{1}{2 \cdot 1 \cdot 3} + \frac{1}{2^2 \cdot 2! \cdot 5} - \frac{1}{2^3 \cdot 3! \cdot 7} \right] \approx 0.3413.$$

Review Exercises for Chapter 8

13. $a_n = \sqrt{n+1} - \sqrt{n}$

$$\lim_{n\to\infty} = \lim_{n\to\infty}\left[\sqrt{n+1} - \sqrt{n}\right]$$

$$= \lim_{n\to\infty}\left[(\sqrt{n+1} - \sqrt{n})\frac{\sqrt{n+1}+\sqrt{n}}{\sqrt{n+1}+\sqrt{n}}\right]$$

$$= \lim_{n\to\infty}\frac{1}{\sqrt{n+1}+\sqrt{n}} = 0$$

The sequence converges to 0.

21. (a) $\displaystyle\sum_{n=1}^{\infty}\frac{(-1)^{n+1}}{(2n)!}$

n	5	10	15	20	25
S_n	0.4597	0.4597	0.4597	0.4597	0.4597

(b) The graph is shown in the figure.

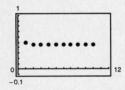

(c) Observe that $\displaystyle\lim_{n\to\infty} a_n = \lim_{n\to\infty}\frac{1}{(2n)!} = 0$ and

$$a_{n+1} = \frac{1}{[2(n+1)]!} < \frac{1}{(2n)!} = a_n$$

for all n. Therefore, by the Alternating Series Test, the series converges.

25. $\displaystyle\sum_{n=0}^{\infty}\left(\frac{1}{2^n} - \frac{1}{3^n}\right)$

Write the series as the difference of two geometric series.

$$\sum_{n=0}^{\infty}\left(\frac{1}{2^n} - \frac{1}{3^n}\right) = \sum_{n=0}^{\infty}\left(\frac{1}{2}\right)^n - \sum_{n=0}^{\infty}\left(\frac{1}{3}\right)^n$$

Since these two geometric series have the values $a = 1$, $r = \frac{1}{2}$, and $r = \frac{1}{3}$, respectively, it follows that

$$\sum_{n=0}^{\infty}\left(\frac{1}{2^n} - \frac{1}{3^n}\right) = \frac{1}{1-(1/2)} - \frac{1}{1-(1/3)}$$

$$= 2 - \frac{3}{2} = \frac{1}{2}.$$

37. $\displaystyle\sum_{n=1}^{\infty}\frac{1}{\sqrt{n^3+2n}}$

$$\sum_{n=1}^{\infty}\frac{1}{\sqrt{n^3+2n}} < \sum_{n=1}^{\infty}\frac{1}{\sqrt{n^3}} = \sum_{n=1}^{\infty}\frac{1}{n^{3/2}}$$

Since

$$\sum_{n=1}^{\infty}\frac{1}{n^{3/2}}$$

is a p-series with $p > 1$, it converges. Therefore, by the Comparison Test, the given series converges.

49. $\displaystyle\sum_{n=0}^{\infty}\left(\frac{x}{10}\right)^n$

The series is geometric with common ratio $r = x/10$. Therefore, it converges only if

$$\left|\frac{x}{10}\right| < 1 \implies -10 < x < 10.$$

53. $\displaystyle\sum_{n=0}^{\infty} n!(x-2)^n$

Since

$$\lim_{n\to\infty}\left|\frac{u_{n+1}}{u_n}\right| = \lim_{n\to\infty}\left|\frac{(n+1)!(x-2)^{n+1}}{n!(x-2)^2}\right|$$

$$= \lim_{n\to\infty}|(n+1)(x-2)| = \infty$$

for all $x \neq 2$, it follows that the radius of convergence is $R = 0$, the series will converge only at $x = 2$.

57. Since $3^x = e^{\ln(3^x)} = e^{x(\ln 3)}$, we can substitute $x(\ln 3)$ for x in the series

$$e^x = 1 + x + \frac{x^2}{2!} + \frac{x^3}{3!} + \frac{x^4}{4!} + \cdots$$

to obtain

$$3^x = e^{x(\ln 3)}$$

$$= 1 + (\ln 3)x + \frac{(\ln 3)^2 x^2}{2!} + \frac{(\ln 3)^3 x^3}{3!} + \frac{(\ln 3)^4 x^4}{4!} + \cdots$$

$$= \sum_{n=0}^{\infty} \frac{(x \ln 3)^n}{n!}.$$

65. $\displaystyle\sum_{n=0}^{\infty} \frac{1}{2^n n!}$

$$e^x = 1 + x + \frac{x^2}{2!} + \frac{x^3}{3!} + \frac{x^4}{4!} + \cdots = \sum_{n=0}^{\infty} \frac{x^n}{n!}$$

Substituting $x = \frac{1}{2}$ into the series yields

$$\sum_{n=0}^{\infty} \frac{(1/2)^n}{n!} = \sum_{n=0}^{\infty} \frac{1}{2^n n!} = e^{1/2} \approx 1.6487.$$

73. $1 + \dfrac{2}{3}x + \dfrac{4}{9}x^2 + \dfrac{8}{27}x^3 + \cdots$

is a geometric series with $r = 2x/3$. Therefore, it converges if

$$\left|\frac{2x}{2}\right| < 1 \implies -\frac{3}{2} < x < \frac{3}{2}.$$

Using the formula for the sum of a convergent geometric series we have

$$1 + \frac{2}{3}x + \frac{4}{9}x^2 + \frac{8}{27}x^3 + \cdots = \sum_{n=0}^{\infty} \left(\frac{2x}{3}\right)^n$$

$$= \frac{1}{1 - (2x/3)} = \frac{3}{3 - 2x}, \quad -\frac{3}{2} < x < \frac{3}{2}.$$

81.
$$y = \sum_{n=0}^{\infty} \frac{(-1)^n x^{2n}}{4^n (n!)^2}$$

$$y' = \sum_{n=1}^{\infty} \frac{(-1)^n (2n) x^{2n-1}}{4^n (n!)^2} = \sum_{n=0}^{\infty} \frac{(-1)^{n+1}(2n + 2)x^{2n+1}}{4^{n+1}[(n + 1)!]^2}$$

$$y'' = \sum_{n=0}^{\infty} \frac{(-1)^{n+1}(2n + 2)(2n + 1)x^{2n}}{4^{n+1}[(n + 1)!]^2}$$

$$x^2 y'' + xy' + x^2 y = \sum_{n=0}^{\infty} \frac{(-1)^{n+1}(2n + 2)(2n + 1)x^{2n+2}}{4^{n+1}[(n + 1)!]^2} + \sum_{n=0}^{\infty} \frac{(-1)^{n+1}(2n + 2)x^{2n+2}}{4^{n+1}[(n + 1)!]^2} + \sum_{n=0}^{\infty} \frac{(-1)^n x^{2n+2}}{4^n (n!)^2}$$

$$= \sum_{n=0}^{\infty} \left[\frac{(-1)^{n+1}(2n + 2)(2n + 1)}{4^{n+1}[(n + 1)!]^2} + \frac{(-1)^{n+1}(2n + 2)}{4^{n+1}[(n + 1)!]^2} + \frac{(-1)^n}{4^n (n!)^2} \right] x^{2n+2}$$

$$= \sum_{n=0}^{\infty} \left[\frac{(-1)^{n+1} 4(n + 1)^2}{4^{n+1}[(n + 1)!]^2} - \frac{(-1)^{n+1}}{4^n (n!)^2} \right] x^{2n+2}$$

$$= \sum_{n=0}^{\infty} \left[\frac{(-1)^{n+1}}{4^n (n!)^2} - \frac{(-1)^{n+1}}{4^n (n!)^2} \right] x^{2n+2} = 0$$

CHAPTER 9
Conics, Parametric Equations, and Polar Coordinates

CHAPTER 9
Conics, Parametric Equations, and Polar Coordinates

Section 9.1   Conics and Calculus
Solutions to Selected Odd-Numbered Exercises

9. $y^2 = -6x$

Write the equation in standard form.

$$(y - k)^2 = 4p(x - h)$$
$$y^2 = -6x$$
$$(y - 0)^2 = 4\left(-\tfrac{3}{2}\right)(x - 0)$$

Thus, $k = 0$, $h = 0$, and $p = -\tfrac{3}{2}$. It follows that:

Vertex, (h, k): $(0, 0)$

Focus, $(h + p, k)$: $\left(-\tfrac{3}{2}, 0\right)$

Directrix, $(x = h - p)$: $x = \tfrac{3}{2}$

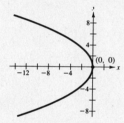

13. $y^2 - 4y - 4x = 0$

Write the equation in standard form.

$$(y - k)^2 = 4p(x - h)$$
$$y^2 - 4y = 4x$$
$$y^2 - 4y + 4 = 4x + 4$$
$$(y - 2)^2 = 4(1)(x + 1)$$

Thus, $h = -1$, $k = 2$, and $p = 1$.

Vertex, (h, k): $(-1, 2)$

Focus, $(h + p, k)$: $(0, 2)$

Directrix, $(x = h - p)$: $x = -2$

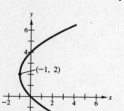

21. The vertex and focus of the parabola are $(3, 2)$ and $(1, 2)$, respectively. Since the vertex and focus lie on a horizontal line, the axis of this parabola must be horizontal and its standard form is

$$(y - k)^2 = 4p(x - h).$$

Since the vertex is at $(3, 2)$, $h = 3$ and $k = 2$. Furthermore, the *directed distance* from the focus to the vertex is

$$p = 1 - 3 = -2.$$

Thus,

$$(y - 2)^2 = 4(-2)(x - 3)$$
$$y^2 - 4y + 4 = -8x + 24$$
$$y^2 - 4y + 8x - 20 = 0.$$

27. Since the axis of the parabola is vertical, the standard form is

$$(x - h)^2 = 4p(y - k).$$

For this exercise, a more convenient form for this equation is

$$y = ax^2 + bx + c.$$

—CONTINUED—

27. —CONTINUED—

Substituting the values of the coordinates of the points $(0, 3)$, $(3, 4)$, and $(4, 11)$ into this equation, yields the following three equations.

(i) $3 = a(0)^2 + b(0) + c$

(ii) $4 = a(3)^2 + b(3) + c$

(iii) $11 = a(4)^2 + b(4) + c$

Simplification yields:

(i) $3 = c$

(ii) $4 = 9a + 3b + c \implies 1 = 9a + 3b$

(iii) $11 = 16a + 4b + c \implies 8 = 16a + 4b$

Solving (ii) and (iii) for a and b yields:

(ii) $9a + 3b = 1 \implies 9a \left| + 3b = 1 \right.$

(iii) $4a + b = 2 \implies 12a \left| + 3b = 6 \right.$

$$-3a \left| = -5 \right.$$

$$a = \tfrac{5}{3}$$

$$b = -\tfrac{14}{3}$$

Finally,

$$y = ax^2 + bx + c$$

$$y = \tfrac{5}{3}x^2 - \tfrac{14}{3}x + 3$$

$$3y = 5x^2 - 14x + 9$$

$$5x^2 - 14x - 3y + 9 = 0.$$

35. The position of the receiver is the x-intercept of the tangent line to the curve $y = x - x^2$ that passes through the point $(-1, 1)$ (see figure). The point of tangency has coordinates $(x, y) = (x, x - x^2)$. The slope of the tangent line can be calculated in two ways. First, use the formula for the slope of a line through two points and obtain

$$m = \frac{(x - x^2) - 1}{x - (-1)} = \frac{x - x^2 - 1}{x + 1}.$$

The slope of the line is also given by $y' = 1 - 2x$. Equating these two expressions for the slope of the tangent line and solving the resulting equation gives the x-coordinate of the point of tangency.

$$\frac{x - x^2 - 1}{x + 1} = 1 - 2x$$

$$x - x^2 - 1 = (1 - 2x)(1 + x)$$

$$x^2 + 2x - 2 = 0 \implies x = \sqrt{3} - 1$$

The value of the derivative at this value of x is

$$y' = 1 - 2\left(\sqrt{3} - 1\right) = 3 - 2\sqrt{3},$$

and the equation of the tangent line is

$$y - 1 = \left(3 - 2\sqrt{3}\right)\left[x - \left(\sqrt{3} - 1\right)\right]$$

$$y = \left(3 - 2\sqrt{3}\right)x + 2\left(2 - \sqrt{3}\right).$$

Since the receiver is located at the x-intercept of the line, set $y = 0$ and solve for x to obtain

$$x_0 = x = \frac{2\sqrt{3}}{3}.$$

The distance between the base of the hill and the receiver is

$$\frac{2\sqrt{3}}{3} - 1 \approx 0.155.$$

39. (a) Place the coordinate axes so that the standard form of the equation of the parabola is $x^2 = 4\,py$ (see figure). Since the parabola passes through the point $(60, 20)$,

$$60^2 = 4p(20) \quad \text{or} \quad 45 = p.$$

Therefore,

$$x^2 = 4(45)y \quad \text{or} \quad y = \frac{1}{180}x^2.$$

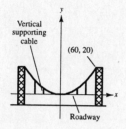

Vertical supporting cable

(60, 20)

Roadway

(b) $y' = \dfrac{1}{90}x$

$$s = \int_{-60}^{60} \sqrt{1 + (y')^2}\,dx$$

$$= 2\int_{0}^{60} \sqrt{1 + \left(\frac{x}{90}\right)^2}\,dx$$

$$= \frac{1}{45}\int_{0}^{60} \sqrt{90^2 + x^2}\,dx$$

$$= \frac{1}{90}\left[x\sqrt{90^2 + x^2} + 90^2\ln\left(x + \sqrt{90^2 + x^2}\right) \right]_0^{60}$$

$$= 10\left[2\sqrt{13} + 9\ln\left(\frac{2 + \sqrt{13}}{3}\right) \right] \approx 128.4 \text{ meters}$$

41. $x^2 + 4y^2 = 4$

Write the equation in standard form.

$$x^2 + 4y^2 = 4$$

$$\frac{x^2}{4} + y^2 = 1$$

$$\frac{(x - 0)^2}{2^2} + \frac{(y - 0)^2}{1^2} = 1$$

$$\frac{(x - h)^2}{a^2} + \frac{(y - k)^2}{b^2} = 1$$

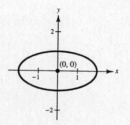

Thus, $h = 0$, $k = 0$, $a = 2$, $b = 1$, and $c = \sqrt{2^2 - 1^2} = \sqrt{3}$.

Center, (h, k): $(0, 0)$

Foci, $(h \pm c, k)$: $(\pm\sqrt{3}, 0)$

Vertices, $(h \pm a, k)$: $(\pm 2, 0)$

$$e = \frac{c}{a} = \frac{\sqrt{3}}{2}$$

45. $9x^2 + 4y^2 + 36x - 24y + 36 = 0$

Write the equation in standard form.

$$9x^2 + 4y^2 + 36x - 24y + 36 = 0$$

$$9x^2 + 36x + 4y^2 - 24y = -36$$

$$9(x^2 + 4x + 4) + 4(y^2 - 6y + 9) = -36 + 36 + 36$$

$$9(x + 2)^2 + 4(y - 3)^2 = 36$$

$$\frac{(x + 2)^2}{4} + \frac{(y - 3)^2}{9} = 1$$

$$\frac{(x + 2)^2}{2^2} + \frac{(y - 3)^2}{3^2} = 1$$

$$\frac{(x - h)^2}{b^2} + \frac{(y - k)^2}{a^2} = 1$$

—CONTINUED—

45. —CONTINUED—

Thus, $h = -2$, $k = 3$, $a = 3$, $b = 2$, and $c = \sqrt{3^2 - 2^2} = \sqrt{5}$.

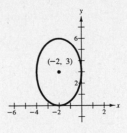

Center, (h, k): $(-2, 3)$

Foci, $(h, k \pm c)$: $\left(-2, 3 \pm \sqrt{5}\right)$

Vertices, $(h, k \pm a)$: $(-2, 3 \pm 3)$

$$e = \frac{c}{a} = \frac{\sqrt{5}}{3}$$

53. Vertices: $(3, 1)$, $(3, 9)$

Minor axis length: 6

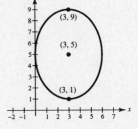

Since the vertices lie on a vertical line (see figure), the standard form of the equation of the ellipse is

$$\frac{(x - h)^2}{b^2} + \frac{(y - k)^2}{a^2} = 1.$$

Since the center of the ellipse is the midpoint of the line segment connecting the vertices, we have

$$(h, k) = \left(\frac{3 + 3}{2}, \frac{1 + 9}{2}\right) = (3, 5).$$

The length of the major axis is $2a = 8$, and therefore, $a = 4$. Furthermore, since the length of the minor axis is 6, $2b = 6$ or $b = 3$. Finally, the equation is

$$\frac{(x - 3)^2}{9} + \frac{(y - 5)^2}{16} = 1.$$

59. Since the height of the fireplace arch is 2 feet, $b = 2$ (see figure). Since the width of the arch is 5 feet, $a = \frac{5}{2}$. Therefore,

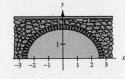

$$c^2 = a^2 - b^2 = \frac{25}{4} - 4 = \frac{9}{4} \implies c = \frac{3}{2}.$$

Hence, the tacks should be placed $c = 1.5$ feet from the center, and the length of the string should be $2a = 5$ feet.

69. $\dfrac{x^2}{4} + \dfrac{y^2}{1} = 1$

(a) Using the symmetry of the region and trigonometric substitution to evaluate the integral, we have

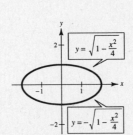

$$\text{Area} = 4\int_0^2 \sqrt{1 - \frac{x^2}{4}}\, dx$$

$$= 2\int_0^2 \sqrt{4 - x^2}\, dx$$

$$= \left[x\sqrt{4 - x^2} + 4\arcsin\frac{x}{2}\right]_0^2$$

$$= 4\arcsin 1 = 2\pi.$$

—CONTINUED—

69. —CONTINUED—

(b) Using the disc method and the symmetry of the region, yields

$$V = 2\pi\int_0^2\left(\sqrt{1 - \frac{x^2}{4}}\right)^2 dx = \frac{\pi}{2}\int_0^2(4 - x^2)\,dx = \frac{\pi}{2}\left[4x - \frac{1}{3}x^3\right]_0^2 = \frac{8\pi}{3}.$$

$$\sqrt{1 + \left(\frac{dy}{dx}\right)^2} = \sqrt{1 + \left(\frac{x}{2\sqrt{4 - x^2}}\right)^2} = \frac{\sqrt{16 - 3x^2}}{2\sqrt{4 - x^2}} = \frac{\sqrt{16 - 3x^2}}{4y}$$

$$S = 2(2\pi)\int_0^2 y\sqrt{1 + \left(\frac{dy}{dx}\right)^2}\,dx$$

$$= 4\pi\int_0^2 y\frac{\sqrt{16 - 3x^2}}{4y}\,dx$$

$$= \pi\int_0^2\sqrt{4^2 - \left(\sqrt{3}x\right)^2}\,dx$$

$$= \frac{\pi}{2\sqrt{3}}\left[\sqrt{3}x\sqrt{16 - 3x^2} + 16\arcsin\left(\frac{\sqrt{3}x}{4}\right)\right]_0^2$$

$$= \frac{2\pi}{9}\left(9 + 4\sqrt{3}\pi\right) \approx 21.48$$

(c) Using the Shell Method and the symmetry of the region we have

$$V = 2(2\pi)\int_0^2 x\left(\frac{1}{2}\sqrt{4 - x^2}\right)dx = 2\pi\left(-\frac{1}{2}\right)\int_0^2(4 - x^2)^{1/2}(-2x)\,dx = -\frac{2\pi}{3}\left[(4 - x^2)^{3/2}\right]_0^2 = \frac{16\pi}{3}.$$

Since $x = 2\sqrt{1 - y^2}$, the derivative of x with respect to y is

$$\frac{dx}{dy} = \frac{-2y}{\sqrt{1 - y^2}}.$$

Therefore,

$$\sqrt{1 + \left(\frac{dx}{dy}\right)^2} = \sqrt{1 + \frac{4y^2}{1 - y^2}} = \frac{\sqrt{1 + 3y^2}}{\sqrt{1 - y^2}} = \frac{2\sqrt{1 + 3y^2}}{x}$$

and

$$S = 2(2\pi)\int_0^1 x\sqrt{1 + \left(\frac{dx}{dy}\right)^2}\,dy$$

$$= 4\pi\int_0^1 x\frac{2\sqrt{1 + 3y^2}}{x}\,dy$$

$$= 8\pi\int_0^1\sqrt{1 + 3y^2}\,dy$$

$$= \frac{8\pi}{2\sqrt{3}}\left[\sqrt{3}y\sqrt{1 + 3y^2} + \ln\left|\sqrt{3}y + \sqrt{1 + 3y^2}\right|\right]_0^1$$

$$= \frac{4\pi}{3}\left[6 + \sqrt{3}\ln\left(2 + \sqrt{3}\right)\right] \approx 34.69.$$

75.
$$y^2 - \frac{x^2}{4} = 1$$

$$\frac{(y-0)^2}{1^2} - \frac{(x-0)^2}{2^2} = 1$$

$$\frac{(y-k)^2}{a^2} - \frac{(x-h)^2}{b^2} = 1$$

Thus, $h = 0$, $k = 0$, $a = 1$, $b = 2$, and $c = \sqrt{1^2 + 2^2} = \sqrt{5}$.

Center, (h, k): $(0, 0)$

Vertices, $(h, k \pm a)$: $(0, \pm 1)$

Foci, $(h, k \pm c)$: $\left(0, \pm\sqrt{5}\right)$

Finally, the asymptotes are given by

$$y = k \pm \frac{a}{b}(x - h) = \pm\frac{x}{2}.$$

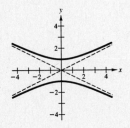

79. $9x^2 - y^2 - 36x - 6y + 18 = 0$

Write the equation in standard form.

$$9x^2 - y^2 - 36x - 6y + 18 = 0$$

$$9x^2 - 36x - (y^2 + 6y) = -18$$

$$9(x^2 - 4x + 4) - (y^2 + 6y + 9) = -18 + 36 - 9$$

$$9(x - 2)^2 - (y + 3)^2 = 9$$

$$\frac{(x-2)^2}{1^2} - \frac{(y+3)^2}{3^2} = 1$$

Thus, $h = 2$, $k = -3$, $a = 1$, $b = 3$, and $c = \sqrt{1^2 + 3^2} = \sqrt{10}$.

Center, (h, k): $(2, -3)$

Vertices, $(h \pm a, k)$: $(1, -3)$ and $(3, -3)$

Foci, $(h \pm c, k)$: $\left(2 \pm \sqrt{10}, -3\right)$

Finally, the asymptotes are

$$y = k \pm \frac{b}{a}(x - h) = -3 \pm \frac{3}{1}(x - 2)$$

$$y = 3x - 9 \text{ and } y = -3x + 3.$$

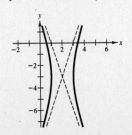

77. $\dfrac{(x-1)^2}{4} - \dfrac{(y+2)^2}{1} = 1$

$$\frac{(x-h)^2}{a^2} - \frac{(y+k)^2}{b^2} = 1$$

Thus, $h = 1$, $k = -2$, $a = 2$, $b = 1$, and $c = \sqrt{4 + 1} = \sqrt{5}$.

Center, (h, k): $(1, -2)$

Vertices, $(h \pm a, k)$: $(-1, -2), (3, -2)$

Foci, $(h \pm c, k)$: $\left(1 \pm \sqrt{5}, -2\right)$

Finally, the asymptotes are

$$y = k \pm \frac{b}{a}(x - h) = -2 \pm \frac{1}{2}(x - 1)$$

$$y = \frac{1}{2}x - \frac{5}{2} \text{ and } y = -\frac{1}{2}x - \frac{3}{2}.$$

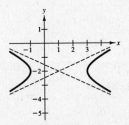

87. Vertices: $(-1, 0), (1, 0)$

Asymptotes: $y = \pm 3x$

Since the vertices lie on a horizontal line, the standard form is

$$\frac{(x-h)^2}{a^2} - \frac{(y-k)^2}{b^2} = 1.$$

The center of the hyperbola lies at the midpoint of the line segment connecting the vertices. Thus,

$$(h, k) = \left(\frac{1 + (-1)}{2}, \frac{0 + 0}{2}\right) = (0, 0)$$

and we have $h = 0$ and $k = 0$. Since the asymptotes are of the form

$$y = k \pm \frac{b}{a}(x - h) = \pm 3x$$

we have

$$\pm\frac{b}{a} = \pm 3 \text{ or } b = 3a.$$

The distance from the center to the vertices is $a = 1$. Therefore, $b = 3$ and the equation is

$$\frac{x^2}{1^2} - \frac{y^2}{3^2} = 1 \text{ or } x^2 - \frac{y^2}{9} = 1.$$

95. The points $(2, 2)$ and $(10, 2)$ are the foci of the hyperbola. From the figure and the distance formula we have

$$d_1 - d_2 = 6$$

$$\sqrt{(x - 2)^2 + (y - 2)^2} - \sqrt{(x - 10)^2 + (y - 2)^2} = 6.$$

Isolate the radicals one at a time, square each member of the resulting equation, and simplify.

$$\sqrt{(x - 2)^2 + (y - 2)^2} = 6 + \sqrt{(x - 10)^2 + (y - 2)^2}$$

$$(x - 2)^2 + (y - 2)^2 = 36 + 12\sqrt{(x - 10)^2 + (y - 2)^2} + (x - 10)^2 + (y - 2)^2$$

$$(x - 2)^2 - (x - 10)^2 - 36 = 12\sqrt{(x - 10)^2 + (y - 2)^2}$$

$$4x - 33 = 3\sqrt{(x - 10)^2 + (y - 2)^2}$$

$$16x^2 - 264x + 1089 = 9(x^2 - 20x + 100 + y^2 - 4y + 4)$$

$$7x^2 - 9y^2 - 84x + 36y + 153 = 0$$

$$7(x^2 - 12x) - 9(y^2 - 4y) = -153$$

$$7(x - 6)^2 - 9(y - 2)^2 = 63$$

$$\frac{(x - 6)^2}{9} - \frac{(y - 2)^2}{7} = 1$$

101. Let (x, y) be a point on the hyperbola and the line through the points $(0, 10)$ and $(10, 0)$. The equation of the line is $y = 10 - x$. Substituting this expression for y into the equation of the hyperbola yields

$$\frac{x^2}{36} - \frac{y^2}{64} = 1$$

$$\frac{x^2}{36} - \frac{(10 - x)^2}{64} = 1$$

$$16x^2 - 9(10 - x)^2 = 576$$

$$7x^2 + 180x - 1476 = 0.$$

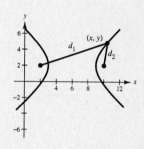

Using the Quadratic Formula yields

$$x = \frac{-180 \pm \sqrt{180^2 - 4(7)(-1476)}}{2(7)} = \frac{-180 \pm 192\sqrt{2}}{14} = \frac{-90 \pm 96\sqrt{2}}{7}.$$

Choosing the positive value for x, we have

$$x = \frac{-90 \pm 96\sqrt{2}}{7} \approx 6.538 \quad \text{and} \quad y = 10 - x = \frac{160 - 96\sqrt{2}}{7} \approx 3.462.$$

105.
$$\frac{x^2}{a^2} + \frac{2y^2}{b^2} = 1 \implies \frac{2y^2}{b^2} = 1 - \frac{x^2}{a^2}, \, c^2 = a^2 - b^2$$

$$\frac{x^2}{a^2 - b^2} - \frac{2y^2}{b^2} = 1 \implies \frac{2y^2}{b^2} = \frac{x^2}{a^2 - b^2} - 1$$

$$1 - \frac{x^2}{a^2} = \frac{x^2}{a^2 - b^2} - 1 \implies 2 = x^2\left(\frac{1}{a^2} + \frac{1}{a^2 - b^2}\right)$$

$$x^2 = \frac{2a^2(a^2 - b^2)}{2a^2 - b^2} \implies x = \pm\frac{\sqrt{2}a\sqrt{a^2 - b^2}}{\sqrt{2a^2 - b^2}} = \pm\frac{\sqrt{2}ac}{\sqrt{2a^2 - b^2}}$$

$$\frac{2y^2}{b^2} = 1 - \frac{1}{a^2}\left(\frac{2a^2c^2}{2a^2 - b^2}\right) \implies \frac{2y^2}{b^2} = \frac{b^2}{2a^2 - b^2}$$

$$y^2 = \frac{b^4}{2(2a^2 - b^2)} \implies y = \pm\frac{b^2}{\sqrt{2}\sqrt{2a^2 - b^2}}$$

—**CONTINUED**—

105. —CONTINUED—

There are four points of intersection: $\left(\dfrac{\sqrt{2}ac}{\sqrt{2a^2 - b^2}}, \pm\dfrac{b^2}{\sqrt{2}\sqrt{2a^2 - b^2}}\right), \left(-\dfrac{\sqrt{2}ac}{\sqrt{2a^2 - b^2}}, \pm\dfrac{b^2}{\sqrt{2}\sqrt{2a^2 - b^2}}\right)$

$$\frac{x^2}{a^2} + \frac{2y^2}{b^2} = 1 \implies \frac{2x}{a^2} + \frac{4yy'}{b^2} = 0 \implies y'_e = -\frac{b^2 x}{2a^2 y}$$

$$\frac{x^2}{a^2 - b^2} - \frac{2y^2}{b^2} = 1 \implies \frac{2x}{c^2} - \frac{4yy'}{b^2} = 0 \implies y'_h = \frac{b^2 x}{2c^2 y}$$

At $\left(\dfrac{\sqrt{2}ac}{\sqrt{2a^2 - b^2}}, \dfrac{b^2}{\sqrt{2}\sqrt{2a^2 - b^2}}\right)$, the slopes of the tangent lines are:

$$y'_e = \frac{-b^2\left(\dfrac{\sqrt{2}ac}{\sqrt{2a^2 - b^2}}\right)}{2a^2\left(\dfrac{b^2}{\sqrt{2}\sqrt{2a^2 - b^2}}\right)} = -\frac{c}{a} \quad \text{and} \quad y'_h = \frac{b^2\left(\dfrac{\sqrt{2}ac}{\sqrt{2a^2 - b^2}}\right)}{2c^2\left(\dfrac{b^2}{\sqrt{2}\sqrt{2a^2 - b^2}}\right)} = \frac{a}{c}$$

Since the slopes are negative reciprocals, the tangent lines are perpendicular. Similarly, the curves are perpendicular at the other three points of intersection.

Section 9.2   Plane Curves and Parametric Equations

7. $x = t^3, y = \dfrac{t^2}{2}$

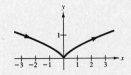

t	$-\frac{3}{2}$	-1	$-\frac{1}{2}$	0	$\frac{1}{2}$	1	$\frac{3}{2}$
x	$-\frac{27}{8}$	-1	$-\frac{1}{8}$	0	$\frac{1}{8}$	1	$\frac{27}{8}$
y	$\frac{9}{8}$	$\frac{1}{2}$	$\frac{1}{8}$	0	$\frac{1}{8}$	$\frac{1}{2}$	$\frac{9}{8}$

By plotting these points in the order of increasing t and using the continuity of the parametric equation, we obtain the curve shown in the figure.

Since $x = t^3$ and $y = t^2/2$, we have

$$t = x^{1/3} \quad \text{or} \quad y = \frac{1}{2}(x^{1/3})^2 = \frac{1}{2}x^{2/3}.$$

15. $x = 3\cos\theta, y = 3\sin\theta$

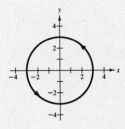

t	0	$\frac{\pi}{4}$	$\frac{\pi}{2}$	$\frac{3\pi}{4}$	π	$\frac{5\pi}{4}$	$\frac{3\pi}{2}$	$\frac{7\pi}{4}$
x	3	$\frac{3\sqrt{2}}{2}$	0	$-\frac{3\sqrt{2}}{2}$	-3	$-\frac{3\sqrt{2}}{2}$	0	$\frac{3\sqrt{2}}{2}$
y	0	$\frac{3\sqrt{2}}{2}$	3	$\frac{3\sqrt{2}}{2}$	0	$-\frac{3\sqrt{2}}{2}$	-3	$-\frac{3\sqrt{2}}{2}$

By plotting these points in order of increasing θ and by using the continuity of the parametric equations, we obtain the curve shown in the figure.

To eliminate the parameter, use the identity $\sin^2\theta + \cos^2\theta = 1$.

$$x^2 + y^2 = 9\cos^2\theta + 9\sin^2\theta$$
$$= 9(\cos^2\theta + \sin^2\theta) = 9$$

The graph of the equation is a circle centered at the origin with radius 3.

23. $x = 4 \sec \theta$, $y = 3 \tan \theta$

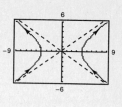

θ	$-\dfrac{\pi}{3}$	$-\dfrac{\pi}{6}$	0	$\dfrac{\pi}{6}$	$\dfrac{\pi}{3}$	$\dfrac{\pi}{2}$	$\dfrac{2\pi}{3}$	$\dfrac{5\pi}{6}$	π	$\dfrac{7\pi}{6}$	$\dfrac{4\pi}{3}$
x	8	$\dfrac{8}{\sqrt{3}}$	4	$\dfrac{8}{\sqrt{3}}$	8	undef.	-8	$-\dfrac{8}{\sqrt{3}}$	-4	$-\dfrac{8}{\sqrt{3}}$	-8
y	$-3\sqrt{3}$	$-\sqrt{3}$	0	$\sqrt{3}$	$3\sqrt{3}$	undef.	$-3\sqrt{3}$	$-\sqrt{3}$	0	$\sqrt{3}$	$3\sqrt{3}$

By plotting these points in order of increasing θ we obtain the curve shown in the figure.

Since the parametric equations involve secants and tangents, use the identity $\sec^2 \theta - \tan^2 \theta = 1$. Therefore,

$$\frac{x}{4} = \sec \theta \text{ and } \frac{y}{3} = \tan \theta \quad \text{or} \quad \frac{x^2}{16} - \frac{y^2}{9} = \sec^2 \theta - \tan^2 \theta = 1.$$

The graph of this equation is a hyperbola centered at the origin with vertices $(\pm 4, 0)$.

29. By eliminating the parameter in each part of this exercise, we get $y = 2x + 1$. The graphs differ in their orientation and domains. In each case the derivatives of $x = f(t)$ and $y = g(t)$ are continuous and not simultaneously zero. Therefore, the graphs are smooth.

(a) $x = t$
$-\infty < x < \infty$

$y = 2t + 1$
$-\infty < y < \infty$

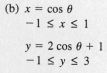

(b) $x = \cos \theta$
$-1 \le x \le 1$

$y = 2 \cos \theta + 1$
$-1 \le y \le 3$

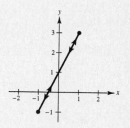

(c) $x = e^{-t}$
$x > 0$

$y = 2e^{-t} + 1$
$y > 1$

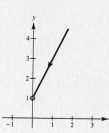

(d) $x = e^t$
$x > 0$

$y = 2e^t + 1$
$y > 1$

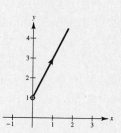

37. Since the parametric equations involve sines and cosines, use the identity $\sin^2 \theta + \cos^2 \theta = 1$.

$$x = h + a \cos \theta \qquad y = k + b \sin \theta$$

$$\frac{x - h}{a} = \cos \theta \qquad \frac{y - k}{b} = \sin \theta$$

$$\frac{(x - h)^2}{a^2} = \cos^2 \theta \qquad \frac{(y - k)^2}{b^2} = \sin^2 \theta$$

$$\frac{(x - h)^2}{a^2} + \frac{(y - k)^2}{b^2} = \cos^2 \theta + \sin^2 \theta = 1$$

$$\frac{(x - h)^2}{a^2} + \frac{(y - k)^2}{b^2} = 1$$

45. The midpoint of the line segment joining the vertices is the center of the hyperbola. Therefore, $(h, k) = (0, 0)$. The transverse axis of the hyperbola is horizontal with $a = 4$ and $c = 5$. Since $b^2 = c^2 - a^2$, we have $b = 3$. Using the result of Exercise 38 a set of parametric equations are

$$x = h + a \sec \theta \qquad y = k + b \tan \theta$$

$$x = 4 \sec \theta, \qquad y = 3 \tan \theta.$$

53. $x = \theta - \frac{3}{2} \sin \theta, \; y = 1 - \frac{3}{2} \cos \theta$

Select the parametric equations mode of your graphing utility and enter the parametric equations. The parametric mode may require the parameter t rather than θ. A graph of the curve is shown in the figure. The curve is smooth for all values of θ.

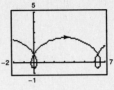

69. $x = (v_0 \cos \theta)t, \; y = h + (v_0 \sin \theta)t - 16t^2$

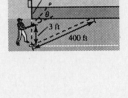

(a) Since the ball is hit 3 feet above the ground, $h = 3$. Since the initial velocity is 100 miles per hour we have

$$v_0 = 100 \text{ mi/hr} = \frac{100(5280)}{3600} = \frac{440}{3} \text{ ft/sec.}$$

Therefore,

$$x = \left(\frac{440}{3} \cos \theta\right)t, \; y = 3 + \left(\frac{440}{3} \sin \theta\right)t - 16t^2.$$

(b) When $\theta = 15°$, the parametric equations are

$$x = \left(\frac{440}{3} \cos 15°\right)t, \; y = 3 + \left(\frac{440}{3} \sin 15°\right)t - 16t^2,$$

and the graph of the curve is given in the figure. It is not a home run, since the ball drops to the ground in front of the fence which is 400 feet from the home plate.

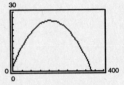

(c) When $\theta = 23°$, the parametric equations are

$$x = \left(\frac{440}{3} \cos 23°\right)t, \; y = 3 + \left(\frac{440}{3} \sin 23°\right)t - 16t^2,$$

and the graph of the curve is given in the figure. By using the trace feature of the graphing utility we observe that $y \approx 32$ feet when $x = 400$ feet. Therefore, it is a home run.

(d) For the hit to be a home run, y must be at least 10 when x is 400. When $x = 400$, we have

$$\left(\frac{440}{3} \cos \theta\right)t = 400 \implies t = \frac{1200}{440 \cos \theta}.$$

Substituting this expression for t into the parametric equation for y when $y = 10$ yields

$$y = 3 + \left(\frac{440}{3} \sin \theta\right)t - 16t^2$$

$$10 = 3 + \left(\frac{440}{3} \sin \theta\right) \cdot \frac{1200}{440 \cos \theta} - 16\left(\frac{1200}{440 \cos \theta}\right)^2$$

$$0 = -7 + 400 \tan \theta - 16\left(\frac{30}{11} \sec \theta\right)^2.$$

Using the root finding capabilities of a graphing utility yields the root

$$\theta \approx 0.3383 \text{ radians} = 0.3383\left(\frac{180}{\pi}\right) \approx 19.4°.$$

Section 9.3   Parametric Equations and Calculus

1. Since $x = 2t$ and $y = 3t - 1$, you have

$$\frac{dy}{dx} = \frac{dy/dt}{dx/dt} = \frac{3}{2} \text{ for all values of } t$$

and

$$\frac{d^2y}{dx^2} = \frac{d[dy/dx]/dt}{dx/dt} = \frac{0}{2} = 0 \text{ for all values of } t.$$

7. Since $x = 2 + \sec\theta$ and $y = 1 + 2\tan\theta$, we have

$$\frac{dy}{dx} = \frac{dy/d\theta}{dx/d\theta} = \frac{2\sec^2\theta}{\sec\theta\tan\theta} = \frac{2\sec\theta}{\tan\theta} = 2\csc\theta$$

$$\frac{d^2y}{dx^2} = \frac{d[dy/dx]/d\theta}{dx/d\theta} = \frac{-2\csc\theta\cot\theta}{\sec\theta\tan\theta} = -2\cot^3\theta.$$

At $\theta = \pi/6$,

$$\frac{dy}{dx} = 2\csc\frac{\pi}{6} = 2(2) = 4$$

$$\frac{d^2y}{dx^2} = -2\cot^3\frac{\pi}{6} = -2\left(\sqrt{3}\right)^3 = -6\sqrt{3}.$$

11. $x = 2\cot\theta,\ y = 2\sin^2\theta$

$$\frac{dy}{dx} = \frac{dy/d\theta}{dx/d\theta} = \frac{4\sin\theta\cos\theta}{-2\csc^2\theta} = -2\sin^3\theta\cos\theta$$

At the point $\left(-\dfrac{2}{\sqrt{3}}, \dfrac{3}{2}\right)$, $\theta = \dfrac{2\pi}{3}$. When $\theta = \dfrac{2\pi}{3}$,

$$\frac{dy}{dx} = -2\sin^3\frac{2\pi}{3}\cos\frac{2\pi}{3} = -2\left(\frac{\sqrt{3}}{2}\right)^3\left(-\frac{1}{2}\right) = \frac{3\sqrt{3}}{8}.$$

Therefore, the equation of the tangent line is

$$y - \frac{3}{2} = \frac{3\sqrt{3}}{8}\left(x + \frac{2}{\sqrt{3}}\right)$$

$$3\sqrt{3}x - 8y + 18 = 0.$$

At the point $(0, 2)$, $\theta = \dfrac{\pi}{2}$. When $\theta = \dfrac{\pi}{2}$,

$$\frac{dy}{dx} = -2\sin^3\frac{\pi}{2}\cos\frac{\pi}{2} = -2(1)^3(0) = 0.$$

Therefore, the equation of the tangent line is

$$y - 2 = 0(x - 0)$$

$$y - 2 = 0.$$

At the point $\left(2\sqrt{3}, \dfrac{1}{2}\right)$, $\theta = \dfrac{\pi}{6}$. When $\theta = \dfrac{\pi}{6}$,

$$\frac{dy}{dx} = -2\sin^3\frac{\pi}{6}\cos\frac{\pi}{6} = -2\left(\frac{1}{2}\right)^3\left(\frac{\sqrt{3}}{2}\right) = -\frac{\sqrt{3}}{8}.$$

Therefore, the equation of the tangent line is

$$y - \frac{1}{2} = -\frac{\sqrt{3}}{8}\left(x - 2\sqrt{3}\right)$$

$$\sqrt{3}x + 8y - 10 = 0.$$

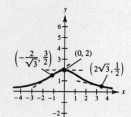

21. $x = 1 - t,\ y = t^3 - 3t$

Since $\dfrac{dy}{dx} = \dfrac{dy/dt}{dx/dt} = \dfrac{3t^2 - 3}{-1} = 3 - 3t^2$, the horizontal tangents occur when

$$3 - 3t^2 = 0 \quad \text{or} \quad t = \pm 1.$$

The corresponding points are $(0, -2)$, and $(2, 2)$. Since dy/dx is never undefined, there are no points of vertical tangency.

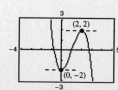

31.
$$x = e^{-t} \cos t$$

$$\frac{dx}{dt} = -e^{-t}(\sin t + \cos t)$$

$$\left(\frac{dx}{dt}\right)^2 = e^{-2t}(\sin^2 t + 2 \sin t \cos t + \cos^2 t) = e^{-2t}(1 + \sin 2t)$$

$$y = e^{-t} \sin t$$

$$\frac{dy}{dt} = e^{-t}(\cos t - \sin t)$$

$$\left(\frac{dy}{dt}\right)^2 = e^{-2t}(\cos^2 t - 2 \sin t \cos t + \sin^2 t) = e^{-2t}(1 - \sin 2t)$$

$$\left(\frac{dx}{dt}\right)^2 + \left(\frac{dy}{dt}\right)^2 = 2e^{-2t}$$

Therefore,

$$s = \int_0^{\pi/2} \sqrt{\left(\frac{dx}{dt}\right)^2 + \left(\frac{dy}{dt}\right)^2}\, dt = \int_0^{\pi/2} \sqrt{2e^{-2t}}\, dt$$

$$= -\sqrt{2} \int_0^{\pi/2} e^{-t}(-1)\, dt = -\sqrt{2}\left[e^{-t}\right]_0^{\pi/2} = \sqrt{2}(1 - e^{-\pi/2}) \approx 1.12.$$

37. $x = a \cos^3 \theta,\; y = a \sin^3 \theta$

Using $dx/d\theta = -3a \cos^2 \theta \sin \theta$, $dy/d\theta = 3a \sin^2 \theta \cos \theta$, and the symmetry of the graph, the perimeter is given by

$$s = 4 \int_0^{\pi/2} \sqrt{\left(\frac{dx}{d\theta}\right)^2 + \left(\frac{dy}{d\theta}\right)^2}\, d\theta = 4 \int_0^{\pi/2} \sqrt{9a^2 \cos^4 \theta \sin^2 \theta + 9a^2 \sin^4 \theta \cos^2 \theta}\, d\theta$$

$$= 4 \int_0^{\pi/2} \sqrt{9a^2 \sin^2 \theta \cos^2 \theta(\cos^2 \theta + \sin^2 \theta)}\, d\theta$$

$$= 4 \int_0^{\pi/2} 3a \sin \theta \cos \theta\, d\theta = 12a\left[\frac{\sin^2 \theta}{2}\right]_0^{\pi/2} = 6a.$$

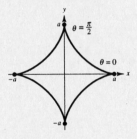

49. $x = a \cos^3 \theta,\; y = a \sin^3 \theta$

$$\frac{dx}{d\theta} = -3a \cos^2 \theta \sin \theta,\; \frac{dy}{d\theta} = 3a \sin^2 \theta \cos \theta$$

Using the integral for surface area and the symmetry of the graph shown in the figure, we have

$$S = 2\pi \int_a^b g(\theta) \sqrt{\left(\frac{dx}{d\theta}\right)^2 + \left(\frac{dy}{d\theta}\right)^2}\, d\theta$$

$$= 4\pi \int_0^{\pi/2} a \sin^3 \theta \sqrt{9a^2 \cos^4 \theta \sin^2 \theta + 9a^2 \sin^4 \theta \cos^2 \theta}\, d\theta$$

$$= 12a^2 \pi \int_0^{\pi/2} \sin^4 \theta \cos \theta\, d\theta$$

$$= 4 \frac{12a^2 \pi}{5}\left[\sin^5 \theta\right]_0^{\pi/2} = \frac{12}{5}\pi a^2.$$

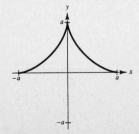

51. The sphere is formed by revolving a circle of radius r about the x-axis. We represent the circle by

$$x = f(\phi) = r \cos \phi \qquad y = g(\phi) = r \sin \phi$$

$$\frac{dx}{d\phi} = -r \sin \phi \qquad \frac{dy}{d\phi} = r \cos \phi.$$

From the integrals of Theorem 9.9 and the accompanying figure, we have

$$S = 2\pi \int_0^\theta g(\phi) \sqrt{\left(\frac{dx}{d\phi}\right)^2 + \left(\frac{dy}{d\phi}\right)^2}\, d\phi$$

$$= 2\pi \int_0^\theta r \sin \phi \sqrt{r^2 \sin^2 \phi + r^2 \cos^2 \phi}\, d\phi$$

$$= 2\pi r^2 \int_0^\theta \sin \phi\, d\phi = -2\pi r^2\left[\cos \phi\right]_0^\theta$$

$$= 2\pi r^2(1 - \cos \theta).$$

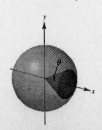

57. $x = 2 \sin^2 \theta,\ y = 2 \sin^2 \theta \tan \theta$

From Exercise 52, we have the following convergent improper integral:

$$A = \int_0^2 y\,dx = \int_0^{\pi/2} y \frac{dx}{d\theta}\,d\theta$$

$$= \int_0^{\pi/2} 2\sin^2\theta \tan\theta(4\sin\theta\cos\theta)\,d\theta$$

$$= 8\int_0^{\pi/2} \sin^4\theta\,d\theta$$

$$= 8\int_0^{\pi/2} \left(\frac{1 - \cos 2\theta}{2}\right)^2 d\theta$$

$$= 2\int_0^{\pi/2} (1 - 2\cos 2\theta + \cos^2 2\theta)\,d\theta$$

$$= 2\int_0^{\pi/2} \left(1 - 2\cos 2\theta + \frac{1 + \cos 4\theta}{2}\right)d\theta$$

$$= 2\int_0^{\pi/2} \left(\frac{3}{2} - 2\cos 2\theta + \frac{1}{2}\cos 4\theta\right)d\theta$$

$$= 2\left[\frac{3}{2}\theta - \sin 2\theta + \frac{1}{8}\sin 4\theta\right]_0^{\pi/2} = 2\left(\frac{3\pi}{4}\right) = \frac{3\pi}{2}$$

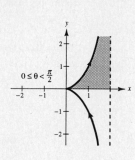

67. (a) Let d_1 be the distance between point P and the plane flying 375 miles per hour (see figure). Therefore,

$$d_1 = 150 - 375t$$

where t is time in hours. If (x_1, y_1) are the coordinates of the position of this plane, then we obtain the required parametric equations in the following manner.

$$\cos 70° = \frac{x_1}{150 - 375t} \implies x_1 = \cos 70°(150 - 375t)$$

$$\sin 70° = \frac{y_2}{150 - 375t} \implies y_1 = \sin 70°(150 - 375t)$$

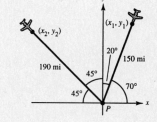

Let d_2 be the distance between the point P and the plane flying 450 miles per hour. Therefore,

$$d_2 = 190 - 450t$$

where t is time in hours. If (x_2, y_2) are the coordinates of the position of this plane, then we obtain the required parametric equations in the following manner.

$$\cos 45° = \frac{-x_2}{190 - 450t} \implies x_2 = -\cos 45°(190 - 450t)$$

$$\sin 45° = \frac{y_2}{190 - 450t} \implies y_2 = \sin 45°(190 - 450t)$$

(b) The distance d between the planes is

$$d = \sqrt{(x_2 - x_1)^2 + (y_2 - y_1)^2}$$

$$= \sqrt{[-\cos 45°(190 - 450t) - \cos 70°(150 - 375t)]^2 + [\sin 45°(190 - 450t) - \sin 70°(150 - 375t)]^2}.$$

(c) The graph of the distance function of part (b) is shown in the figure. Using the capabilities of finding extrema on a graphing utility yields a minimum separation of the planes of 7.59 miles when $t \approx 0.4145$. This meets regulations requiring a minimum separation of at least 3 miles.

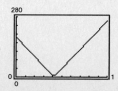

Section 9.4 Polar Coordinates and Polar Graphs

5. Using the conversion from polar to rectangular coordinates, you have

$$x = r \cos \theta = \sqrt{2} \cos(2.36) \approx -1.004$$

$$y = r \sin \theta = \sqrt{2} \sin(2.36) \approx 0.996.$$

Therefore the rectangular coordinates are $(1.004, 0.996)$.

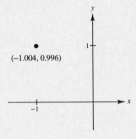

13. First,

$$r = \pm\sqrt{x^2 + y^2} = \pm\sqrt{(-3)^2 + 4^2} = \pm 5$$

$$\tan \theta = -\tfrac{4}{3} \text{ and } \arctan\left(-\tfrac{4}{3}\right) \approx -0.9273.$$

Since $(-3, 4)$ lies in the second quadrant, let $\theta = \pi - 0.9273 = 2.214$. Thus one polar representation is

$$(5, 2.214) \quad (r > 0, 0 \le \theta < 2\pi).$$

To obtain the second representation, change the sign of r and increase θ by π radians to obtain

$$(-5, 5.356) \quad (r < 0, 0 \le \theta < 2\pi).$$

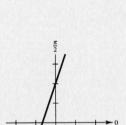

25. Since $x = r \cos \theta$ and $y = r \sin \theta$, we have

$$3x - y + 2 = 0$$

$$3(r \cos \theta) - r \sin \theta + 2 = 0$$

$$r(3 \cos \theta - \sin \theta) = -2$$

$$r = \frac{-2}{3 \cos \theta - \sin \theta}.$$

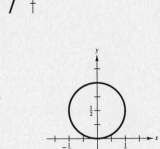

31. Since

$$r^2 = x^2 + y^2 \text{ and } y = r \sin \theta,$$

we begin by multiplying both members of the polar equation by r to obtain

$$r = \sin \theta$$

$$r^2 = r \sin \theta$$

$$x^2 + y^2 = y$$

$$x^2 + y^2 - y = 0$$

$$x^2 + \left(y^2 - y + \tfrac{1}{4}\right) = \tfrac{1}{4}$$

$$x^2 + \left(y - \tfrac{1}{2}\right)^2 = \left(\tfrac{1}{2}\right)^2.$$

Therefore, the graph is a circle of radius $\tfrac{1}{2}$ centered at $\left(0, \tfrac{1}{2}\right)$.

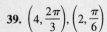

39. $\left(4, \dfrac{2\pi}{3}\right), \left(2, \dfrac{\pi}{6}\right)$

$$d = \sqrt{r_1^2 + r_2^2 - 2r_1 r_2 \cos(\theta_1 - \theta_2)}$$

$$= \sqrt{4^2 + 2^2 - 2(4)(2) \cos\left(\frac{2\pi}{3} - \frac{\pi}{6}\right)} = \sqrt{20 - 16 \cos \frac{\pi}{2}} = 2\sqrt{5} \approx 4.5$$

45. $r = 3(1 - \cos \theta)$

(a) The graph is shown in the figure.

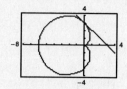

(b) The graph of the tangent line when $\theta = \pi/2$ is shown on the graph of part (a),

(c) $f(\theta) = 3(1 - \cos \theta)$

$f'(\theta) = 3 \sin \theta$

$$\frac{dy}{dx} = \frac{f'(\theta) \sin \theta + f(\theta) \cos \theta}{f'(\theta) \cos \theta - f(\theta) \sin \theta}$$

$$= \frac{(3 \sin \theta) \sin \theta + 3(1 - \cos \theta) \cos \theta}{(3 \sin \theta) \cos \theta - 3(1 - \cos \theta) \sin \theta}$$

$$= \frac{\sin^2 \theta + \cos \theta - \cos^2 \theta}{\sin \theta (2 \cos \theta - 1)}$$

$$= \frac{(1 + 2 \cos \theta)(1 - \cos \theta)}{\sin \theta (2 \cos \theta - 1)}$$

When $\theta = \pi/2$, the slope is

$$\frac{dy}{dx} = \frac{(1 + 0)(1 - 0)}{1(0 - 1)} = -1.$$

49. Since $f(\theta) = 1 + \sin \theta$ and $f'(\theta) = \cos \theta$, we have

$$\frac{dy}{dx} = \frac{f'(\theta) \sin \theta + f(\theta) \cos \theta}{f'(\theta) \cos \theta - f(\theta) \sin \theta}$$

$$= \frac{\cos \theta \sin \theta + (1 + \sin \theta) \cos \theta}{\cos^2 \theta - (1 + \sin \theta) \sin \theta}$$

$$= \frac{\cos \theta (1 + 2 \sin \theta)}{1 - \sin \theta - 2 \sin^2 \theta}$$

$$= \frac{\cos \theta (1 + 2 \sin \theta)}{(1 - 2 \sin \theta)(1 + \sin \theta)}.$$

Since $dy/dx = 0$ when $\cos \theta (1 + 2 \sin \theta) = 0$ or $\theta = \pi/2, 7\pi/6, 11\pi/6$, there are horizontal tangents at the points

$$\left(2, \frac{\pi}{2}\right), \left(\frac{1}{2}, \frac{7\pi}{6}\right), \text{ and } \left(\frac{1}{2}, \frac{11\pi}{6}\right).$$

Since dy/dx is undefined when $(1 - 2 \sin \theta)(1 + \sin \theta) = 0$ or $\theta = \pi/6, 5\pi/6, 3\pi/2$, there are vertical tangents at the points

$$\left(\frac{3}{2}, \frac{\pi}{6}\right) \text{ and } \left(\frac{3}{2}, \frac{5\pi}{6}\right).$$

59. The equation has the form $r = a \cos(n\theta) = 2 \cos 3\theta$ where n is odd. This means the graph is a *rose curve* with $n = 3$ petals. The curve has polar axis symmetry. The relative extrema of r are $(2, 0)$, $(-2, \pi/3)$, and $(2, 2\pi/3)$. Since $r = 0$ and $dr/d\theta \neq 0$ when $\theta = \pi/6$, $\theta = \pi/2$, and $\theta = 5\pi/6$, we have tangents at the pole for these values of θ.

θ	0	$\dfrac{\pi}{12}$	$\dfrac{\pi}{6}$	$\dfrac{\pi}{3}$	$\dfrac{\pi}{2}$	$\dfrac{2\pi}{3}$
r	2	$\sqrt{2}$	0	-2	0	2

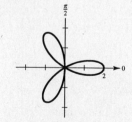

67. $r = 2\theta$

Spiral of Archimedes

Symmetric to $\theta = \pi/2$

θ	0	$\dfrac{\pi}{4}$	$\dfrac{\pi}{2}$	$\dfrac{3\pi}{4}$	π	$\dfrac{5\pi}{4}$	$\dfrac{3\pi}{2}$
r	2	$\dfrac{\pi}{2}$	π	$\dfrac{3\pi}{2}$	2π	$\dfrac{5\pi}{2}$	3π

Tangent at the pole: $\theta = 0$

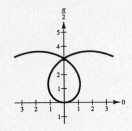

79. Before using the graphing utility to obtain the graph of the polar equation, observe that the equation has the form

$$r^2 = a^2 \sin 2\theta = 4 \sin 2\theta$$

and is a *leminscate* with symmetry with respect to the pole. (If your graphing utility doesn't have a polar mode, but does have a parametric mode, you can sketch the graph of $r = f(\theta)$ by writing the equation as $x = f(\theta) \cos \theta$ and $y = f(\theta) \sin \theta$.) The entire graph will be traced for $0 \le \theta \le \pi$. Also, the tangents at the pole are $\theta = 0$ and $\theta = \pi/2$ since $r = 0$ and $dr/d\theta \ne 0$ at these values.

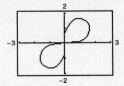

81. The graph of the polar equation obtained by a graphing utility is shown in the figure. To locate the vertical asymptote, note that

$$r \to -\infty \text{ as } \theta \to \frac{\pi^-}{2} \qquad \text{and} \qquad r \to \infty \text{ as } \theta \to \frac{\pi^+}{2}.$$

To see that the vertical asymptote is located at $x = -1$, write

$$r = 2 - \frac{1}{\cos \theta} = 2 - \frac{r}{r \cos \theta} = 2 - \frac{r}{x}$$

$$rx = 2x - r$$

$$r(1 + x) = 2x$$

$$r = \frac{2x}{1 + x}.$$

Thus $r \to \pm\infty$ as $x \to -1$.

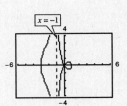

97. $f(\theta) = r = \dfrac{6}{1 - \cos \theta}$

$$f'(\theta) = \frac{-6 \sin \theta}{(1 - \cos \theta)^2}$$

From the definition of the angle ψ between the radial line and the tangent line (see Exercise 92), we have

$$\tan \psi = \left| \frac{f(\theta)}{f'(\theta)} \right| = \left| \frac{6/(1 - \cos \theta)}{-6 \sin \theta/(1 - \cos \theta)^2} \right| = \left| \frac{1 - \cos \theta}{-\sin \theta} \right|.$$

At $\theta = 2\pi/3$,

$$\tan \psi = \left| \frac{1 - (-1/2)}{-\sqrt{3}/2} \right| = \left| \frac{3/2}{-\sqrt{3}/2} \right| = \sqrt{3}.$$

Therefore, $\psi = \pi/3$. The graph is shown in the figure.

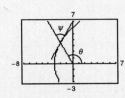

Section 9.5   Area and Arc Length in Polar Coordinates

3. The graph of $r = 2 \cos 3\theta$ is shown in the figure. Use symmetry to find the area of one petal.

$$A = 2\left[\frac{1}{2}\int_0^{\pi/6} (2 \cos 3\theta)^2 \, d\theta\right]$$

$$= \int_0^{\pi/6} 4\left(\frac{1 + \cos 6\theta}{2}\right) d\theta = 2\left[\theta + \frac{1}{6}\sin 6\theta\right]_0^{\pi/6} = \frac{\pi}{3}$$

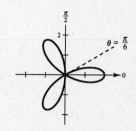

9. Solving the equation $r = 0$ to find the tangents at the pole we have,

$$1 + 2 \cos \theta = 0$$

$$\cos \theta = -\frac{1}{2}$$

$$\theta = \frac{2\pi}{3} \text{ or } \frac{4\pi}{3}.$$

Therefore, the lower half of the inner loop is generated when θ is in the interval $2\pi/3 \le \theta \le \pi$. From the symmetry of the graph in the figure, the area of the inner loop is given by

$$2\int_{2\pi/3}^{\pi} \frac{1}{2} r^2 \, d\theta = \int_{2\pi/3}^{\pi} (1 + 2 \cos \theta)^2 \, d\theta$$

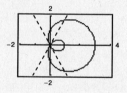

$$= \int_{2\pi/3}^{\pi} (1 + 4 \cos \theta + 4 \cos^2 \theta) \, d\theta$$

$$= \int_{2\pi/3}^{\pi} \left(1 + 4 \cos \theta + 4 \frac{1 + \cos 2\theta}{2}\right) d\theta$$

$$= \int_{2\pi/3}^{\pi} (3 + 4 \cos \theta + 2 \cos 2\theta) \, d\theta$$

$$= \left[3\theta + 4 \sin \theta + \sin 2\theta\right]_{2\pi/3}^{\pi}$$

$$= \pi - \frac{3\sqrt{3}}{2} = \frac{2\pi - 3\sqrt{3}}{2}.$$

11. From the symmetry of the graph given in the figure, the area of the region inside the outer loop is

$$2\int_0^{2\pi/3} \frac{1}{2} r^2 \, d\theta = \int_0^{2\pi/3} (1 + 2 \cos \theta)^2 \, d\theta$$

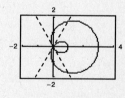

$$= \left[3\theta + 4 \sin \theta + \sin 2\theta\right]_0^{2\pi/3} \qquad \text{From Exercise 9}$$

$$= 2\pi + 2\sqrt{3} - \frac{\sqrt{3}}{2} = 2\pi + \frac{3\sqrt{3}}{2}.$$

From Exercise 9, we see that the area of the inner loop is given by $\pi - \left(3\sqrt{3}/2\right)$. Finally, the area of the region between the two loops is

$$A = \left(2\pi + \frac{3\sqrt{3}}{2}\right) - \left(\pi - \frac{3\sqrt{3}}{2}\right) = \pi + 3\sqrt{3}.$$

17. From Section 9.4 we know the graph of $r = 4 - 5 \sin \theta$ is a limaçon and the graph of $r = 3 \sin \theta$ is a circle (see figure). Solving the two equations simultaneously, yields

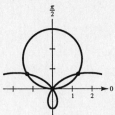

$$4 - 5 \sin \theta = 3 \sin \theta$$

$$8 \sin \theta = 4$$

$$\sin \theta = \frac{1}{2}$$

$$\theta = \frac{\pi}{6}, \frac{5\pi}{6} \quad (0 \le \theta < 2\pi).$$

From these values we obtain the points $(3/2, \pi/6)$ and $(3/2, 5\pi/6)$. To test additional points of intersection, replace r by $-r$ and θ by $\pi + \theta$ in $r = 4 - 5 \sin \theta$ to obtain

$$-r = 4 - 5 \sin(\pi + \theta) = 4 + 5 \sin \theta.$$

Solving this equation simultaneously with $r = 3 \sin \theta$, yields

$$-4 - 5 \sin \theta = 3 \sin \theta$$

$$8 \sin \theta = -4$$

$$\sin \theta = -\frac{1}{2}$$

$$\theta = \frac{7\pi}{6}, \frac{11\pi}{6}.$$

The corresponding points are $(-3/2, 7\pi/6)$ and $(-3/2, 11\pi/6)$. However, these two points coincide with the previous two points. Finally, observe that both curves pass through the pole. Hence there are three points of intersection, $(3/2, \pi/6)$, $(3/2, 5\pi/6)$, and $(0, 0)$ as seen in the figure.

21. From Section 9.4, we know the graph of $r = 4 \sin 2\theta$ is a rose curve with 4 petals and is symmetric to the polar axis, the vertical axis and the pole. Also, the graph of $r = 2$ is a circle of radius 2 centered at the pole. Solving the two equations simultaneously, yields

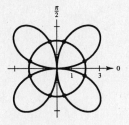

$$4 \sin 2\theta = 2$$

$$\sin 2\theta = \frac{1}{2}$$

$$2\theta = \frac{\pi}{6}, \frac{5\pi}{6}$$

$$\theta = \frac{\pi}{12}, \frac{5\pi}{12}.$$

Therefore, the points of intersection for one petal are $(2, \pi/12)$ and $(2, 5\pi/12)$. By symmetry, the other points of intersection are $(2, 7\pi/12)$, $(2, 11\pi/12)$, $(2, 13\pi/12)$, $(2, 17\pi/12)$, $(2, 19\pi/12)$, and $(2, 23\pi/12)$.

27. From the graph we see that we need only consider the region common to both curves in one petal of the rose curve and multiply the result by four. From Exercise 21 the points of intersection on the first petal occur when $\theta = \pi/12$ and $5\pi/12$. There are three subregions within one petal with the following bounds.

(a) for $0 \le \theta \le \dfrac{\pi}{12}$, $r = 4 \sin 2\theta$

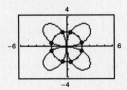

(b) for $\dfrac{\pi}{12} \le \theta \le \dfrac{5\pi}{12}$, $r = 2$

(c) for $\dfrac{5\pi}{12} \le \theta \le \dfrac{\pi}{2}$, $r = 4 \sin 2\theta$

—CONTINUED—

27. —CONTINUED—

Therefore, the area within one petal is

$$A = \int_0^{\pi/12} \frac{1}{2}(4\sin 2\theta)^2 \, d\theta + \int_{\pi/12}^{5\pi/12} \frac{1}{2}(2)^2 \, d\theta + \int_{5\pi/12}^{\pi/2} \frac{1}{2}(4\sin 2\theta)^2 \, d\theta.$$

By the symmetry of the petal, the first and third integrals are equal. Thus,

$$A = 2\int_0^{\pi/12} \frac{1}{2}(4\sin 2\theta)^2 \, d\theta + \int_{\pi/12}^{5\pi/12} \frac{1}{2}(2)^2 \, d\theta$$

$$= 16\int_0^{\pi/12} \sin^2 2\theta \, d\theta + 2\int_{\pi/12}^{5\pi/12} d\theta$$

$$= 8\int_0^{\pi/12} (1 - \cos 4\theta) \, d\theta + 2\int_{\pi/12}^{5\pi/12} d\theta$$

$$= 8\left[\theta - \frac{1}{4}\sin 4\theta\right]_0^{\pi/12} + \left[2\theta\right]_{\pi/12}^{5\pi/12}$$

$$= \frac{2\pi}{3} - \frac{2\sqrt{3}}{2} + \frac{2\pi}{3} = \frac{4\pi}{3} - \sqrt{3}.$$

Finally, multiplying by 4, we obtain the total area of $\frac{4}{3}(4\pi - 3\sqrt{3})$.

33. The graphs of $r = a(1 + \cos\theta)$ and $r = a\cos\theta$ are symmetric to the polar axis (see figure). Therefore, double the area of the top half of the cardiod and subtract the area of the circle of radius $a/2$.

$$A = 2\left[\frac{1}{2}\int_0^\pi [a(1 + \cos\theta)]^2\right] - \frac{\pi a^2}{4}$$

$$= a^2\int_0^\pi (1 + 2\cos\theta + \cos^2\theta) \, d\theta - \frac{\pi a^2}{4}$$

$$= a^2\int_0^\pi \left(1 + 2\cos\theta + \frac{1 + \cos 2\theta}{2}\right) d\theta - \frac{\pi a^2}{4}$$

$$= a^2\int_0^\pi \left(\frac{3}{2} + 2\cos\theta + \frac{1}{2}\cos 2\theta\right) d\theta - \frac{\pi a^2}{4}$$

$$= a^2\left[\frac{3}{2}\theta + 2\sin\theta + \frac{1}{4}\sin 2\theta\right]_0^\pi - \frac{\pi a^2}{4}$$

$$= \frac{3\pi a^2}{2} - \frac{\pi a^2}{4} = \frac{5\pi a^2}{4}$$

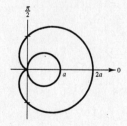

43. $r = 1 + \sin\theta, \, 0 \le \theta \le 2\pi$

$$\frac{dr}{d\theta} = \cos\theta$$

The graph of the polar equation is symmetric to $\theta = \pi/2$. Therefore, double the arc length for $-\pi/2 \le \theta \le \pi/2$.

$$s = \int_\alpha^\beta \sqrt{r^2 + \left(\frac{dr}{d\theta}\right)^2} \, d\theta$$

$$= 2\int_{-\pi/2}^{\pi/2} \sqrt{(1 + \sin\theta)^2 + (\cos\theta)^2} \, d\theta$$

$$= 2\int_{-\pi/2}^{\pi/2} \sqrt{1 + 2\sin\theta + \sin^2\theta + \cos^2\theta} \, d\theta$$

$$= 2\int_{-\pi/2}^{\pi/2} \sqrt{2(1 + \sin\theta)} \, d\theta$$

$$= 2\sqrt{2}\int_{-\pi/2}^{\pi/2} \frac{\cos\theta}{\sqrt{1 - \sin\theta}} \, d\theta$$

$$= -4\sqrt{2}\left[\sqrt{1 - \sin\theta}\right]_{-\pi/2}^{\pi/2}$$

$$= -4\sqrt{2}\left(0 - \sqrt{2}\right) = 8$$

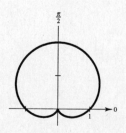

47.
$$s = \int_\alpha^\beta \sqrt{[f(\theta)]^2 + [f'(\theta)]^2}\, d\theta$$

$$= \int_\pi^{2\pi} \sqrt{\left(\frac{1}{\theta}\right)^2 + \left(\frac{-1}{\theta^2}\right)^2}\, d\theta$$

$$= \int_\pi^{2\pi} \frac{1}{\theta^2}\sqrt{\theta^2 + 1}\, d\theta$$

$$= \left[-\frac{\sqrt{\theta^2 + 1}}{\theta} + \ln\left|\theta + \sqrt{\theta^2 + 1}\right|\right]_\pi^{2\pi}$$

$$= \frac{2\sqrt{\pi^2 + 1} - \sqrt{4\pi^2 + 1}}{2\pi} + \ln\left|\frac{2\pi + \sqrt{4\pi^2 + 1}}{\pi + \sqrt{\pi^2 + 1}}\right| \approx 0.7112$$

53.
$$S = 2\pi\int_\alpha^\beta f(\theta)\cos\theta\sqrt{[f(\theta)]^2 + [f'(\theta)]^2}\, d\theta$$

$$= 2\pi\int_0^{\pi/2} e^{a\theta}(\cos\theta)\sqrt{(e^{a\theta})^2 + (ae^{a\theta})^2}\, d\theta$$

$$= 2\pi\int_0^{\pi/2} \cos\theta e^{a\theta}\sqrt{e^{2a\theta}(1 + a^2)}\, d\theta$$

$$= 2\pi\sqrt{1 + a^2}\int_0^{\pi/2} \cos\theta e^{2a\theta}\, d\theta$$

$$= 2\pi\sqrt{1 + a^2}\left[\frac{e^{2a\theta}}{4a^2 + 1}(2a\cos\theta + \sin\theta)\right]_0^{\pi/2} \quad \text{(Integration by Parts)}$$

$$= \frac{2\pi\sqrt{1 + a^2}}{4a^2 + 1}(e^{\pi a} - 2a)$$

Section 9.6   Polar Equations of Conics and Kepler's Laws

13. From the form of the equation we have

$$r = \frac{-1}{1 - \sin\theta} = \frac{ed}{1 - e\sin\theta}.$$

We can conclude that the graph of the equation is a parabola ($e = 1$). Sketch the left half of the parabola by plotting points in the table. Then using symmetry with respect to $\theta = \pi/2$, sketch the right half.

θ	$-\dfrac{\pi}{2}$	$-\dfrac{\pi}{3}$	$-\dfrac{\pi}{4}$	$-\dfrac{\pi}{6}$	0	$\dfrac{\pi}{6}$	$\dfrac{\pi}{4}$	$\dfrac{\pi}{3}$	$\dfrac{\pi}{2}$
r	-0.50	-0.54	-0.59	-0.67	-1	-2	-3.41	-7.46	Undefined

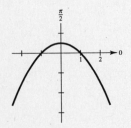

21. To determine the type of conic, rewrite the equation as

$$r = \frac{3}{2 + 6\sin\theta} = \frac{3/2}{1 + 3\sin\theta} = \frac{ed}{1 - e\sin\theta}.$$

From this form we can conclude that the graph is a hyperbola with $e = 3$. Since $r = f(\sin\theta)$, its graph is symmetric to $\theta = \pi/2$. Therefore, the entries of the table are solution points of the equation for the right half of the hyperbola. Plot these points and sketch the right half and then sketch the other half by symmetry.

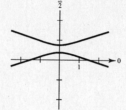

θ	0	$\dfrac{\pi}{6}$	$\dfrac{\pi}{3}$	$\dfrac{\pi}{2}$	$\dfrac{7\pi}{6}$	$\dfrac{4\pi}{3}$	$\dfrac{3\pi}{2}$
r	1.500	0.600	0.417	0.125	-3.000	-0.417	-0.750

35. Since the directrix is horizontal and above the pole (see figure), choose an equation of the form

$$r = \frac{ed}{1 + e\sin\theta}.$$

Moreover, since the eccentricity of the ellipse is $\frac{1}{2}$ and the directed distance from the focus to the directrix is $d = 1$, we have the equation

$$r = \frac{1/2}{1 + (1/2)\sin\theta} = \frac{1}{2 + \sin\theta}.$$

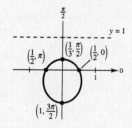

39. Since the directrix is horizontal and below the pole (see figure), choose an equation of the form

$$r = \frac{ed}{1 - e\sin\theta}.$$

Moreover, since the eccentricity of a parabola is $e = 1$ and the distance from the focus to the directrix is $d = 2$, we have the equation

$$r = \frac{2}{1 - \sin\theta}.$$

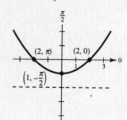

49. The polar equation for the hyperbola

$$\frac{x^2}{a^2} - \frac{y^2}{b^2} = 1 \text{ is } r^2 = \frac{-b^2}{1 - e^2\cos^2\theta} \quad \text{(see Exercise 46).}$$

For the hyperbola $\dfrac{x^2}{9} - \dfrac{y^2}{16} = 1$, we have

$$a = 3, b = 4, c = \sqrt{a^2 + b^2} = 5, \text{ and } e = \frac{c}{a} = \frac{5}{3}.$$

Therefore, $r^2 = \dfrac{-16}{1 - (25/9)\cos^2\theta} = \dfrac{-144}{9 - 25\cos^2\theta}.$

Review Exercises for Chapter 9

9. $3x^2 + 2y^2 - 12x + 12y + 29 = 0$

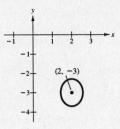

Write the standard form of the equation.

$$3x^2 + 2y^2 - 12x + 12y + 29 = 0$$

$$3(x^2 - 4x) + 2(y^2 + 6y) = -29$$

$$3(x^2 - 4x + 4) + 2(y^2 + 6y + 9) = -29 + 12 + 18$$

$$3(x - 2)^2 + 2(y + 3)^2 = 1$$

$$\frac{(x - 2)^2}{\left(1/\sqrt{3}\right)^2} + \frac{(y + 3)^2}{\left(1/\sqrt{2}\right)^2} = 1$$

The graph is an ellipse where $h = 2, k = -3, a = 1/\sqrt{2}, b = 1/\sqrt{3}$, and $c = \sqrt{(1/2) - (1/3)} = 1/\sqrt{6}$.

Center, (h, k): $(2, -3)$

Vertices, $(h, k \pm a)$: $\left(2, -3 \pm \dfrac{\sqrt{2}}{2}\right)$

Foci, $(h, k \pm c)$: $\left(2, -3 \pm \dfrac{\sqrt{6}}{6}\right)$

11. Since the directrix $(x = -3)$ is vertical, the axis of the parabola is horizontal and its standard form is

$$(y - k)^2 = 4p(x - h).$$

Since the vertex $(0, 2)$ is to the right of the directrix, the parabola opens to the right and $p > 0$. Therefore, $p = 3$ (the distance between the vertex and directrix), $h = 0$, and $k = 2$.

$$(y - 2)^2 = 4(3)(x - 0)$$

$$y^2 - 4y + 4 = 12x$$

$$y^2 - 4y - 12x + 4 = 0$$

19. (a) The volume of the tank is the area of the ellipse times the length of the tank.

$$\frac{x^2}{16} + \frac{y^2}{9} = 1 \implies x = \frac{4}{3}\sqrt{9 - y^2}$$

Using the symmetry of the ellipse (see figure) and trigonometric substitution to evaluate the integral, the area is

$$A = 4\int_0^3 \frac{4}{3}\sqrt{9 - y^2}\, dy$$

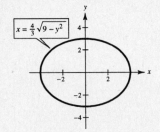

$$= \frac{16}{3}\left(\frac{1}{2}\right)\left[y\sqrt{9 - y^2} + 9 \arcsin \frac{y}{3}\right]_0^3 = 12\pi.$$

Therefore, the volume of the tank is

$$16(12\pi) = 192\pi \text{ cubic feet.}$$

—CONTINUED—

19.—CONTINUED—

(b) The force against a representative rectangle of length $2x$ is

$$\Delta F = (\text{density})(\text{depth})(\text{area})$$

$$= 62.4(3 - y)(2x\,\Delta y) = 62.4(3 - y)(2)\left(\frac{4}{3}\sqrt{9 - y^2}\right)\Delta y = 166.4(3 - y)\sqrt{9 - y^2}\,\Delta y.$$

The total force is

$$F = 166.4\int_{-3}^{3}(3 - y)(9 - y^2)^{1/2}\,dy = 166.4(3)\int_{-3}^{3}\sqrt{9 - y^2}\,dy - 166.4\int_{-3}^{3}3y\sqrt{9 - y^2}\,dy.$$

The second integral is 0 since the integrand is an odd function and it is evaluated over the interval $[-3, 3]$. The first integral is the area of a semicircle of radius 3. Therefore,

$$F = 166.4(3)\left(\frac{1}{2}\right)(\pi)(3^2) \approx 7057.3 \text{ pounds.}$$

(c) The truck will be carrying $\frac{3}{4}$ of its total capacity when the water covers $\frac{3}{4}$ of the area of a cross section of the tank. One-half of this area will be below the major axis and $\frac{1}{4}$ above the major axis of the ellipse

$$\frac{x^2}{16} + \frac{y^2}{9} = 1.$$

One-fourth the total area is 3π [see part (a)]. From the figure we have

$$\int_0^h 2\left(\frac{4}{3}\sqrt{9 - y^2}\right)dy = 3\pi$$

$$\int_0^y \sqrt{3^2 - y^2}\,dy = \frac{9\pi}{8}$$

$$\frac{1}{2}\left[y\sqrt{9 - y^2} + 9\arcsin\frac{y}{3}\right]_0^h = \frac{9\pi}{8}$$

$$h\sqrt{9 - h^2} + 9\arcsin\frac{h}{3} = \frac{9\pi}{4}.$$

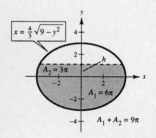

Find h such that $f(h) = \sqrt{9 - h^2} + \arcsin\left(\frac{h}{3}\right) - \frac{9\pi}{4} - 0.$

Using the root finding capabilities of a graphing utility yields $h \approx 1.212$ and therefore the total height of the water in the tank is $3 + 1.212 = 4.212$ feet.

(d) To approximate the surface area of the tank, multiply the perimeter of an elliptical cross section by the length of the tank and add the area of the ends of the tank [see part (a)]. Using rectangular coordinates yields the following improper integral. (Observe that the integrand is undefined when $x = 4$.)

$$\frac{x^2}{16} + \frac{y^2}{9} = 1$$

$$y = \frac{3}{4}\sqrt{16 - x^2}$$

$$\frac{dy}{dx} = \frac{-3x}{4\sqrt{16 - x^2}}$$

$$s = 4\int_0^4 \sqrt{1 + \left(\frac{dy}{dx}\right)^2}\,dx = 4\int_0^4 \sqrt{1 + \frac{9x^2}{16(16 - x^2)}}\,dx$$

If we use parametric equations to represent the ellipse, we have

$$x = 4\cos\theta \quad \text{and} \quad y = 3\sin\theta.$$

Using symmetry and the integration capabilities of a graphing utility we approximate the perimeter of an elliptical cross section of the tank.

$$S = 4\int_0^{\pi/2}\sqrt{\left(\frac{dx}{d\theta}\right)^2 + \left(\frac{dy}{d\theta}\right)^2}\,d\theta = 4\int_0^{\pi/2}\sqrt{(-4\sin\theta)^2 + (3\cos\theta)^2}\,d\theta = 4\int_0^{\pi/2}\sqrt{16\sin^2\theta + 9\cos^2\theta}\,d\theta \approx 22.1$$

Therefore, $S \approx 22.1(16) + 2(12\pi) \approx 429$ square feet.

27. (a) Since $x = 3 + 2 \cos \theta$ and $y = 2 + 5 \sin \theta$, we have

$$\frac{dy}{dx} = \frac{dy/d\theta}{dx/d\theta} = \frac{5 \cos \theta}{-2 \sin \theta} = -\frac{5}{2} \cot \theta.$$

The points of horizontal tangency occur at $\theta = \pi/2$ and $\theta = 3\pi/2$. Substituting these values of θ into the set of parametric equations yields the points $(3, 7)$ and $(3, -3)$.

(b) To eliminate the parameter, consider the identity $\sin^2 \theta + \cos^2 \theta = 1$ and write the parametric equations in the form

$$\frac{x - 3}{2} = \cos \theta, \quad \frac{y - 2}{5} = \sin \theta.$$

By squaring and adding these equations, we obtain

$$\frac{(x - 3)^2}{4} + \frac{(y - 2)^2}{25} = \sin^2 \theta + \cos^2 \theta = 1$$

which is an equation for the ellipse centered at $(3, 2)$ with vertices at $(3, -3)$ and $(3, 7)$.

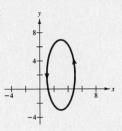

35. Semi-major axis: $a = 4$
Semi-minor axis: $b = 3$
Center: $(h, k) = (-3, 4)$

The parametric equations for an ellipse with horizontal major axis is

$$x = h + a \cos \theta = -3 + 4 \cos \theta$$

$$y = k + b \sin \theta = 4 + 3 \sin \theta.$$

41. $x = r(\cos \theta + \theta \sin \theta) \quad y = r(\sin \theta - \theta \cos \theta)$

$$\frac{dx}{d\theta} = r\theta \cos \theta \qquad \frac{dy}{d\theta} = r\theta \sin \theta$$

$$s = \int_0^{\pi} \sqrt{\left(\frac{dx}{d\theta}\right)^2 + \left(\frac{dy}{d\theta}\right)^2} \, d\theta$$

$$= \int_0^{\pi} \sqrt{(r\theta \cos \theta)^2 + (r\theta \sin \theta)^2} \, d\theta$$

$$= r \int_0^{\pi} \theta \, d\theta = r \left[\frac{\theta^2}{2}\right]_0^{\pi} = \frac{1}{2}\pi^2 r$$

51. First, replace $\cos 2\theta$ by $2 \cos^2 \theta - 1$ and $\sec \theta$ by $1/\cos \theta$ to obtain

$$r = 4 \cos 2\theta \sec \theta = 4(2 \cos^2 \theta - 1)\frac{1}{\cos \theta}$$

$$r \cos \theta = 8 \cos^2 \theta - 4.$$

Now since $x = r \cos \theta$ and $r^2 = x^2 + y^2$, we have

$$x = 8\left(\frac{x^2}{r^2}\right) - 4 = \frac{8x^2 - 4(x^2 + y^2)}{x^2 + y^2}$$

$$x^3 + xy^2 = 4x^2 - 4y^2$$

$$(4 + x)y^2 = (x^2)(4 - x)$$

$$y^2 = x^2\left(\frac{4 - x}{4 + x}\right).$$

55. Since $r^2 = x^2 + y^2$ and $\theta = \arctan(y/x)$, we have

$$x^2 + y^2 = a^2\left(\arctan \frac{y}{x}\right)^2$$

$$r^2 = a^2\theta^2.$$

67. Given the polar equation $r^2 = 4 \sin^2 2\theta$, we have $r = \pm 2 \sin 2\theta$.

Type of curve: Rose curve with four petals
Symmetry: With respect to the pole, polar axis, and $\theta = \pi/2$.
Extrema of r: $(\pm 2, \pi/4), (\pm 2, 3\pi/4)$
Tangents at the pole: $\theta = 0, \pi/2$

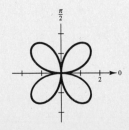

θ	0	$\dfrac{\pi}{12}$	$\dfrac{\pi}{8}$	$\dfrac{\pi}{6}$	$\dfrac{\pi}{4}$	$\dfrac{\pi}{3}$	$\dfrac{3\pi}{8}$	$\dfrac{5\pi}{12}$	$\dfrac{\pi}{2}$
r	0	± 1	$\pm\sqrt{2}$	$\pm\sqrt{3}$	± 2	$\pm\sqrt{3}$	$\pm\sqrt{2}$	± 1	0

75. (a) The graph has polar axis symmetry and the tangents at the pole are $\theta = \pi/3$ and $\theta = -\pi/3$.

(b) To find the points of vertical or horizontal tangency, note that $f'(\theta) = 2 \sin \theta$ and find dy/dx as follows:

$$\frac{dy}{dx} = \frac{f'(\theta) \sin \theta + f(\theta) \cos \theta}{f'(\theta) \cos \theta - f(\theta) \sin \theta}$$

$$= \frac{2 \sin^2 \theta + (1 - 2 \cos \theta) \cos \theta}{2 \sin \theta \cos \theta - (1 - 2 \cos \theta) \sin \theta}$$

$$= \frac{2 \sin^2 \theta + \cos \theta - 2 \cos^2 \theta}{4 \sin \theta \cos \theta - \sin \theta}$$

$$= \frac{2(1 - \cos^2 \theta) + \cos \theta - 2 \cos^2 \theta}{\sin \theta(4 \cos \theta - 1)}$$

$$= \frac{2 + \cos \theta - 4 \cos^2 \theta}{\sin \theta(4 \cos \theta - 1)}$$

The graph has horizontal tangents when $dy/dx = 0$, and this occurs when

$$-4 \cos^2 \theta + \cos \theta + 2 = 0$$

$$\cos \theta = \frac{-1 \pm \sqrt{1 + 32}}{-8} = \frac{1 \mp \sqrt{33}}{8}.$$

When $\cos \theta = \left(1 \mp \sqrt{33}\right)/8$,

$$r = 1 - 2\left(\frac{1 \mp \sqrt{33}}{8}\right) = \frac{3 \pm \sqrt{33}}{4}.$$

Therefore, the points of horizontal tangency are

$$\left(\frac{3 - \sqrt{33}}{4}, \arccos\left[\frac{1 + \sqrt{33}}{8}\right]\right) \approx (-0.686, 0.586)$$

$$\left(\frac{3 - \sqrt{33}}{4}, -\arccos\left[\frac{1 + \sqrt{33}}{8}\right]\right) \approx (-0.686, -0.568)$$

$$\left(\frac{3 + \sqrt{33}}{4}, \arccos\left[\frac{1 - \sqrt{33}}{8}\right]\right) \approx (2.186, 2.206)$$

$$\left(\frac{3 + \sqrt{33}}{4}, -\arccos\left[\frac{1 - \sqrt{33}}{8}\right]\right) \approx (2.186, -2.206).$$

The graph has vertical tangents when

$$\sin \theta(4 \cos \theta - 1) = 0$$

$$\theta = 0, \pi, \text{ or } \pm\arccos\frac{1}{4}.$$

When $\cos \theta = 1/4$, $r = 1 - 2(1/4) = 1/2$. Thus, the points of vertical tangency are

$$(-1, 0), \ (3, \pi), \text{ and } \left(\frac{1}{2}, \pm\arccos\frac{1}{4}\right) \approx (0.5, \pm1.318).$$

(c) The graph is shown in the figure.

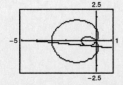

87. The graphs of the polar equations are shown in the figure. To find the area of the common interior of the two curves, we must first find their points of intersection by solving the two equations simultaneously.

$$4 \cos \theta = 2$$

$$\cos \theta = \frac{1}{2} \implies \theta = \pm \frac{\pi}{3}$$

Using symmetry and the integration capabilities of a graphing utility, we have

$$A = 2\left[\frac{1}{2}\int_0^{\pi/3} 2^2 \, d\theta + \frac{1}{2}\int_{\pi/3}^{\pi/2} (4 \cos \theta)^2 \, d\theta\right] = \frac{8\pi - 6\sqrt{3}}{3} \approx 4.91.$$

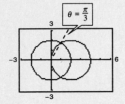

93. Parabola

Vertex: $(2, \pi)$

Focus: $(0, 0)$

Since the vertex is 2 units to the left of the pole, the parabola opens to the right, is symmetric to the polar axis, the distance between the focus and directrix is $d = 4$, and the eccentricity is $e = 1$. The polar equation of the parabola is

$$r = \frac{ed}{1 \pm e \cos \theta} = \frac{4}{1 - \cos \theta}.$$

Appendix A.1

7. Since $4^3 = 64$, it follows that $\sqrt[3]{64} = 4$ and is therefore rational.

13. Let $x = 0.297297\ldots$. Since the repeating pattern occurs every three decimal places, multiply both sides of the equation by 1000 and obtain

$$1000x = 297.297297\ldots$$

$$\underline{x = 0.297297\ldots} \quad \text{Subtract}$$

$$999x = 297$$

$$x = \tfrac{297}{999} = \tfrac{11}{37}.$$

27.
$$-4 < \quad 2x - 3 \quad < 4$$
$$-4 + 3 < 2x - 3 + 3 < 4 + 3$$
$$-1 < \quad 2x \quad < 7$$
$$-\tfrac{1}{2} < \quad x \quad < \tfrac{7}{2}$$

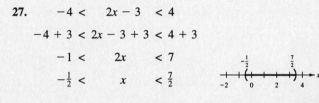

37. $|2x + 1| < 5$
$$-5 < 2x + 1 < 5$$
$$-6 < \quad 2x \quad < 4$$
$$-3 < \quad x \quad < 2$$

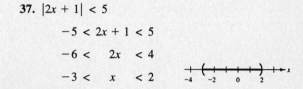

43.
$$x^2 + x - 1 \le 5$$
$$x^2 + x - 6 \le 0$$
$$(x + 3)(x - 2) \le 0$$
Critical numbers are: $x = -3$ and $x = 2$
Solution: $-3 \le x \le 2$

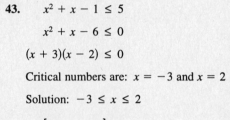

Interval	$-\infty < x < -3$	$-3 < x < 2$	$2 < x < \infty$
Test value	$x = -4$	$x = 0$	$x = 3$
Sign of $x + 3$	$-$	$+$	$+$
Sign of $x - 2$	$-$	$-$	$+$
Sign of $(x + 3)(x - 2)$	$+$	$-$	$+$

47. (a) $a = 126, b = 75$

Directed distance from a to b: $b - a = -51$
Directed distance from b to a: $a - b = 51$
Distance between a and b: $|a - b| = 51$

(b) $a = -126, b = -75$

Directed distance from a to b: $b - a = 51$
Directed distance from b to a: $a - b = -51$
Distance between a and b: $|a - b| = 51$

55. From the figure we have
$$x < \quad 0 \text{ or } \quad x > 4$$
$$x - 2 < -2 \text{ or } x - 2 > 4 - 2 \quad \text{Centered at 2}$$
$$x - 2 < -2 \text{ or } x - 2 > 2.$$

Therefore, the magnitude of $x - 2$ must be greater than 2 or $|x - 2| > 2$.

59. Since the revenue R must be greater than the cost C for the product to return a profit, it follows that
$$R > C$$
$$115.95x > 95x + 750$$
$$115.95x - 95x > 750$$
$$20.95x > 750$$
$$x > 35.7995 \text{ or } x \ge 36 \text{ units.}$$

61. Since $\left| \dfrac{x - 50}{5} \right| \ge 1.645$, we have

$$-\frac{x - 50}{5} \ge 1.645 \qquad \text{or} \qquad \frac{x - 50}{5} \ge 1.645$$

$$-\frac{x - 50}{5}(-5) \le 1.645(-5) \text{ or } \frac{x - 50}{5}(5) \ge 1.645(5)$$

$$x - 50 \le -8.225 \quad \text{or} \quad x - 50 \ge 8.225$$

$$x \le 41.775 \quad \text{or} \quad x \ge 58.225.$$

Therefore the coin will be declared unfair if $x \le 41$ or $x \ge 59$.

73. Case 1: $a > 0, b > 0,$ and $ab > 0.$

$$|ab| = ab = |a||b|$$

Case 2: $a < 0, b < 0,$ and $ab > 0.$

$$|ab| = ab = (-a)(-b) = |a||b|$$

Case 3: $a > 0, b < 0,$ and $ab < 0.$

$$|ab| = -(ab) = a(-b) = |a||b|$$

Case 4: $a < 0, b > 0,$ and $ab < 0.$

$$|ab| = -(ab) = (-a)(b) = |a||b|$$

Appendix A.2

1. (a)

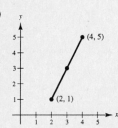

(b) Let $(2, 1) = (x_1, y_1)$ and $(4, 5) = (x_2, y_2).$ Then

$$d = \sqrt{(x_2 - x_1)^2 + (y_2 - y_1)^2}$$

$$= \sqrt{(4 - 2)^2 + (5 - 1)^2} = \sqrt{2^2 + 4^2}$$

$$= \sqrt{20} = 2\sqrt{5}.$$

(c) midpoint $= \left(\dfrac{x_1 + x_2}{2}, \dfrac{y_1 + y_2}{2} \right) = \left(\dfrac{2 + 4}{2}, \dfrac{1 + 5}{2} \right) = (3, 3)$

7. Let $d_1 =$ distance between $(4, 0)$ and $(2, 1).$

$$d_1{}^2 = (2 - 4)^2 + (1 - 0)^2 = (-2)^2 + 1^2 = 5$$

Let $d_2 =$ distance between $(2, 1)$ and $(-1, -5).$

$$d_2{}^2 = (-1 - 2)^2 + (-5 - 1)^2 = (-3)^2 + (-6)^2 = 45$$

Let $d_3 =$ distance between $(4, 0)$ and $(-1, -5).$

$$d_3{}^2 = (-1 - 4)^2 + (-5 - 0)^2 = (-5)^2 + (-5)^2 = 50$$

Since $d_1{}^2 + d_2{}^2 = 5 + 45 = 50 = d_3{}^2,$ the triangle is a right triangle.

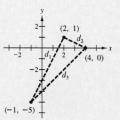

19. Let $d_1 =$ distance between $(-2, 1)$ and $(-1, 0).$ Then

$$d_1 = \sqrt{[-1 - (-2)]^2 + (0 - 1)^2} = \sqrt{1^2 + (-1)^2} = \sqrt{2}.$$

Let $d_2 =$ distance between $(-1, 0)$ and $(2, -2).$ Then

$$d_2 = \sqrt{[2 - (-1)]^2 + (-2 - 0)^2} = \sqrt{3^2 + (-2)^2} = \sqrt{13}.$$

Let $d_3 =$ distance between $(-2, 1)$ and $(2, -2).$ Then

$$d_3 = \sqrt{[2 - (-2)]^2 + (-2 - 1)^2} = \sqrt{4^2 + (-3)^2} = \sqrt{25} = 5.$$

The points $(-2, 1), (-1, 0),$ and $(2, -2)$ lie on a line only if $d_1 + d_2 = d_3.$ Since $\sqrt{2} + \sqrt{13} \approx 5.02 \neq 5,$ the points are *not* collinear.

25. The midpoint of the line segment from (x_1, y_1) to (x_2, y_2) is $\left(\dfrac{x_1 + x_2}{2}, \dfrac{y_1 + y_2}{2} \right).$

The midpoint between (x_1, y_1) and $\left(\dfrac{x_1 + x_2}{2}, \dfrac{y_1 + y_2}{2} \right)$ is

$$\left(\dfrac{x_1 + \dfrac{x_1 + x_2}{2}}{2}, \dfrac{y_1 + \dfrac{y_1 + y_2}{2}}{2} \right) = \left(\dfrac{1}{2} \left(\dfrac{2x_1 + x_1 + x_2}{2} \right), \dfrac{1}{2} \left(\dfrac{2y_1 + y_1 + y_2}{2} \right) \right)$$

$$= \left(\dfrac{3x_1 + x_2}{4}, \dfrac{3y_1 + y_2}{4} \right).$$

—CONTINUED—

25. —CONTINUED—

The midpoint between $\left(\dfrac{x_1 + x_2}{2}, \dfrac{y_1 + y_2}{2}\right)$ and (x_2, y_2) is

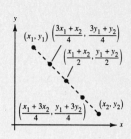

$$\left(\dfrac{\dfrac{x_1 + x_2}{2} + x_2}{2}, \dfrac{\dfrac{y_1 + y_2}{2} + y_2}{2}\right) = \left(\dfrac{1}{2}\left(\dfrac{x_1 + x_2 + 2x_2}{2}\right), \dfrac{1}{2}\left(\dfrac{y_1 + y_2 + 2y_2}{2}\right)\right)$$

$$= \left(\dfrac{x_1 + 3x_2}{4}, \dfrac{y_1 + 3y_2}{4}\right)$$

Thus the three points are

$$\left(\dfrac{3x_1 + x_2}{4}, \dfrac{3y_1 + y_2}{4}\right), \left(\dfrac{x_1 + x_2}{2}, \dfrac{y_1 + y_2}{2}\right), \left(\dfrac{x_1 + 3x_2}{4}, \dfrac{y_1 + 3y_2}{4}\right).$$

33. Let $(2, -1) = (h, k)$ and $r = 4$. Then using the standard form of the equation of a circle, we have

$$(x - h)^2 + (y - k)^2 = r^2$$

$$(x - 2)^2 + [y - (-1)]^2 = 4^2$$

$$x^2 - 4x + 4 + y^2 + 2y + 1 = 16$$

$$x^2 + y^2 - 4x + 2y - 11 = 0.$$

43. $$x^2 + y^2 - 2x + 6y + 10 = 0$$

$$(x^2 - 2x + \underline{\ \ }) + (y^2 + 6y + \underline{\ \ }) = -10$$

$$(x^2 - 2x + 1) + (y^2 + 6y + 9) = -10 + 1 + 9$$

$$(x - 1)^2 + (y + 3)^2 = 0$$

The only solution point of the equation is $(1, -3)$.

51. $$x^2 + y^2 - 4x + 2y + 1 \le 0$$

$$(x^2 - 4x + 4) + (y^2 + 2y + 1) \le -1 + 4 + 1$$

$$(x - 2)^2 + (y + 1)^2 \le 4$$

Therefore, the inequality is satisfied by the set of all points lying on the boundary and in the interior of the circle with center $(2, -1)$ and radius 2.

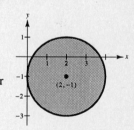

61. For simplicity, assume the semicircle is centered at the origin with a radius r (see figure). If (a, b) is a point on the semicircle, then it must satisfy the equation $a^2 + b^2 = r^2$. To verify that the angle at (a, b) is a right angle, it is sufficient to show that $d_1^2 + d_2^2 = d_3^2$.

$$d_1^2 = [a - (-r)]^2 + (b - 0)^2$$

$$d_2^2 = (a - r)^2 + (b - 0)^2$$

$$d_1^2 + d_2^2 = (a^2 + 2ar + r^2 + b^2) + (a^2 - 2ar + r^2 + b^2)$$

$$= 2a^2 + 2b^2 + 2r^2$$

$$= 2(a^2 + b^2) + 2r^2$$

$$= 2r^2 + 2r^2 = 4r^2 = (2r)^2 = d_3^2$$

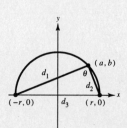

Appendix A.3

7. Since $180° = \pi$ radians, it follows that 1 radian $= 180°/\pi$.

(a) $\dfrac{3\pi}{2}$ radians $= \left(\dfrac{3\pi}{2}\right)\left(\dfrac{180°}{\pi}\right) = 270°$

(c) $-\dfrac{7\pi}{12}$ radians $= \left(-\dfrac{7\pi}{12}\right)\left(\dfrac{180°}{\pi}\right) = -105°$

(b) $\dfrac{7\pi}{6}$ radians $= \left(\dfrac{7\pi}{6}\right)\left(\dfrac{180°}{\pi}\right) = 210°$

(d) -2.367 radians $= (-2.367)\left(\dfrac{180°}{\pi}\right) \approx -135.619°$

11. (a) From the figure we have $x = 3$, $y = 4$, and $r = \sqrt{x^2 + y^2} = 5$.

$$\sin\theta = \frac{y}{r} = \frac{4}{5} \qquad \csc\theta = \frac{r}{y} = \frac{5}{4}$$

$$\cos\theta = \frac{x}{r} = \frac{3}{5} \qquad \sec\theta = \frac{r}{x} = \frac{5}{3}$$

$$\tan\theta = \frac{y}{x} = \frac{4}{3} \qquad \cot\theta = \frac{x}{y} = \frac{3}{4}$$

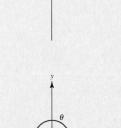

(b) From the figure we have $x = -12$, $y = -5$, and $r = \sqrt{x^2 + y^2} = 13$.

$$\sin\theta = \frac{y}{r} = -\frac{5}{13} \qquad \csc\theta = \frac{r}{y} = -\frac{13}{5}$$

$$\cos\theta = \frac{x}{r} = -\frac{12}{13} \qquad \sec\theta = \frac{r}{x} = -\frac{13}{12}$$

$$\tan\theta = \frac{y}{x} = \frac{5}{12} \qquad \cot\theta = \frac{x}{y} = \frac{12}{5}$$

17. Using the fact that

$$\cos\theta = \frac{4}{5},$$

construct the figure and obtain

$$\cot\theta = \frac{4}{y} = \frac{4}{\sqrt{25 - 16}} = \frac{4}{3}.$$

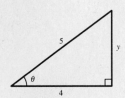

21. (a) The angle $225°$ is in Quadrant III and the reference angle is $225° - 180° = 45°$. Therefore,

$$\sin 225° = -\sin 45° = -\frac{\sqrt{2}}{2}$$

$$\cos 225° = -\cos 45° = -\frac{\sqrt{2}}{2}$$

$$\tan 225° = \tan 45° = 1.$$

(b) The angle $-225°$ is in Quadrant II and the reference angle is $225° - 180° = 45°$. Therefore,

$$\sin(-225°) = \sin 45° = \frac{\sqrt{2}}{2}$$

$$\cos(-225°) = -\cos 45° = -\frac{\sqrt{2}}{2}$$

$$\tan(-225°) = -\tan 45° = -1.$$

(c) The angle $5\pi/3$ is in Quadrant IV and the reference angle is $2\pi - (5\pi/3) = \pi/3$. Therefore,

$$\sin\frac{5\pi}{3} = -\sin\frac{\pi}{3} = -\frac{\sqrt{3}}{2}$$

$$\cos\frac{5\pi}{3} = \cos\frac{\pi}{3} = \frac{1}{2}$$

$$\tan\frac{5\pi}{3} = -\tan\frac{\pi}{3} = -\sqrt{3}.$$

(d) The angle $11\pi/6$ is in Quadrant IV and the reference angle is $2\pi - (11\pi/6) = \pi/6$. Therefore,

$$\sin\frac{11\pi}{6} = -\sin\frac{\pi}{6} = -\frac{1}{2}$$

$$\cos\frac{11\pi}{6} = \cos\frac{\pi}{6} = \frac{\sqrt{3}}{2}$$

$$\tan\frac{11\pi}{6} = -\tan\frac{\pi}{6} = -\frac{\sqrt{3}}{3}.$$

27. (a) $\cos \theta = \dfrac{\sqrt{2}}{2}$

To solve the equation, we realize that the cosine is positive in Quadrants I and IV and that

$$\cos \frac{\pi}{4} = \frac{\sqrt{2}}{2}.$$

Therefore, the reference angle is $\pi/4$ and the required angles are

$$\theta = \frac{\pi}{4} \text{ and } \theta = 2\pi - \frac{\pi}{4} = \frac{7\pi}{4}.$$

(b) $\cos \theta = -\dfrac{\sqrt{2}}{2}$

To solve the equation, we realize that the cosine is negative in Quadrants II and III and that

$$\cos \frac{\pi}{4} = \frac{\sqrt{2}}{2}.$$

Therefore, the reference angle is $\pi/4$ and the required angles are

$$\theta = \pi - \frac{\pi}{4} = \frac{3\pi}{4} \text{ and } \theta = \pi + \frac{\pi}{4} = \frac{5\pi}{4}.$$

33. $\tan^2 \theta - \tan \theta = 0, \ 0 \le \theta < 2\pi$

$$\tan \theta (\tan \theta - 1) = 0$$

If $\tan \theta = 0$, then $\theta = 0$ or $\theta = \pi$. If $\tan \theta - 1 = 0$, then $\tan \theta = 1$ and $\theta = \pi/4$ or $\theta = 5\pi/4$. Thus for

$$0 \le \theta < 2\pi,$$

there are four solutions:

$$\theta = 0, \frac{\pi}{4}, \pi, \frac{5\pi}{4}.$$

39. In one minute the plane travels

$$(60 \text{ sec})(275 \text{ ft/sec}) = 16{,}500 \text{ ft.}$$

This distance is the approximate length of the hypotenuse of a right triangle whose side opposite the angle of magnitude $18°$ is the altitude h of the plane. Therefore,

$$\sin 18° \approx \frac{h}{16{,}500}$$

$$16{,}500 \sin 18° \approx h$$

$$h \approx 5100 \text{ ft.}$$

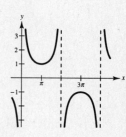

47. Since the period of $y = \sec x$ is 2π, the period of $y = \sec 5x$ is $\dfrac{2\pi}{5}$.

55. The graph of $y = \csc(x/2)$ has the following characteristics:

Period: $\dfrac{2\pi}{1/2} = 4\pi$

Vertical asymptote: $x = 2n\pi$, n an integer

Using the basic shape of the graph of the cosecant function, we sketch one period of the function on the interval $[0, 4\pi]$, following the pattern

minimum: $(\pi, 1)$

maximum: $(3\pi, -1)$.

67. $S = 58.3 + 32.5 \cos \dfrac{\pi t}{6}$

The graph of the sales function and the horizontal line $S = 75$ are shown in the figure. Using the capabilities of a graphing utility to find the points of intersection of two graphs, we find that the graphs intersect when $t = 1.97$ and $t = 10.03$. Therefore, sales exceed 75,000 during the months of January, November, and December.

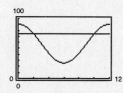

Appendix E

3. From the equations
$$x^2 - 10xy + y^2 + 1 = 0$$
$$Ax^2 + Bxy + Cy^2 + Dx + Ey + F = 0$$
we have $A = 1$, $B = -10$, $C = 1$, $D = 0$, $E = 0$, and $F = 1$. Thus

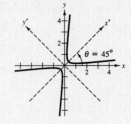

$$\cot 2\theta = \frac{A - C}{B} = 0 \quad \text{or} \quad 2\theta = \frac{\pi}{2} \quad \text{and} \quad \theta = \frac{\pi}{4}.$$

Therefore, $\sin \theta = \cos \theta = \dfrac{\sqrt{2}}{2}$ and

$$x = x' \cos \theta - y' \sin \theta = \frac{\sqrt{2}}{2}x' - \frac{\sqrt{2}}{2}y'$$

$$y = x' \sin \theta + y' \cos \theta = \frac{\sqrt{2}}{2}x' + \frac{\sqrt{2}}{2}y'.$$

Substitution into $x^2 - 10xy + y^2 + 1 = 0$ yields

$$\left(\frac{\sqrt{2}}{2}x' - \frac{\sqrt{2}}{2}y'\right)^2 - 10\left(\frac{\sqrt{2}}{2}x' - \frac{\sqrt{2}}{2}y'\right)\left(\frac{\sqrt{2}}{2}x' + \frac{\sqrt{2}}{2}y'\right) + \left(\frac{\sqrt{2}}{2}x' + \frac{\sqrt{2}}{2}y'\right)^2 + 1 = 0.$$

After expanding and combining like terms we have

$$-4(x')^2 + 6(y')^2 + 1 = 0 \quad \text{or} \quad \frac{(x')^2}{1/4} - \frac{(y')^2}{1/6} = 1.$$

7. From the equations
$$5x^2 - 2xy + 5y^2 - 12 = 0$$
$$Ax^2 + Bxy + Cy^2 + Dx + Ey + F = 0$$
we have $A = 5$, $B = -2$, $C = 5$, $D = 0$, $E = 0$, and $F = -12$. Thus

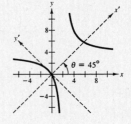

$$\cot 2\theta = \frac{A - C}{B} = 0 \quad \text{or} \quad 2\theta = \frac{\pi}{2} \quad \text{and} \quad \theta = \frac{\pi}{4}.$$

Therefore, $\sin \theta = \cos \theta = \dfrac{\sqrt{2}}{2}$ and

$$x = x' \cos \theta - y' \sin \theta = \frac{\sqrt{2}}{2}x' - \frac{\sqrt{2}}{2}y'$$

$$y = x' \sin \theta + y' \cos \theta = \frac{\sqrt{2}}{2}x' + \frac{\sqrt{2}}{2}y'$$

Substituting into $5x^2 - 2xy + 5y^2 - 12 = 0$ yields

$$5\left(\frac{\sqrt{2}}{2}x' - \frac{\sqrt{2}}{2}y'\right)^2 - 2\left(\frac{\sqrt{2}}{2}x' - \frac{\sqrt{2}}{2}y'\right)\left(\frac{\sqrt{2}}{2}x' + \frac{\sqrt{2}}{2}y'\right) + 5\left(\frac{\sqrt{2}}{2}x' + \frac{\sqrt{2}}{2}y'\right)^2 - 12 = 0.$$

After expanding and combining like terms we have

$$4(x')^2 + 6(y')^2 - 12 = 0$$

$$\frac{(x')^2}{3} + \frac{(y')^2}{2} = 1.$$

11. From the equations

$$9x^2 + 24xy + 16y^2 + 90x - 130y = 0$$

$$Ax^2 + Bxy + Cy^2 + Dx + Ey + F = 0$$

we have $A = 9$, $B = 24$, $C = 16$, $D = 90$, $E = -130$, and $F = 0$. Thus

$$\cot 2\theta = \frac{A - C}{B} = \frac{9 - 16}{24} = \frac{-7}{24}.$$

From the identity

$$\cot 2\theta = \frac{\cot^2 \theta - 1}{2 \cot \theta},$$

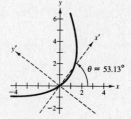

we have

$$\frac{\cot^2 \theta - 1}{2 \cot \theta} = \frac{-7}{24}$$

$$24 \cot^2 \theta - 24 = -14 \cot \theta$$

$$12 \cot^2 \theta + 7 \cot \theta - 12 = 0$$

$$(4 \cot \theta - 3)(3 \cot \theta + 4) = 0$$

$$\cot \theta = \frac{3}{4} \text{ or } -\frac{4}{3}.$$

Since $0 < \theta < 90°$, choose $\cot \theta = \frac{3}{4}$ and therefore, $\theta \approx 53.13°$. Since $\cot \theta = \frac{3}{4}$, $\sin \theta = \frac{4}{5}$ and $\cos \theta = \frac{3}{5}$. Therefore, using the equations

$$x = x' \cos \theta - y' \sin \theta = \frac{3}{5}x' - \frac{4}{5}y'$$

$$y = x' \sin \theta + y' \cos \theta = \frac{4}{5}x' + \frac{3}{5}y'$$

and

$$9x^2 + 24xy + 16y^2 + 90x - 130y = 0,$$

we have

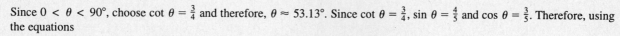

$$9\left(\frac{3}{5}x' - \frac{4}{5}y'\right)^2 + 24\left(\frac{3}{5}x' - \frac{4}{5}y'\right)\left(\frac{4}{5}x' + \frac{3}{5}y'\right) + 16\left(\frac{4}{5}x' + \frac{3}{5}y'\right)^2 + 90\left(\frac{3}{5}x' - \frac{4}{5}y'\right) - 130\left(\frac{4}{5}x' + \frac{3}{5}y'\right) = 0.$$

After expanding and combining like terms we have

$$25(x')^2 - 50x' - 150y' = 0$$

$$(x')^2 - 2x' - 6y' = 0$$

$$(x' - 1)^2 = 4\left(\frac{3}{2}\right)\left(y' + \frac{1}{6}\right).$$

17. From the equations

$$32x^2 + 50xy + 7y^2 = 52$$

$$Ax^2 + Bxy + Cy^2 + Dx + Ey + F = 0,$$

we have $A = 32$, $B = 50$, $C = 7$, $D = 0$, $E = 0$, and $F = -52$. Therefore,

$$\cot 2\theta = \frac{A - C}{B} = \frac{1}{2}$$

$$\theta = \frac{1}{2} \operatorname{arccot} \frac{1}{2} \approx 31.72°.$$

Use the Quadratic Formula to solve for y in terms of x.

$$7y^2 + (50x)y + (32x^2 - 52) = 0$$

$$y = \frac{-50x \pm \sqrt{(50x)^2 - 4(7)(32x^2 - 52)}}{2(7)}$$

$$= \frac{-50x \pm 2\sqrt{401x^2 + 364}}{2(7)}$$

$$= \frac{-25x \pm \sqrt{401x^2 + 364}}{7}$$

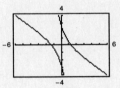

Use a graphing utility to graph the two resulting equations and obtain the hyperbola shown in the figure.

25. From the equations

$$x^2 + 4xy + 4y^2 - 5x - y - 3 = 0$$

$$Ax^2 + Bxy + Cy^2 + Dx + Ey + F = 0$$

we have $A = 1$, $B = 4$, $C = 4$, $D = -5$, $E = -1$, and $F = -3$. The value of the discriminant is

$$B^2 - 4AC = 4^2 - 4(1)4 = 0$$

and the curve is a parabola.